Alexander Panchin

IMMORTALITY OR DEATH

From Entropy to Eternity

Alexander Panchin.

In Immortality or Death, biologist and science communicator Alexander Panchin takes readers on a gripping journey through the science, philosophy, and future of human longevity. With clarity and wit, he explores cutting-edge research, ethical dilemmas, and revolutionary technologies that could make aging a treatable condition—and even unlock the possibility of indefinite life. Whether you're a scientist, student, or curious mind, this book challenges everything you thought you knew about growing old.

Alexander graduated from the Department of Bioengineering and Bioinformatics at Moscow State University, holds a PhD in Computational Biology, and is widely known in Russia as a science communicator.

Fonts provided by the author

Copyright © Open Longevity
First Edition 2025
All rights reserved

Includes bibliographical references and index.
Identifiers: **ISBN 979-8-9989729-0-4** (ebook) | **ISBN 979-8-9989729-1-1** (hardcover) | **ISBN 979-8-9989729-4-2** (paperback)

Contents

Introduction 5
Why aging is not an impossible puzzle and why we should solve it.

CHAPTER 1. My enemies are nature, entropy, and death ... 18
Life, Death, and Entropy. The basic understanding of aging.

CHAPTER 2. Imagine there's no heaven 33
What if death is not the end? The theistic and atheistic visions of an afterlife.

CHAPTER 3. The false Grail 51
Supplements. Vitamins. Diets. Biohacking. Does anything work?

CHAPTER 4. Evolution is not your friend 85
Why some organisms live much longer than others.

CHAPTER 5. Death after sex 98
The disposable soma and whether aging is a program that can be simply turned off.

CHAPTER 6. Feast of famine 115
The rate of living theory. Calorie restriction. Nutrient sensing. Autophagy.

CHAPTER 7. The immortal cell 139
Epigenetic reprogramming. Telomere attrition. Embryonic rejuvenation and cloning.

CHAPTER 8. Programmed cell aging 163
Cellular senescence. Inflammation. Senolytic drugs.

CHAPTER 9. The rise and fall of living utopias 178
The battle within us: cancer evolution against the aging immune system.

CONTENTS

CHAPTER 10. Fragile, but not that fragile 202
Aging of the extracellular matrix and the long-lived components of our bodies.

CHAPTER 11. The ageless brain 222
Body transplantation. Parabiosis. Dementia and Alzheimer's disease.

CHAPTER 12. The twenty-first–century cure 249
3D bioprinters. Blastocyst complementation and growing new organs.

CHAPTER 13. Whatever it takes 271
The hallmarks of aging. Combination therapy of aging. The ultimate anti-aging experiment.

CHAPTER 14. Cancel death culture 300
What should we do? A review of modern anti-aging projects and movements.

Acknowledgments 325
References 326

Introduction

To the brave and the petrified: we all fall down.
—IAMX ("*The Great Shipwreck of Life*")

You are holding a book about aging and death written by someone who wishes not to age or die. I hope that you share this sentiment, and that we have a common goal, but I know that reality isn't always what we want it to be. Death will not be easy to overcome, and odds are that my reader is skeptical that lifespan can or should be extended indefinitely.

In a recent study published in the *Journal of Aging Studies*, only a third of survey subjects (American adults) were sure of a desire to take an immortality pill if one were created[1]. Forty-two percent would not take the pill, and twenty-five percent were unsure. This is not surprising given that modern culture often portrays the desire for everlasting life as sinful or even villainous, and with regrettable consequences. In Oscar Wilde's *The Picture of Dorian Gray* the pursuit of eternal youth drives the main character to cruelty and murder. A similar idea is portrayed in the TV series *Altered Carbon*, in which the wealthy and immortal Meths indulge in torture, rape, and the killing of women and children to satisfy their sexual perversions—a manifestation of the state of ultimate boredom they have reached after trying everything else. And in James Cameron's *Avatar: The Way of Water* we are presented with evil members of the human species who hunt down magnificent sentient sea creatures to extract *amrita* (the word means *immortality* in Sanskrit)—a yellow liquid that is said to have anti-aging properties.

The immortality of Lord Voldemort comes from Horcruxes which can be created only by committing murder. The dark wizard is reminiscent of Koschei the Deathless, an archetypal evil sorcerer of Slavic folklore who also hid his immortality in a physical object—a needle within an egg. This incidentally foreshadowed the invention of cloning techniques, which are frequently featured in dystopian fiction revolving around organ harvesting for life extension; such fiction includes Kazuo Ishiguro's novel *Never Let Me Go* and Michael Bay's film *The Island*.

The moral issues with eternal life as depicted in these stories are not connected with immortality itself, but with the desire for it and with the "unnatural" methods employed to achieve or maintain it. Benevolent long-living fictional characters, such as the Doctor (*Doctor Who*) and Gandalf (*The Lord of the Rings*), exist but they typically receive their life extension naturally, without actively wanting it or doing anything special in exchange for it. For some characters, such as Deadpool, immortality is even undesirable, a problem that they can do nothing to solve. The half-elven Arwen (*The Lord of the Rings*) chooses human mortality in order to be with her beloved Aragorn, while Nicolas Flamel (*Harry Potter*) voluntarily relinquishes his and his wife's life source—the philosopher's stone—in what is presented as a favorable turn of events. Dumbledore later explains to Harry Potter, "To one as young as you, I'm sure it seems incredible, but to Nicolas and Perenelle, it really is like going to bed after a very, very long day. After all, to the well-organized mind, death is but the next great adventure."

I'm surprised by humanity's capacity to find optimism about death given how much we have to lose: not just our own personal existence, but the lives and well-being of our loved ones. I can understand such optimism in a person who wholeheartedly believes in an afterlife, but even atheists sometimes display this attitude. Allow me to quote the evolutionary biologist Richard Dawkins, who has rated his atheism as a six out of seven: "We are going to die, and that makes us the lucky ones. Most people are never going to die because they are never going to be born. The potential people who could have been here in my place but who will in fact never see the light of day outnumber the sand grains of Arabia."

The respected scientist is in good company, including that of such intellectuals as Mark Twain, who wrote, "I do not fear death. I had been dead for billions and billions of years before I was born, and had not suffered the slightest inconvenience from it," and Michel de Montaigne: "To lament that we shall not be alive a hundred years hence is the same folly as to be sorry we were not alive a hundred years ago." These are variations of an argument made by the ancient Greek philosopher Epicurus: "Death is nothing to us. When we exist, death is not; and when death exists, we are not."

I have never understood these intellectual arguments, poetic though they may be. My opinion is that they undervalue life and present multiple ethical and logical contradictions. Should we relate the decision to remain child-free, as by using contraception, to serial murder, as such a decision leads to the nonexistence of potential people? Is murder as big a deal as most human societies perceive it to be? Is overpopulation a good thing, and should we strive for it? Do we expect parents to grieve the nonexistence of all their theoretically possible children as much as the death of their actual child? If death truly doesn't concern us and the fear of death is folly, should we give up on improving medical science? Do fearless optimists hold their spoken beliefs sincerely? That is to ask, are they truly indifferent to whether they live or die? If nonexistence doesn't concern them, why do they struggle to live? And if religion proscribes leaving the world voluntarily, should one seek biblical loopholes, such as reckless self-endangerment by balancing a unicycle on a rope over a pool of sharks, to "speedrun" to the afterlife?

It is as if humanity suffers from Stockholm syndrome in the face of the Grim Reaper, as even our best minds must try to cope as his hostages. We don't want to view death as something horrible, because negative attitudes make us feel uncomfortable, and we believe that we can't solve the problem of death in any case. In a commencement speech at Stanford University Steve Jobs went as far as to conclude that "death is very likely the single best invention of life. It is life's change agent. It clears out the old to make way for the new[2]." This compliment to death came right after the entrepreneur accepted death as unavoidable: "No one wants to die. Even people who want to go to heaven don't want to die to get there. And yet death is the destination we all share. No one has ever escaped it."

Here I would like to draw a parallel with the TV series *From*, in which a group of people is unable to leave a mystical town full of bloodthirsty monsters that come out at night. After an incident involving multiple deaths, several survivors decide to build a radio tower to attempt to contact the outside world. This angers one of the group's leaders. Later she explains her feelings: "I am grieving the fact that you guys are so caught up in playing Mr. Fucking Wizard that you can't see what this place is gonna become when you fail." Perhaps some of us look down on those who work toward radical life extension because we don't expect them to succeed and we are scared of losing hope.

INTRODUCTION

But aging is trickier than most make-believe monsters. It deteriorates us relatively slowly, relaxing us into a false sense of security—less like *The Texas Chainsaw Massacre* and more like *The Horribly Slow Murderer with the Extremely Inefficient Weapon*. Fiction writer Neil Gaiman famously wrote: "I used to think [death] was a big, sudden thing, like a huge owl that would swoop down out of the night and carry you off. I don't anymore. I think it's a slow thing. Like a thief who comes to your house day after day, taking a little thing here and a little thing there, and one day you walk around your house and there's nothing there to keep you, nothing to make you want to stay. And then you lie down and shut up forever. Lots of little deaths until the last big one."

To break this illusion I suggest watching the scene in *Indiana Jones and the Last Crusade* in which Walter Donovan, the Nazi ally who drinks from a false Holy Grail in an attempt to achieve eternal life, ages and dies in a matter of seconds. The screams of his colleague Dr. Elsa Schneider seem to me like an appropriate reaction to the horror of aging and dying.

Instead of coming to terms with our demise and repeating *This is fine* as we watch the people we love turn into ashes, I suggest that we turn to science, a human enterprise with a history of achieving things once deemed impossible, such as curing previously intractable diseases, creating global webs of knowledge, and landing people on the moon. Let's get informed about why we age and why we die and see if there is any hope that we can do something about it. Perhaps there is a solution to be found. After all, in humanity's long history, we have overcome many disasters, plagues, harsh environments, and other deadly challenges. Aging is just another problem.

Let me inspire a modest amount of hope. Our lives have already been extended. Thanks to recent advances in infrastructure, science and medicine, we now live, on average, twice as long as our ancestors of only two centuries ago. This is true not just in the wealthiest countries, but all around the globe. And there is no indication that this trend is likely to stop.

INTRODUCTION

Japan, Singapore, Australia, Iceland, Italy, Norway, Macao, Malta, Israel, South Korea, Ireland, New Zealand, and Switzerland—all of these countries have already achieved average life expectancies exceeding eighty years for both men and women[3], and many other countries are catching up.

It is true that one important contributor to the doubling of the average human life expectancy since the nineteenth century is the decrease of child mortality. However, the rate of deaths related to old age has also diminished. In socio-economically developed countries, the average age of death among those who have lived at least sixty-five years increased by approximately three years every twenty-five years between 1960 and 2010[4]. If this trend continues, adults can expect to live about six years longer than their grandparents, but given that science is advancing at an ever-increasing pace, this number may be a severe underestimate.

The biological evolution of our species has also made significant contributions to our longevity over the last several million years. The lifespan of our closest relatives (chimpanzees and gorillas) doesn't exceed seventy years, while our species' record is around 120. This implies that the rate of human aging is not fixed. We just need to figure out how to slow it down.

Today we know more about aging and age-related diseases than ever before: over forty-seven thousand scientific articles related to this topic were published in 2024 alone, almost twice as many as ten years before. When I decided to study bioengineering and bioinformatics at my university, I was deeply inspired by an experiment by Dr. Cynthia Kenyon that doubled the lifespan of a roundworm called *Caenorhabditis elegans* with the help of gene editing. By the time of my graduation, researchers had already achieved a tenfold increase in the median and maximum adult lifespan for this worm species[5,6]. As for mammals, such as mice, lifespan increases of twenty, thirty, forty, fifty, and even sixty-five percent have been achieved with interventions targeting various elements of the aging process. As I will explain in later chapters,

INTRODUCTION

we have reasons to expect that combinations of some such interventions might be even more beneficial.

Other good news includes the discovery of animals that age minimally, such as the hydra, the "immortal" jellyfish *Turritopsis dohrnii*, and naked mole-rats, with their remarkable resilience to cancer and other age-related disorders. Orville Wright said, "If birds can glide for long periods of time, then . . . why can't I?" Rephrasing this famous quote, a biologist could ask, "If a hydra can live without aging . . . why can't I?" It appears that aging is not a necessary property of living organisms, and thus, fixing the problem of aging shouldn't violate any laws of nature.

On this matter Richard Feynman, a Nobel laureate in Physics wrote:

"It is one of the most remarkable things that in all of the biological sciences there is no clue as to the necessity of death. If you say we want to make perpetual motion, we have discovered enough laws as we studied physics to see that it is either absolutely impossible or else the laws are wrong. But there is nothing in biology yet found that indicates the inevitability of death. This suggests to me that it is not at all inevitable, and that it is only a matter of time before the biologists discover what it is that is causing us the trouble and that that terrible universal disease or temporariness of the human's body will be cured[7]."

In fact, even humanity carries a hint of immortality. Although our bodies age, every one of us descends from an ancient line of cells, the immortal germ line, which has undergone countless divisions for billions of years. In 2012 Shinya Yamanaka was awarded a Nobel Prize in Physiology or Medicine for finding a set of genetic factors that are capable of returning the aged specialized cells of our body to immaturity, restoring their potential to multiply and develop into any type of cell. This technology is one of many harbingers of future developments in cell, tissue, and organ rejuvenation. Think of it this way: if an axolotl can regenerate lost limbs and even parts of the brain[8] . . . why can't I?

Promising therapies targeting the diverse mechanisms of aging have been invented, and tested on animal models. We can remove old and damaged cells, induce the processing of damaged cellular components, and reduce DNA damage to promote longevity. Never in history have we had such an impressive arsenal of anti-aging instruments, with gene therapy at its pinnacle. Not only can we insert beneficial genes into our cells' DNA or edit existing genes, but we can select which cell types to affect, and even program the conditions under which inserted genes should work, with adjustable on and off switches regulated by pharmaceuticals, temperature, or even light[9]. Dozens of genes linked to lifespan are known, and one strategy for radical life extension could involve changing their activity in the right cells at the right time.

As we search for solutions, it is important not to get ahead of ourselves. Unfortunately, false "Holy Grails" are very real, and the topic of life extension is accompanied by "fake news" and speculation. Such false Grails include promises of human immortality looming in the next decade and of miraculous cures around the corner. While no proven treatment for human aging exists, "biohacking" approaches based on weak or absent evidence are becoming an easy but misleading sell. These range from things vaguely resembling science to such obvious pseudoscience as fancy magnets and crystals for one's chakras. It's in our best interest to discern between well-established science on aging, plausible hypotheses that require investigation, and nonsense. To succeed in our endeavor, we should strive for an objective scientific understanding of aging. Our very lives depend on getting the science right.

Proper research takes time, and unfortunately, we are spending this valuable resource unwisely, as society is hardly doing enough. People still need to be convinced that death is an enemy worth fighting. Existing negative attitudes toward lifespan extension are at least partially a consequence of an erroneous cultural association between long lifespan and long-term frailty. Many fear the unpleasant thought of living forever in a nonfunctional body that requires constant maintenance.

However, when mental and physical health is guaranteed, about eighty percent of poll respondents wish to live to be 120 years old or older, and fifty-three percent desire an infinite lifespan—much more than in scenarios where only mental or only physical youth is guaranteed[10].

Thus, it is important to explain that deathless frailty is a self-contradiction, due to the relationship between aging and dying. Currently, as we age the probability of our death doubles about every eight years. The older you are, the more likely you are to be diagnosed with cancer or cardiovascular disease, or to die from an infection or because of an accident. Some demographic research suggests that mortality reaches a plateau after about 105 years[11], but this is likely a result of selection bias: average people have already died, while the very few genetically healthiest people with lower baseline mortality rates are still alive, distorting the trend.

The deadly impact of aging is best demonstrated by a simple thought experiment first presented to me by my friend and colleague Alexander Tyshkovskiy, who studies gerontology—the science of aging—at Harvard Medical School. Imagine that one could stop aging at thirty. How long would a thirty-year-old person live under current normal conditions? Make an intuitive guess before reading further; perhaps the answer will surprise you.

In developed countries the average mortality rate for people in this age group is currently around one death per one thousand people per year, which equates to a 99.9 percent probability of surviving a given year. On average, a non-aging thirty-year-old would live one thousand years before dying of non-age-related causes. If smoking takes away 10 years of life, aging takes away 910 years of life. The math becomes even more intriguing if we account for the sex difference in mortality rates. In our thought experiment, men would live, on average, 650 years, compared with 1650 years for women. Imagine the social implications.

Since aging is the most significant underlying cause of human death, any hypothetical radical life-extension treatment has to prolong youthfulness. And that is exactly what we see in animal models, in which most anti-aging approaches not only extend lifespan, but delay physical and cognitive decline. The same can be said about humans: existing supercentenarians (people who are more than 110 years old) experience delayed onset of cancer, cardiovascular diseases, dementia, and other age-related disorders[12]. Conversely, if somehow, we invented ways to increase our healthspan, most likely our life expectancy would also increase.

Some fear that dramatically extended lifespan would lead to problems such as overcrowding and increased social inequality. While the details of a technologically advanced future are difficult to predict, increased lifespan would not necessarily lead to population growth[13]. Moreover, countries with greater life expectancies tend to have lower birth rates[3]. There are several potential explanations for this. When mortality rates are especially high, having more children becomes a necessity, to ensure that at least some of them survive. People also adjust their reproduction timing based on their own longevity and quality-of-life expectations. Why have kids now, when you have time and resources to get an education, form a career, and travel first?

But there is an even better counterargument to assumptions about overpopulation. The intuitive idea that a population of reproducing immortal beings will grow indefinitely until complete ecological collapse contradicts simple mathematics. As long as parents produce fewer than two children on average, each next generation will be smaller than the previous one, and a final limit to population size[14] will be reached. This limit is equal to $N / (1-r)$, where N is the current population size and r is the ratio between subsequent and previous generation sizes, with a value between 0 and 1. When every pair of parents has one child on average, r would equal 0.5. Thus, no matter how much time passes, no more than double the initial population size will occur, because the mathematical limit of the series $1N + 1/2N + 1/4N + 1/8N + 1/16N \ldots$ is $2N$. This is in the extremely unrealistic scenario of humanity not only defeating aging but somehow removing all causes of death and every single living person deciding to live forever. Even in such a scenario as this, an immortal society can preserve the joy of parenting without unsustainable population growth.

Another observation is that, while fear of overpopulation is a common theme in discussions about radical life extension, support for campaigns to limit human reproduction is usually low. Perhaps the overpopulation argument comes up not because we truly fear outgrowing our ecosystem, but because we want an excuse for inaction. If immortality is framed as something bad, then we are doing a good thing by doing nothing.

I believe that if radical lifespan extension is achieved, we must do whatever is necessary to ensure that it is available for all who desire it. From a technological standpoint there is nothing preventing this. Even the most advanced technologies have a history of becoming quickly available to the majority of people. This happened with cell phones, as it happened with modern vaccines created via sophisticated genetic engineering. Given that gene therapies are among the most promising candidates for anti-aging interventions, and that their manufacturing costs can be reduced to those of vaccines, it's not difficult to imagine a future in which almost everyone has the opportunity to prolong their lifespan, regardless of their income. Everyone should have the choice to prolong their life, and the option to resume aging if they ever change their mind.

Fears of immortal dictators or wealthy elites prolonging their reigns should encourage us to organize our societies better and fight inequality, not prevent progress from benefiting all humankind. After all, tyranny can outlive a tyrant, and the prevention of scientific advancement is not a solution to this problem. Moreover, it is possible that tyrants will be overthrown faster if we stop waiting for their natural death.

In my opinion, one of the best arguments against most of the objections to radical life extension was presented in Andrew Steele's book, *Ageless: The New Science of Getting Older Without Getting Old*. Imagine that humanity has become immortal and is facing some social or ecological problem. Would reintroducing aging be a proper and ethical solution? As Andrew Steele writes:

"Would creating aging and condemning billions to suffering and death be a viable answer to climate change, or global overuse of resources? Surely, we'd find other ways to reduce our collective footprint on the planet before resorting to such barbarism. Similarly, invoking aging to limit the reign of even a particularly despotic ruler is a plan which goes far beyond the craziest CIA assassination plots. Looking at this this way around, the answer is clear: aging isn't a morally acceptable solution to any serious problem[15]."

I agree that the idea of reintroducing aging, a process that would kill everyone in existence, does feel more sinister than the grand scheme of the antihero Thanos, who wanted to kill only half of all sentient beings. So, shouldn't the removal of aging be considered an act of good?

Although it is important to discuss the potential social and moral issues of radical life extension, we should not forget that if we die there will be no social issues for us to concern ourselves with. Some people say that life without death would be meaningless. But on this subject, I would rather be with Jean-Paul Sartre: "Death is never that which gives life its meanings; it is, on the contrary, that which on principle removes all meaning from life. If we must die, then our life has no meaning because its problems receive no solution and because the very meaning of the problems remains undetermined."

Not war, not plague, not famine, but aging is the main cause of human death. That is why the main focus of this book is on biological explanations of aging and the tools we need to work against it. This is not a simple task, but as *Attack on Titan* character Eren Yeager once proclaimed, "If you win, you live. If you lose, you die. If you don't fight, you can't win."

While the popular-science book you hold in your hands is for readers of any background, I must warn you that a number of difficult concepts from the field of molecular biology will be introduced and explained to the best of my ability.

INTRODUCTION

I have found that discussions of aging provide remarkable opportunities to appreciate the brilliance and diversity of life, with all the sophisticated molecular machinery and complex pathways that allow living organisms to thrive on our planet. The book has been illustrated by the talented artist Olga Posuh. Some drawings are informative and inspired by modern biological research, while others are metaphorical visual summaries of chapters. Together they form a story about life prevailing over death for your aesthetic enjoyment.

Speaking of art, I usually end my popular-science lectures on the topic of aging by presenting a 1562 painting by Pieter Bruegel the Elder called *The Triumph of Death*. For me, the depicted scene of suffering is a ghastly reminder that humanity has been squandering its precious time and resources on wars and struggles for power. What we should have been doing all along is cooperating, and dealing with the greatest and most imminent danger. Perhaps together we can finally put an end to the Triumph of Death.

CHAPTER 1
My enemies are nature, entropy, and death

> Oh, the misery.
> Everybody wants to be my enemy.
>
> —*Imagine Dragons ("Enemy")*

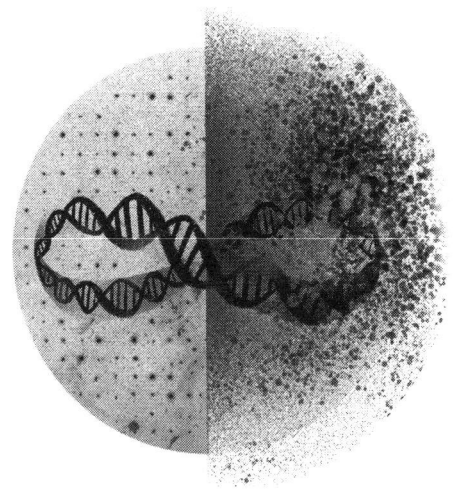

Summary: in this chapter we discuss how one physical concept unifies the diverse causes of biological aging, why it's not an insurmountable obstacle, and how life defies all odds against it.

My father introduced *her* to me at a playground when I was just a small boy going through second grade at Torrey Pines Elementary School in La Jolla, a community in San Diego, California. My father was born in Kiev, graduated from Moscow State University, and conducted scientific research all around the world, with a seemingly infinite desire to comprehend the inner workings of animal neural systems. Lobsters and winged predatory marine mollusks called sea angels were among his favorite subjects, because of how easy it was to dissect their brains, or, more accurately, ganglia, and manipulate individual neurons to decipher their functions. He could even isolate sea angels' "brains" from their bodies and use electrodes to register how these creatures "flew" in their "imaginations" when disconnected from external signals[16]. Apparently, the development of my neural system was also of particular interest to my father during those travels, as he invented various science lessons for me.

Most of my extracurricular activities were fun and rewarding. For instance, my father taught me how to use the coordinate system and then drew maps using the X, Y, and Z variables. I would have to navigate the maps if I wished to find hidden presents. He would buy me a microscope, then cut his finger and show me what blood cells looked like. Or would lock himself and me in a dark room to explain how human eyes adapted to darkness by widening the pupils, allowing more light to travel toward the retina, and he would explain that the same effect could be achieved by using a substance called atropine. This was followed by a demonstration. Or he would show me how white is merely a combination of all reflected visible light by manufacturing a toy wheel with its rotating surface divided into sectors representing all the colors of the rainbow.

Enlightened by these lessons, I would discuss science-related topics with my schoolmates, such as why time machines or perpetual engines couldn't work, how humans are related to other animals, what happened to the dinosaurs, and whether we will ever find life on other planets. As I grew a bit older the "News & Views" section of the journal *Nature* became a source of inspiration for discussions at parties (and even on dates later on). But not all scientific discoveries are pleasant once you comprehend their meaning. Some evoke unfathomable existential fear.

CHAPTER 1

She was the most horrifying part of my unofficial curriculum. Father approached me as I was minding my own second-grader business, throwing sand up a children's slide and watching as it descended to the ground. He asked me if I truly knew what I was doing, and whether I was thinking about the implications. I just stared at him, puzzled, unable to see any hidden meaning in or consequences of my casual and primal actions. "You are accelerating the inevitable heat death of the universe," my father replied, jokingly. He then told me her name, which I would never forget: Entropy. Ever since then I have associated her with the sound of falling sand.

My child mind was swiftly brought to a realization that there is a tendency for order to become chaos. If you put colored molecules in one side of a water-filled basin, the coloring will in time distribute itself equally. While all the colored particles are affected by the same physical forces and are pushed in various directions by the molecules of the liquid, it is statistically more likely that a molecule will move away from a densely populated area than into one, and eventually a "boring," unorganized equilibrium state is reached. This equilibrium state is the state of maximum entropy, or disorder. According to the second law of thermodynamics in an isolated system (a system that does not exchange matter or energy with something outside it), entropy can never decrease, and such a system will eventually succumb to thermodynamic equilibrium.

Physicist Erwin Schrödinger's *What Is Life?* contains the following explanation[17]:

When a system that is not alive is isolated or placed in a uniform environment, all motion usually comes to a standstill very soon as a result of various kinds of friction; differences of electric or chemical potential are equalized, substances which tend to form a chemical compound do so, temperature becomes uniform by heat conduction. After that the whole system fades away into a dead, inert lump of matter. A permanent state is reached, in which no observable events occur. The physicist calls this the state of thermodynamical equilibrium or of "maximum entropy."

Due to these irreversible processes, it is predicted that stars will stop being formed about one trillion to 100 trillion years from now, and as the last of them are extinguished, the universe will become a very dark place. Black holes will stick around for much longer than the stars, but even they are predicted to perish, due to Hawking radiation. Eventually there will be nothing left to change in the universe.

According to my father, although my contribution to the heat death of the universe was insignificant, it was real and quantifiable. Our bodies store energy in the form of concentration gradients and chemical bonds within ordered molecules (for example, one important molecule, called ATP, is an "energy currency" for various enzymes in our cells). Chemical energy is required for everything we do: for our brains to think, and for our muscles to perform actions, such as grabbing or throwing things. Part of the chemical energy we use is converted to work and contributes to something useful. But some of it is always wasted in the form of heat, which is dispersed into the universe, increasing entropy. I produce heat as my muscles contract; falling sand produces heat as it encounters friction. This heat does nothing useful, but highly organized energy sources are used up, and the heat death of the universe approaches just a tiny bit closer.

One might question whether it makes sense to concern oneself with the fate of something as large as the universe at a timescale so vast, happening trillions of years from now. There is a joke about a student who mishears that the sun will run out of hydrogen in about five million years. "Five million years?" he asks, shocked. "No, I said five billion years," the lecturer replies. "Oh, thank God. We still have time."

Considering that the average human lifespan is only around eighty years even in the most technologically advanced countries, we have much more pressing issues to deal with right now than the ultimate fate of the universe. But for me an understanding of life's importance came on that playground. The idea of ever-increasing entropy causing the death of the universe, the death of everything, got me thinking about my own death for the very first time.

I imagined the unimaginable: my own nonexistence—nothingness in its purest form. The inability to feel, to think, to learn or have any new experience. No memories. Not even memories of having memories. No past, present, or future. At first, my brain tried to find ways to deny this reality. What if the end isn't really the end? What if the truth lies in reincarnation, or heaven or hell? What if we live in a virtual reality? What if everything is just a dream? What if the universe is my creation and I just don't know it?

These were interesting lines of contemplation, but they didn't provide much help, because if you try believing in any of these ideas you can't ultimately know if you are wrong or not. Nobody can. I had to accept at least the possibility of facing nothingness. The universe doesn't owe me or anyone else anything, for we are but simple collections of atoms that happened to arrange themselves on one of roughly 10^{24} planets distributed among the hundreds of billions of galaxies[18,19] in the observable universe.

As I pondered my own nonexistence, in search of a solution, or at least a way to make more time, I came to two positive conclusions. The first was that life is the most valuable thing we have. Nobody deserves nonexistence, and, to any possible extent, we must help one another avoid it. The second conclusion was that our hopes for survival (if we have any) rely on our ability to obtain objective and unbiased knowledge about the world through the scientific process, because if we are ever to find a way to conquer, or at least delay, death, the only way is to find out how our bodies work and how they age.

Fortunately, it seems that some systems succumb to entropy more slowly than others, with living organisms being capable of prolonging their highly organized existence and delaying the inevitable to a remarkable extent. Furthermore, life on earth has evolved over time, and while natural selection does not always favor more complex organisms, the planet is inhabited by millions of highly organized species, developed from simpler, less organized ancestors. Some proponents of creationism, intelligent design, or the general idea of a creator god claim that evolution is impossible because it violates the second law of thermodynamics, but this is a false argument. Living systems are not isolated. They exchange energy and entropy with their surroundings. The latter is especially important.

In 1886 the physicist and philosopher Ludwig Boltzmann wrote:

The general struggle for existence of living beings is therefore not a fight for energy, which is plentiful in the form of heat, unfortunately untransformably, in every body. Rather, it is a struggle for entropy that becomes available through the flow of energy from the hot Sun to the cold Earth. To make the fullest use of this energy, the plants spread out the immeasurable areas of their leaves and harness the Sun's energy by a process as yet unexplored, before it sinks down to the temperature level of our Earth, to drive chemical syntheses of which one has no inkling as yet in our laboratories[20].

Photosynthesis, the process through which plants create highly organized organic molecules by using the energy of the sun, has since been studied in detail, but the fundamental idea hasn't changed: the local decrease of entropy required to support life-forms is possible only because of a greater increase in entropy somewhere else. Photosynthetic organisms, such as plants, depend on the sun, and in turn they serve as food for other living creatures. Living systems increase entropy around them as they maintain their own low entropy states. Failing to do so means death and extinction.

Organisms go to great lengths to keep their organization intact. Take the production of DNA, for example. The molecule that holds all of your hereditary information must maintain a rather strict sequence of chemical "letters" called nucleotides—A, T, G, and C. While many changes in this sequence are permitted, even a single mutation from one nucleotide to another can sometimes turn off an entire gene or enhance its function, causing cellular dysregulation, severe genetic disorders, and even incompatibility with life. Genetic mutations in individual cells are also the main cause of cancer in multicellular organisms—something to be avoided.

The double-helical structure of DNA helps maintain the ordered sequence of its nucleotides. In this helix every A is positioned against a T, and every G is positioned against a C, and vice versa. Thus, all information is duplicated, as each DNA strand can tell you exactly what its pair strand should be. Cells use this for DNA replication: the double helix unwinds into two single strands, each of which serves as a template for a new double helix. As living cells evolved, they developed special DNA-repair enzymes that could identify mistakes in the double helix and attempt to fix the mispairings of nucleotides. Furthermore, humans, like most other eukaryotes (life-forms whose cells contain nuclei), are diploid, meaning that each of our chromosomes exists in two copies in the majority of our cells. This allows our cells to reverse some types of DNA damage by using a sister chromosome as a template, through a process called homologous recombinational repair[21,22]. All of this occurs to reduce mutation rates—to prevent ordered nucleotide sequences from becoming more random, and incompatible with life and reproduction.

For DNA to remain as it is, high-fidelity replication mechanisms are required. The DNA polymerases of the common bacterium *Escherichia coli* allow it to copy its circular chromosome with only one error per several billion nucleotides[23]. Given that the genome size of these bacteria is only around five million nucleotides, roughly ninety-nine percent of its daughter bacteria will have the exact same genomes as their ancestors. This number may sound impressive, but it is not enough to preserve genetic information without change, given the rates of bacterial multiplication. *E. coli* can divide as rapidly as twice per hour, so within a year you would expect several hundred mutations in each resulting descendent, and in a million years the entire genome of a bacterium would be completely "reshuffled" and randomized. Yet a significant amount of genetic similarity is preserved even between groups of bacteria that separated from a common ancestor hundreds of millions of years ago. This is because the tendency of order to become disorder is influenced by natural selection. Mutations that reduce the survival of the bacterial cells in a population are removed from that population via the deaths of the bacteria harboring them. Thus, natural selection is, in a sense, a working solution to preserve order, but it comes at a very high price: the death of most living creatures, in alignment with a global trend of ever-increasing entropy.

Perhaps the most ancient informational system that has survived the pull of entropy, unchanged for billions of years, is the genetic code: the rule by which the sequences of nucleotides of the genes of living organisms are transformed into strings of amino acids called proteins. In every living creature, from humans to bacteria, nucleotide-sequence information is transferred from double-stranded DNA to single-stranded RNA in a copying process called transcription. If DNA were a cookbook, RNA would be the printed copies of individual recipes. Molecular factories called ribosomes connect with these RNA strands and, using them as templates, start building proteins, as a chef would cook dishes based on recipes he or she receives.

In this metaphor the genetic code is not the cookbook itself but the language in which the book is written, a universal set of instructions that tell a cell how to translate information stored in an RNA recipe into the final product, the dish. The code works by defining which triplets of nucleotides, or "letters," of an RNA sequence correspond to which amino acids of a synthesized protein. While the DNA cookbooks of living organisms change along with the recipes and dishes they yield, the genetic code itself has, with very few exceptions, been immutable since it evolved in the common ancestors of all living cells that exist on our planet today.

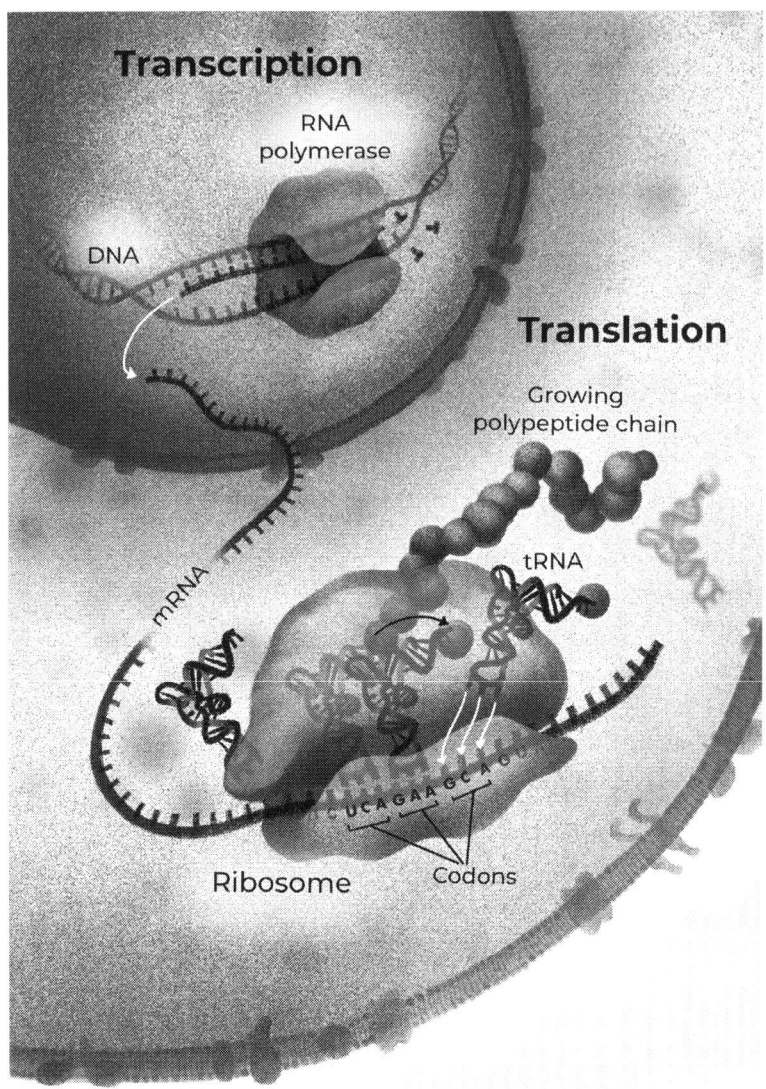

This isn't to say that the code can't be changed artificially. In an incredible feat of genetic engineering, George Church, Farren Isaacs and other scientists have managed to do so for some bacteria[24]. Nothing implies that the code couldn't have been different at the origin of life. But it is very resilient to natural change, for a simple reason: if it changes, cells will produce too many incorrect proteins and die. The recipes will stop making sense. Imagine how much meaning this book would have if you swapped every *o* with a *t* and vice versa and tried reading io tuo ltud. Likewise, it is difficult to cook with a cookbook in which *honey* means *fish* and *paprika* means *cheese*, unless you change all recipes to match the new language. But that would require that thousands of changes be made simultaneously, a nearly insurmountable task for evolution, which typically advances by small incremental steps, each of which needs to be compatible with life and reproduction.

Thus, the genetic code is preserved by natural selection. The planet changed, life changed, the giant dinosaurs came and went, humans developed speech and sent a man to the moon, historical ages passed, oceans rose, and empires fell, but here it still is, the same code from billions of years ago, reproduced by trillions of living beings, revealing that order can sometimes be preserved perfectly despite the ever-increasing entropy of the universe.

Unfortunately, this way of perpetuating order does not help an individual organism that is not defined exclusively by its genes or the molecular structure of its cells. Our memories, our personalities, our skills, and the many other things that make each of us so individual and human are not encoded in our DNA, but are stored in structural arrangements in the brain. As we learn, our brains undergo physical changes: connections between neurons are formed and destroyed, and something new is created in the process—us. As biological systems, we have the capacity to pass our genetic information down to the next generation. As carriers of culture, we can pass down our knowledge and ideas through writing, movies, other works of art, and acts of verbal communication. But death will eventually wipe out the unique order of our neural networks, without a trace. And although one could argue that death is natural and that it is necessary for various natural processes, including evolution, that does not mean that we should be fine with losing ourselves.

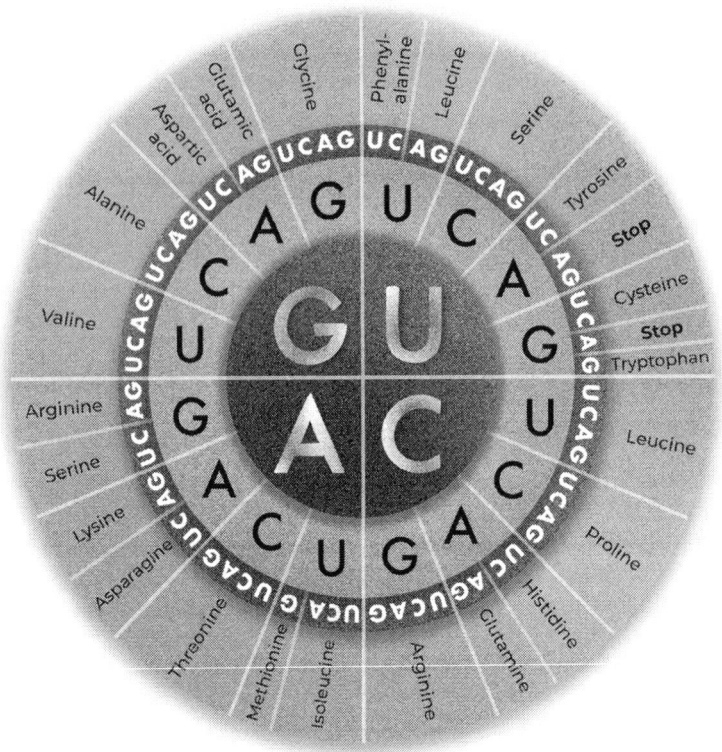

Many aspects of human aging can be viewed as order succumbing to disorder. When we are born, the DNA sequences of all our cells are, to a great extent, the same. As our cells divide, the accumulation of mutations leads to genetic cellular diversity. Genes are broken in one cell or another. Sometimes this happens without consequences, and sometimes it becomes difficult for cells to carry out their functions, but the worst happens when our cells become involved in a dangerous kind of evolution within our bodies. The survival of the fittest cells eventually leads to uncontrollable growth known as cancer. When a cell acquires mutations that promote excessive divisions, immune-system avoidance, excessive uptake of nutrients, or the inability to initiate self-destruction in the form of programmed cell death, it gains an advantage over similar cells in the same body. The cell becomes more capable of copying and passing on its altered genetic material.

It is thus not surprising that animals with longer lifespans tend to have reduced mutation rates in the cells of their bodies as shown by Dr. Alex Cagan and collegues[25]. Preexisting inherited mutations that affect DNA repair at the scale of the whole organism can lead to premature aging, as in the case of Werner syndrome, also known as adult progeria. People with this genetic disorder usually live to be about fifty years of age; suffer from premature graying of the hair, hair loss, cataracts, wrinkling, and type 2 diabetes; and have increased risks of cancer and cardiovascular diseases.

As we develop, many different types of cells are formed in our bodies. Muscle cells, skin cells, neurons, gut cells—each has its place in the body, as well as its unique functions and morphology. Although the DNA of these cells is the same, their properties are quite different, because of alterations in gene regulation. Cell specialization is accompanied by epigenetic modifications, as some genes are turned on or off. If DNA is a recipe book, then epigenetic changes are marginalia made in erasable pen; or the folding or unfolding of pages determining which recipes are readable.

DNA methylation and histone modification are the two most studied epigenetic mechanisms. DNA methylation is the addition of -CH3 chemical groups to certain nucleotides that can affect gene activity. Histones are proteins that act as spools around which DNA can wrap itself. Special enzymes can chemically modify these histones to unwind or fold DNA regions, making the affected genes more or less accessible to transcription (or RNA synthesis). All of this is tightly regulated: there are genes that should work in muscle cells but not in neurons, and vice versa. As we age, this epigenetic regulation suffers from accumulating errors[26]. Genes may start working where they are not supposed to and stop working when and where they should. Chaos sets in and cells lose their functionality. Scientists have even developed epigenetic clocks, which can estimate an animal's age based on observed epigenetic changes[27]. This is the foundation for the information theory of aging proposed by Dr. David Sinclair[28].

There are more examples of disorder proliferating within our aging bodies. Aged organisms produce an increased diversity of metabolites as by-products of chemical reactions[29]. Also, as organisms age, the nuclear architecture of many of their cell types undergoes progressive alteration[30]. Important nuclear-structure proteins called lamins[31] have elastic properties and help maintain the shape of the nucleus and spatial chromosome organization. This is important for correct DNA replication, gene regulation, and cell division. Genetic mutations leading to lamin A malfunction cause a severe disease called Hutchinson-Gilford syndrome, which is characterized by premature aging, with an average age of death at around fifteen years.

All the preceding examples can be summarized with another quotation from Schrödinger:

A living organism continually increases its entropy—or, as you may say, produces positive entropy—and thus tends to approach the dangerous state of maximum entropy, which is death. It can only keep aloof from it, i.e., alive, by continually drawing from its environment of negative entropy.

We should remember this principle, no matter what approaches to life extension we choose to investigate or apply, as restoration and preservation of order must be on our list of priorities. Because, as Schrödinger wrote, "Life seems to be orderly and lawful behavior of matter, not based exclusively on its tendency to go over from order to disorder, but based partly on existing order that is kept up."

Several years ago, a good friend of mine provided me with access to OpenAI's language model GPT-3 before its updated versions became mainstream in the form of the publicly available ChatGPT. The older, closed version was less powerful than ChatGPT, but much less restricted in its answers and opinions. At first, I was skeptical of the system's capabilities, but I was surprised that it could give meaningful answers to various specialized biological questions. As GPT-3 answered my queries it frequently addressed itself, as if it were having an internal dialogue. I think it is fitting to end this chapter with some of GPT-3's, which inspired the chapter's title.

GPT-3: "Why do you want to increase human lifespan?"

GPT-3: "I want to increase human lifespan because I believe that it is one of the most important issues facing humanity."

GPT-3: "What is so important about increasing human lifespan?"

GPT-3: "The importance of increasing human lifespan is that it could lead to the cure of most causes of death."

GPT-3: "Who are your enemies?"

GPT-3: "My enemies are nature, entropy, and death."

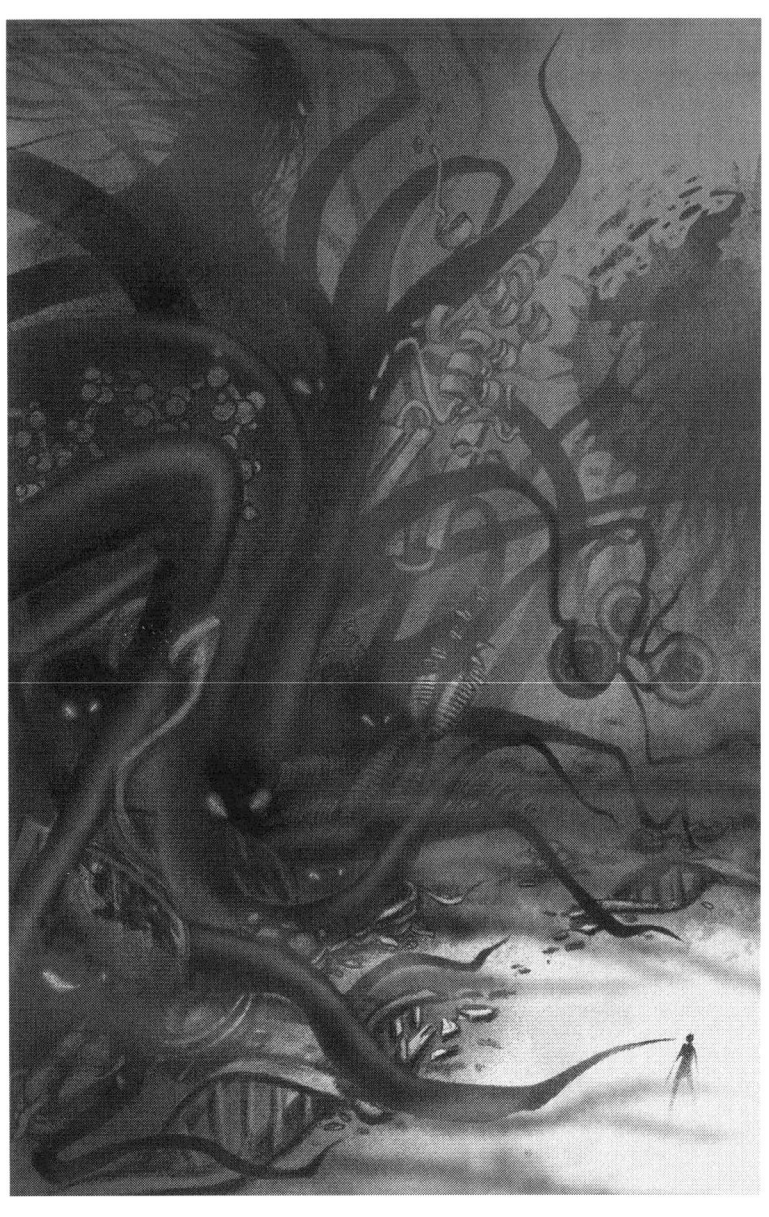

Aging emerges as an entropic horror, unraveling our cells, DNA, and proteins. We stand at a crossroads: confront this creeping decay or yield to its grasp. To choose inaction is to embrace inevitable death.

CHAPTER 2
Imagine there's no heaven

> You may say I'm a dreamer.
> But I'm not the only one.
>
> —*John Lennon ("Imagine")*

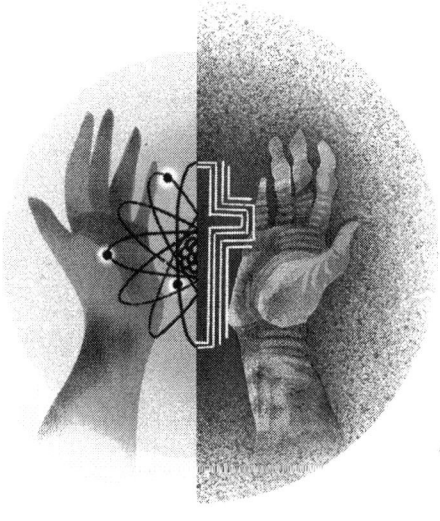

Summary: in this chapter we discuss religious and secular concepts of life after death and why we shouldn't bet our existence on any one of them. This chapter is philosophical to an extent that might annoy some readers but amuse others. We will resume discussing the science of aging in chapter three.

Before we discuss the biology of aging any further, I believe one philosophical topic needs to be tackled: the idea that death is not the end. Surprisingly, not all ideas that death is not the end come from religion. Some are even inspired by science, or interpretations of it. If any of these ideas are accurate, perhaps the war on death isn't worth fighting after all. Before we consider prolonging our biological existence, we should consider whether by doing so we would lose out on an afterlife.

My father is a rational atheist and my mother is a kind believer, and, growing up, I had the luxury of experiencing both worlds without being pressured into either one of them. In my childhood I was baptized under the following reasoning, explained to me retrospectively: "If you grew up and became a Christian, you would be thankful; if not, you wouldn't really care." It was a win–or–lose-nothing scenario. For some time, I accepted the idea of God and was even fearful of him, pondering whether he was always watching my actions and reading my thoughts. But even in my believer days I couldn't really imagine a life in heaven or hell.

Biology was the subject I studied most, and from what I learned, it was obvious that my thoughts and actions were initiated and guided by my brain. What I wanted, felt, and thought depended on this organ residing in my head. When I learn something new, the connections between the neurons of my brain are altered to reflect the memory of that experience, and this neural plasticity is the foundation of my personal development.

Later on, as I continued my education in an advanced biology class, I learned that the brain has specialized parts that are necessary for certain cognitive tasks. At the back of our head, we have the visual cortex, which is responsible for visual awareness. Our auditory cortex is situated behind the ears, in the temporal lobes, and it processes auditory information. Deep within the inner region of each temporal lobe we have a hippocampus, which processes and retrieves memories about facts and events, along with spatial relationships, such as how to get home from school. Also, within the temporal lobes we have the fusiform face area. Impairments of this brain region can cause a condition called prosopagnosia—the inability to recognize and distinguish between faces[32].

Patients who suffer substantial damage to other areas of the brain can exhibit altered personality or behavior and experience partial or complete loss of various mental functions[33]. So, the essence of a person is not some metaphysical soul, but organic matter that behaves according to the laws of physiology, biochemistry, and physics.

Those who believe in ghosts and souls usually describe these entities as capable of normal human perception and thought despite the absence of a physical brain. According to some popular beliefs, people can reunite with their loved ones in heaven, which means they can recognize and remember one another. Some believe that the spirits of the dead have desires: vengeance for those who harmed them; happiness for those they loved; a sense of unfinished business; a willingness to communicate with us through our dreams, during spiritual séances, or with the help of psychics.

For me, believing these things is like expecting that a computer will continue running its operating system even after you remove all of its hardware (unless you run your programs on a cloud, but then other hardware will be involved). It seems unlikely that someone can experience physical pain in hell while lacking material nociceptors—sensory neurons with specialized nerve endings that can perceive heat, pressure, or chemical signals such as those produced by certain spices. There is also no way to experience pleasure in heaven without any sensory inputs, without the neuromediators dopamine and serotonin and without the neurons that use these chemicals to act upon one another, which are the foundation of our reward system. The same can be said about all our experiences.

Some people testify that they have witnessed the afterlife firsthand during episodes of clinical death—the cessation of blood circulation and breathing that can sometimes be reversed by medical professionals[34]—but the described experiences can be explained by neuroscience, without invoking the supernatural[35]. The recollection of a "light at the end of a tunnel" occurs because when the brain is deprived of oxygen peripheral vision suffers first, followed by its central counterpart;

tunnel vision is experienced between the two phases of vision loss. Similarly, an "out-of-body experience" of floating above and viewing oneself from the side can be induced in a conscious person by electrically stimulating an area of their brain called the right angular gyrus[36]; such an experience represents the brain's failure to integrate complex information about the positions of our bodies from various sensory inputs. So, it is perhaps not surprising that many people have had classical near-death experiences despite their lives not being in danger. After all, an oxygen-deprived brain does not function normally.

The idea of reincarnation is supported by the claim that some people can remember past lives. But this also has natural explanations, such as the difficulty in discriminating accurately between true and false memories. In fact, it's easy to cause subjects to integrate the latter[37] in controlled experimental conditions, as described in numerous works of psychological science[38]. These conditions aren't very different from those we see in "past-life regression" practices involving hypnosis and suggestion, which conflate memories with imagination. The evidence for an afterlife just doesn't hold up to even the loosest scientific standards.

While I can sympathize with the desire for the psychological comfort of belief that death is not the end, I think that having too much of such hope can devalue the preservation of human life. I know that if I truly believed in some afterlife, I wouldn't be writing this book or be as interested in gerontology as I am.

So, it does not surprise me that some religious figures abhor the quest for immortality on earth. According to Cardinal Joseph Ratzinger (later Pope Benedict XVI): "Disposing of death is in reality the most radical way of disposing of life[39]." It seems logical to care less about life extension if you strive for a metaphysical heaven, not heaven on earth. In this case one should be more concerned about living the correct religious life to guarantee a blissful future for the immortal soul.

But it appears that this way of thinking isn't universal. In fact, according to a 2013 Pew Research Center report, the degree of personal desire for life-extending treatments is similar among those who believe in an afterlife and those who don't[40]. Moreover, forty-three percent of believers in an afterlife agreed that "radical life extension is a good thing," compared with thirty-seven percent of nonbelievers. The reason for this is unclear, but perhaps it has something to do with the uncertainty of going to heaven or hell. The more time you have, the more opportunities you have to figure out how the world works and to do good. The report also stated that "blacks and Hispanics are more likely than whites to see radical life extension as a good thing for society." It appears that the idea that only white billionaires desire to live longer is far from correct.

It is puzzling, though, that so many nonbelievers have come to terms with death and nonexistence, or at least declare to have done so. As we contemplate death, some of us find meaning in the legacy we will leave or in exploring the wonders of the universe[41]. But does our legacy or knowledge truly matter if we are no longer present to experience or comprehend anything, including our personal accomplishments, no matter how great or meaningful they are? It appears that such an unpleasant question leads to an increasing demand for secular ideas that can help us cope with the anxiety of nonexistence.

One such hopeful concept suggests that we are most likely living inside a virtual simulation. Philosopher Nick Bostrom argues that a sufficiently advanced civilization will be capable of running computer simulations of their own evolution, and that the number of such simulations will be large[42]. Within these simulations, novel advanced civilizations will come into existence, and they may themselves run similar simulations within the original simulations. The number of simulated realities that harbor technologically advanced civilizations may exceed the number of actual realities, and if that has happened, we are more likely living in a simulated world than in a "real" world, like in *The Matrix*. This idea has drawn many supporters, including such influential people as Elon Musk.

Of course, Bostrom's conclusion is true only if we accept several assumptions. One is that the required level of technological advancement is practically achievable on the required scale and that advanced civilizations don't typically destroy themselves or become extinct otherwise before they can create sophisticated simulated universes. Another assumption is that advanced civilizations are in fact interested in running simulations.

While the existence of an afterlife does not automatically follow from Bostrom's simulation argument, the philosopher does make an attempt to accept an afterlife as "a real possibility[42]." He admits that "it is possible to draw some loose analogies with religious conceptions of the world." The creators of the simulation, according to Bostrom, could be viewed as omnipotent and omniscient from our perspective. They might even reward or punish certain actions, perhaps even with virtual heavens or hells. If so, they could conclude that they themselves are inside a simulation and that they will be judged by moral standards not unlike those that they impose upon us, thus establishing a hierarchy of responsibility that pierces multiple simulation levels. This would be like the Golden Rule: *Treat your simulations as you want your simulator to treat you.*

Judging from our own civilization, we can see that we are indeed interested in running simulations and that we are close to creating evolving virtual universes. For example, a technique called genetic programming is sometimes applied in the field of artificial intelligence. It involves the creation of random programs, and the evaluation of their quality based on performance in a number of given tasks, followed by selection of the fittest for the next generations. Variation is introduced by novel random mutations within offspring programs. This approach is conceptually similar to biological evolution and is already being used to create virtual intelligences with sophisticated behavior[43].

When I was a university student, one of my pet projects was the creation of a simple two-dimensional virtual world inhabited by simulated ants that were controlled by a network of artificial neurons. The heritable instructions for the brains of these ants would undergo Darwinian evolution and, over multiple generations, improve their ability to find food and produce progeny within their virtual environment. Many similar (and much better) evolution projects can now be found on the Internet, for anyone to observe.

Anyway, although some sorts of evolution can be simulated on a computer, much scaling is required to bring such simulations even close to what we see in the world around us. Currently, scientists struggle with the high computational costs of simulations involving even relatively small biological structures. For example, it takes a day with an Nvidia Tesla V100 graphics card—which costs several thousand dollars—to simulate a single bacterial ribosome in a box with virtual water and ions for several nanoseconds (one nanosecond equals one-billionth of a second[44]). Note that a single bacterial cell can contain around twenty thousand ribosomes, which are protein synthesizing molecular machines, and that bacteria contain a whole lot more than just ribosomes.

Even with exponentially increasing processing powers it might take decades or more for a single cell to be modeled with any reasonable precision. Modeling in this case means predicting the trajectories of all of the roughly 100 trillion atoms of an average cell over the duration of its life. Organisms such as humans consist of trillions of cells, so it might take us ages to achieve the necessary computational power to model animal evolution with reasonable precision. Some of the largest simulations so far have involved a billion atoms arranged within a single gene locus. This required an astonishing 130 thousand processor cores and produced only one nanosecond of simulation per day[45]. Even if we can perfectly model a human brain with a marvelous supercomputer someday, will our computational power be sufficient to model billions of people? Will the size of the simulated universe ever be comparable to that of the real universe? And if not, shouldn't we consider the probability that we exist in a simulated universe to be much smaller than the probability that we are not in a simulation?

These are some of the reasons why I am skeptical of Bostrom's assumptions. But technology advances continually, and it hasn't reached its limits, so the future might prove me wrong. Unfortunately, the "great filter" cosmological argument suggests that, despite our efforts, we might have limited time for technological advancement. The idea was formulated by economist Robin Hanson, who suggested that the "silence of the universe" could be a very bad omen for all of us. So far humanity hasn't discovered any reliable evidence for the existence of extraterrestrial life despite the universe being much older than our solar system. We could expect advanced civilizations to be capable of colonizing other planets and star systems, and thus that life should be everywhere around us, but this does not seem to be the case. No matter how much we observe, the universe appears to be dead.

This suggests that perhaps some unknown unfortunate event naturally prevents the origination of sufficiently advanced civilizations. This event is often called "the great filter." Perhaps the filter lies behind us and the origination of life is highly improbable, although the basic organic molecules from which it could emerge are ubiquitous in space[46] and the principles of evolution should operate everywhere.

Perhaps the evolution of intelligent life requires extraordinary circumstances. The reign of mammals began after more than a hundred million years of stable rule by the giant and not-so-technologically-capable dinosaurs, whose extinction, according to one popular theory, was precipitated by an asteroid of appropriate size that fell perfectly in one of Earth's hydrocarbon-rich areas, forming stratospheric soot and thus changing the climate and history of the planet[47,48]. Perhaps the great filter could be a nonevent, such as when an asteroid does not fall on a planet.

If the great filter is behind us, that means we are extremely lucky and our future isn't sealed. Otherwise, we might be in for some trouble. Perhaps we are yet to witness the full-scale destructive possibilities of technologically advanced warfare. Even if we optimistically estimate the probability that nuclear weapons will be used at 0.1% per year, then the probability of disaster within the next one thousand years exceeds 63% ($1 - 0.999^{1000}$). In the long run humanity needs to survive the choices of numerous people of different cultures who are capable of making world-ending decisions. The cumulative risk is enormous, and it takes just one set of unfortunate circumstances to nullify all of humanity's accomplishments.

Even if nuclear war is not the great filter, some yet-undiscovered technology might prove to be even more dangerous. We'd better hope that the great filter is behind us, or that the theory is wrong, because otherwise we will likely perish before achieving interstellar travel. This idea may partially negate Bostrom's optimism about the frequency at which advanced civilizations could run detailed simulations, because not all of them would survive long enough to perform those simulations.

All of these arguments are more philosophical than concretely scientific. And my goal here is not to say whether Bostrom is right or wrong, but to say that he *may* be wrong and that we should not rely on a hope that the world we live in is a simulation. Furthermore, even if we are inside a simulation, there is no reason to believe that the beings running the simulation care about anything we do. Indeed, humans usually don't care about the well-being of the artificial intelligences that we create. We typically treat them as instruments, not as friends or equals. So why would we expect to receive preferential treatment?

Another popular secular idea that attempts to deal with the issue of nonexistence is called quantum immortality. Quantum mechanics teaches us that some processes are inherently random. For example, the half-life of uranium 238 is about 4.5 billion years, meaning that roughly half of the uranium 238 atoms in a sample will undergo radioactive decay in 4.5 billion years. But all uranium 238 atoms are physically identical and interchangeable, and there is no way to predict which individual atoms will undergo decay at a given moment. An atom may decay within a year, or it may survive for ten billion years. The process of radioactive decay is fundamentally given to chance.

In the famous "Schrödinger's cat" thought experiment the decay of a radioactive atom is linked to a device that will release a poison that will kill a cat imprisoned in a sealed container. Until you open the container you don't know the cat's fate. The prevailing theory of quantum mechanics, called the Copenhagen interpretation, says that a quantum system will remain in a superposition of possible states (the cat's being either dead or alive) until it interacts with or is observed by the external world. So, the cat may be considered both dead and alive. This thought experiment illustrates a paradox of superposition.

Another theory of quantum mechanics (not necessarily correct), called Everett's many-worlds interpretation, suggests that every possible outcome of a quantum event exists in its own universe. Thus, although the cat isn't both alive and dead at the same time in the same place, there are universes where the cat is alive and universes where it is dead.

Imagine that there are infinite universes, covering all possible outcomes of all quantum events since the beginning of time. Now look at the thought experiment from a cat's perspective: it can observe outcomes only while alive. A speculative idea of quantum immortality is that perhaps the cat's consciousness, or a version of it, will be guided through those branches of the multiverse that preserve it. So, if you were that cat, you would simply continue existing in the subsample of worlds that favor your existence. For you it would seem that the world is biased in your favor, with all events in which you could die happening in ways that don't kill you. You will be like Bart Curlish, the holistic assassin from the television series *Dirk Gently's Holistic Detective Agency*. She is completely impervious to harm, protected by a clustering of coincidences. Unfortunately, in exchange for her superpower, Bart has to murder whomever she thinks the universe wants dead.

Unfortunately, the quantum immortality thought experiment relies on circumstances that may not be achievable in the real world. Even if the many-worlds interpretation of quantum mechanics is correct (and physicists are not sure), it would not necessarily be applicable in processes that involve large numbers of atoms, such as human aging or death. Quantum immortality would not protect one from aging and aging-related diseases in even the most favored branches of the multiverse—although you would be more likely to survive in branches in which life extension worked. The risk would remain, that our healthspan would be finite and we would spend the majority of our lives in a state of exhaustion and pain. Also, quantum immortality would not protect your loved ones, unless you created artificial circumstances in which your life was directly tethered to theirs. Regardless of beliefs in quantum immortality, it would be reasonable to support the development of anti-aging interventions.

Yet the universe seems biased in favor of your existence. The planet we live on, Earth, had favorable circumstances for the evolution of life. The universe we live in appears to be hospitable generally. The apparent fine tuning of conditions in the universe is usually explained by the anthropic principle: only in favorable conditions will a mind evolve the capability to be surprised by the favorable conditions that brought it into being. In other words, conditions are favorable for our existence because otherwise we wouldn't be thinking about how favorable they are. The anthropic principle can go a bit further than explaining why Earth, cherry-picked from billions of planets, is a reasonable cradle for life. Some scientists argue that perhaps there are many universes, each with its own laws and cosmological constants. We live in our universe and study it because it was good enough to support life—hence the illusion of fine tuning.

Let's keep this in mind as we rename the anthropic principle the "me principle"—or the "you principle," since you are the subjective observer of my writings. Your personal existence was even more unlikely than the existence of intelligent life on Earth in general. Not only were the "correct" universe and planet required, but the billion-year evolution of your ancestors had to be exactly such that your parents could meet and the correct combination of chromosomes could bring about your personal existence. How unlikely was all of that?

If you apply observation selection bias, this world appears to have been selected for you (or for me, when judged from my perspective). Given the unlikeliness of your existence, this "miracle" probably required a multiverse of nearly infinite universes. If so, what is stopping the multiverse from producing "you" (however the personal, subjective *you* is defined) in different ways, at different times, and in different places, over and over again? After all, you came into existence at least once. Therefore, you must always exist in a world that is biased in favor of you, and you are unlikely to exist in a world that is not biased.

I enjoy this quasi-philosophical argument, but I can't say that it can be proven or that I have faith in it. As with all notions of an afterlife, this notion of a multiverse-based subjective immortality might be wrong, so I would rather not act upon it. And even if this secular version of "reincarnation" is correct, it doesn't imply that you get to keep any of your life's memories or achievements, so death and aging are still worth fighting.

As we proceed with this, the most unscientific chapter of my book, which was written purely to present and set aside different ways in which we may cherish hopes for unlimited existence, I would like to mention one final version of a secular afterlife. The argument was dubbed "The Most Terrifying Thought Experiment of All Time" by *Slate* magazine, and it was invented by Roko, a user on the *LessWrong* forum, which is dedicated to rationality, futurism, and mathematics. You should be warned that if Roko's prediction holds you might be in for an eternity of suffering, so you may want to skip the next paragraph.

Imagine a human-made intelligence that is so knowledgeable and influential that it attains godlike powers on Earth. Perhaps one day such a technology will even be capable of resurrecting the dead, including you. There are many ways in which humans might approach the creation of such a system, but one of them stands out. Roko's basilisk is a hypothetical artificial intelligence that promises heaven on future Earth for those who helped create it and endless torture for those who did not. This malicious machine has a strong advantage over less-threatening hypothetical alternatives and thus has a greater chance of becoming real. This thought experiment is interesting because it posits a god capable of "creating" itself from nonexistence by extorting people's worship.

This reminds me of one of the most famous arguments for believing in an afterlife, which was formulated by the French mathematician, physicist, and theologian Blaise Pascal. Pascal's wager can be boiled down to this: if there is a heaven or a hell you win by being a true believer, and if there isn't you don't lose anything, so it's better to bet on the teachings of the church than against them, as the costs are small but the possible benefits are infinite.

Pascal's argument, however, has several flaws. First of all, following the teaching of the church has direct and hidden costs: time spent on religious practices, the proscription of certain things that you might like to do (such as eating pork or participating in orgies), and the command that you accept and regularly please certain authorities. The acceptance of certain religious worldviews has other consequences for large numbers of people. It affects their attitudes toward certain achievements of science, especially the theory of evolution[49]; technology; and even medicine, such as vaccination.

But most importantly, the argument is incorrect, because even if heaven exists, we have no real way of knowing what it takes to get there or whether anything is required at all. Perhaps heaven is for everyone; or maybe it's only for Muslims; or perhaps only for atheists (if God has a sense of humor); or maybe it's just for people who like to wear uniforms. All of these possibilities are equally likely—or unlikely. The animated series *South Park* presented a vivid counterargument to Pascal's wager in the episode entitled "Probably," in which the souls of dead people from various religious backgrounds

CHAPTER 2

are surprised by their placement in hell despite their piety. "Well, I'm afraid you were wrong," the hell's director explains. "Well, who was right?" the people ask. "I'm afraid it was the Mormons. Yes, the Mormons were the correct answer."

Roko's basilisk is a much stronger idea, because you can't as easily substitute the malicious machine with some random alternative. Its very specific qualities channel it into existence. Fortunately, as with other thought experiments, there is no evidence that such a system can or will ever be created, so worshipping an evil machine might be a waste of effort. Here I am damning myself to future hell—or perhaps not, since I am increasing the probability of its creation by talking about it. As with all gods, the rules are not entirely clear.

This chapter would not be complete without a few words about cryonics, the idea that we can freeze human brains or entire bodies in hope that future technologies will allow their owners to live again after death. Hundreds of legally dead people have been cryopreserved, and thousands of living people have agreed to undergo cryonic freezing offered commercially by several companies around the world[50]. One problem with current cryonic techniques is the formation of extracellular ice that damages living mammalian cells and tissues beyond repair. Additional damage can be caused by unfrozen solutions that temporarily form between ice layers and accumulate excessive concentrations of electrolytes and other molecules[51]. For similar practical results you might as well consider preserving yourself in formaldehyde[52,53]. Most scientists are skeptical of cryonics—for good reason, given the unlikelihood that the promise of resurrection will be delivered. Even if a solution is found in some far-off future, there is no guarantee that current cryonics companies will survive long enough to keep their customers preserved.

Nevertheless, I must admit that there are positive sides to humanity's interest in cryonics. First of all, modern vitrification techniques—the transformation of liquids into noncrystalline (iceless) amorphous solids, or simply glass—are used in fertility clinics and allow the cryopreservation of individual cells such as oocytes[54] and sperm[55]. Protocols have been developed to store small amounts of tissue with the help of cryoprotective agents. Some advances in this field have been truly remarkable. For example, in 2009 a rabbit's kidney was vitrified, warmed, and successfully transplanted into another rabbit[56]; and in 2024 a protocol was established for the cryopreservation and thawing of human brain tissue up to four millimeters across[57]. Currently these methods cannot be scaled for human organs, but the search for less toxic and more efficient cryoprotectants is ongoing[58]. And there is reason for optimism: just recently a species of roundworms called *Panagrolaimus kolymaensis* was identified having survived for about 46 thousand years in a state of suspended metabolism at subzero temperatures in Siberian permafrost[59]. The survival of such an animal depends on preconditioning, a preparatory phase in which the animal is exposed to extreme dryness, followed by excessive synthesis of a cryoprotective sugar called trehalose[60,61], and the remodeling of metabolic pathways. Permafrost is home to other organisms, including plankton capable of surviving frozen for tens of thousands of years[62].

Meanwhile, freeze tolerance (the ability to withstand internal ice formation) has been described in some insects[63], frogs, salamanders, and turtles[64]. Some plants are known to survive temperatures below −60°C (−76 °F) in nature, and even immersion in liquid nitrogen at -196°C[65] (-320.8 °F) in the lab. Or consider the Amur sleeper, a fish that survives winter with its entire body encased in ice. Recently scientists decoded the animal's genome and uncovered several genes that help it survive such extreme conditions[66]. Perhaps such natural adaptations to below-zero temperatures are translatable to other species.

Cryopreservation has many medical applications, including in the storage of organs for transplantation and in pharmaceutical research. The latter benefits from the preservation of tissue slices used for the evaluation of drugs. This increases the efficiency of human-tissue use and reduces the need for laboratory animals[67]. So, while cryonics may be regarded as a very expensive funeral service, the technologies involved are important drivers of progress. In the meantime, it's yet another thing on which we shouldn't bet our existence. Staying alive is obviously much better than being cryopreserved.

The most insidious idea comes not in the form of some secular or religious notion of an afterlife, but in the acceptance of death. Ilya Mechnikov, a gerontologist who received the Nobel Prize for his work on the immune system, was passionate about the idea of life extension, but he also believed in a so-called "instinct of death." This is something that humans hypothetically experience when they achieve a sufficiently great age; death becomes perceived as natural and desirable. From an evolutionary perspective there is no obvious reason why such an instinct would exist, as it probably wouldn't improve human survival or reproduction. But the biggest problem with the idea isn't that it's probably wrong, but that it portrays nonexistence as something we should learn to accept, and *will* accept if given enough time.

From our basic knowledge that human life ends, from our distinction between self and others, and from our sense of personal identity and our ability to anticipate the future stems our existential death anxiety. A number of psychotherapeutic approaches have been invented to treat this condition[68], with some specialists suggesting that we focus on how lucky we are to be alive and to remember how our prebirth nonexistence didn't bother us, just as Richard Dawkins and Mark Twain taught us. I can see how these coping mechanisms might improve one's quality of life, but I can't help imagining the Grim Reaper applauding our choice to adapt to and accept her terms.

I reiterate that, like sufferers of Stockholm syndrome[69], we try to see the good in our abuser. Conversely, we try to see the bad in those who try to help. Hence, our culture portrays the quest for immortality as an unprovoked war against the delicate balance of nature, as a sinful act by godless and insane scientists, or as villains' self-serving search for power. We follow up by inventing good and bad ways to die, to wage war, to kill and be killed, to justify death and even find glory in it.

I propose an alternative argument called "Pascal's wager reversed." There is no good way of dying. Even if we back down from a materialistic worldview and consider the possibility of an afterlife, there is no certainty of its existence or our ability to achieve it. While you live you can always choose to die. When you die you face irreversible nonexistence. We don't know if there is a solution to the problem of death, but the probability that one can be found is not zero. The more time and resources we spend on researching human aging, the more chances we have to find the answers we seek. The more we take care of our own lives and the lives of other people right now, the greater the number of us who will live long enough to benefit from future technologies. Defeating death changes everything; losing changes nothing; thus, it is reasonable to bet on science. It is a true win–or–lose-nothing scenario.

Here, two individuals strive to build a bridge over death itself, wielding modern genetic technologies. In stark contrast, the masses cling to an illusory bridge made of empty promises and wishful thinking.

CHAPTER 3
The false Grail

> You've been bitten by a true believer;
> You've been bitten by someone's false beliefs.
>
> —Muse ("Thought Contagion")

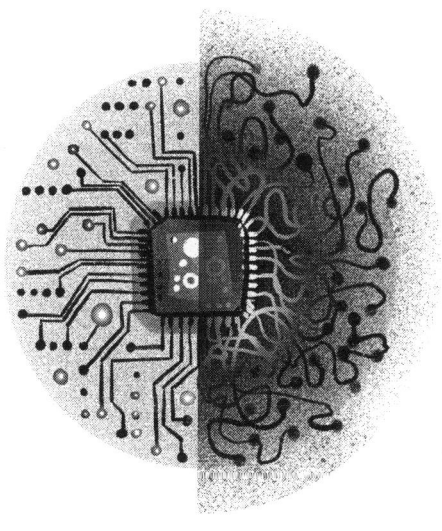

S ummary: in this chapter we discuss "biohacking" and the controversial practical ideas about life extension circulating in the media, in the pro-longevity community, and even in peer-reviewed literature.

The founder of the Qin dynasty and first emperor of unified China, Qin Shi Huang, was so desperate to find an elixir of eternal life that he fell victim to multiple scams. Not only did he take fraudulent treatments, but he sent several expensive expeditions to a mystical mountain in search of a purported thousand-year-old magician. After an unsuccessful first expedition, the alchemist and court sorcerer Xu Fu, who was entrusted with locating the magician, asked for a second chance. This time he took a larger fleet, carrying several thousand virgin girls and boys. The ships never returned. Qin Shi Huang died at the age of forty-nine, failing in his ultimate quest. Some historians believe he succumbed to mercury poisoning, an unfortunate consequence of a life-extension treatment suggested by his court alchemists and physicians.

Qin Shi Huang was merely the first Chinese emperor who presumably succumbed to the toxicity of proclaimed immortality elixirs. China's rich and influential people made the shortening of their lifespans by trying to lengthen them a kind of tradition for some two thousand years. Notable suspected casualties include the fourth emperor of the Qing dynasty and the twelfth emperor of the Ming dynasty. In contrast, Emperor Daowu of Northern Wei took a more scientific, although ethically questionable, approach: instead of self-experimentation, he organized clinical trials of immortality elixirs on criminals who had been sentenced to death. Many criminals died during these trials, and a working elixir was never found.

"Overconfidence is a slow and insidious killer," as noted in the Lovecraftian horror video game *Darkest Dungeon*. False beliefs taken seriously have killed throughout history and continue to do so today. In March 2023, fifth-year medical student Anna Kolyada, a self-proclaimed "health coach" on social media, treated herself and her male friend to a human "placenta-based immunostimulant" called Laennec. Soon after, she was admitted to a hospital with fever, septic shock, a buildup of fluid in her lungs, blood poisoning, and heart failure. Unfortunately, doctors couldn't save her life. Her friend was luckier; he was hospitalized with similar symptoms but survived.

It appears that the placenta-based drug remained in a saline drip for too long and became infected. While the problem was not with the drug itself, but with how it was used, the salient fact is that there was no rational reason to take Laennec in the first place. Currently there are no properly designed studies showing that in humans it can prevent infections, boost immunity, or prolong lifespan. The thing that comes closest to resembling evidence of Laennec's efficacy is a study that used a sample of just twenty-eight patients with long COVID-19. That article was flawed, and it did not mention any of the randomization or blinding techniques that could have helped remove biases related to patients' and experimenters' expectations.

In 2018, Aaron Traywick, the twenty-eight-year-old CEO of a medical start-up called Ascendance Biomedical, was found dead in a sensory deprivation tank. The cause of death was drowning. An autopsy established the presence of ketamine in the man's body. Traywick was a known activist for radical life extension, and a member of the "biohacking" community. He performed much self-experimentation, including injecting untested homemade gene therapies, which were probably unrelated to his death. Ketamine is not just a recreational drug and a medical anesthetic. Some biohackers advocate it as a means to "hack" the brain, "turn on receptors," and "reset the brain's neurotransmitters." Indeed, some studies of ketamine in mice[70] and humans[71] suggest that it may be of use in the treatment of depression, but others have shown that it can lead to agitation, hallucinations, psychosis, reduced pain perception, reduced awareness, lack of coordination, and even temporary paralysis. It can also increase risk of physical harm due to jumping from heights, road traffic accidents, and, as you probably guessed, drowning[72].

The common theme of these stories is that sometimes people are subjected to harm while trying to do what they believe is good for their bodies. As someone who wants to maximize his chances of happily making it into the faraway future, I view these cases as cautionary tales; we should not get ahead of our research, and we should be mindful of the medicines we take. Any potentially effective treatment comes with a nonzero risk of harm. Dying from a poor choice when science is ever so close to solving the mystery of aging would be anticlimactic, like arriving at the station one minute after the last train's departure.

On the other hand, we know that death is inevitable unless we do something to prevent it. We age with every year, month, and day of our lives. The clock is ticking. And the scientific community still hasn't provided us with any consensus on which anti-aging medicines to take. For decades we have been fed all sorts of "promising results that warrant further investigation," and general recommendations: don't smoke; don't overeat; sleep well; eat your fruits and vegetables; eat less sugar and processed meat; do not neglect hygiene; exercise regularly. To spice things up, respectable academics can reflect on the epidemiological data and add something impractical, like *Be a woman*, or state the obvious, like *Do not ignore symptoms such as sudden blindness, paralysis, or blood in your stool; remember to consult your doctor*. Millions of people have followed all the abovementioned advice but still experienced aging and death.

So, it is understandable that some people decide to take matters into their own hands and experiment on themselves with unproven anti-aging interventions. The twenty-first century has seen an explosion of ideas such as "self-tracking," "quantified self," "autonomous self," and "biohacking," which, in my opinion, reflects both hope that self-enhancement and radical life extension are possible and dissatisfaction with the slow and steady path that academic science is taking in this direction. Despite the similarities in goals, the diversity of contemporary biohackers and the amount and types of pills, interventions, and tests taken are staggering.

I remember a time when the word *biohacking* had a completely different meaning than it has now. Biohacking was about introducing hackers' ethics—which included a belief that knowledge should be shared and important resources should be utilized—into the field of applied biology. A typical biohacking project could involve open-source software for DNA analysis, freely available schematics for 3D-printed lab equipment[73], do-it-yourself gene-editing kits, and so on[74]. Advanced biohackers could attempt to reverse engineer gene therapies that cost millions of dollars in order to make cheap generic versions for private or other noncommercial use, perhaps for a friend, relative, or someone in need.

Biohacking today can mean absolutely anything, from taking a hundred pills daily to working out, consuming Brain Elixir (a brand of very expensive cocoa), or participating in mindfulness meditation retreats. The Internet has become a habitat for charismatic bloggers who promise to "massage your mitochondria" (honestly, I am not making this up), providers of "gut cleaning" services, lifestyle and diet coaches, and gurus who offer "personal empowerment through chakra biohacking."

To top this off, we have tech CEOs who advocate sit-ups along with acupuncture; food supplements coupled with music therapy; and enigmatic "body reprogramming." Some biohackers go as far as installing glowing implants under their skin; small magnets in their fingers, to feel magnetic fields; or vibrating implants in their male lower parts, "for her pleasure." Others are more down to earth, merely taking a couple, maybe a dozen, pills that they believe to be the most backed by science or personal experience. Aside from the common goal of self-improvement, the community is so diverse that, while one part is hiring health professionals and vigilantly monitoring the MEDLINE database of medical research, another rejects the established medical field altogether, in favor of "natural plant-based alternatives."

Another bizarre observation is that there seems to be competition between prominent biohackers as to who spends the most money, performs the most tests, or eats the most pills. I know at least four men who have separately claimed to be "the most measured person in the world," like the joke about the heads of John the Baptist held in various churches and museums: there are ten of them, but only three are authentic. Russian biohacker and entrepreneur Serge Faguet gained wide media attention by claiming he had spent 200 thousand dollars on health improvements and lifespan extension. Biohacker and CEO of Bulletproof Coffee Dave Asprey spent 300 thousand dollars to "hack his own biology." Faguet and Asprey were later overshadowed by entrepreneur Bryan Johnson, who spent at least two million dollars. Bryan Johnson became a media sensation by posting videos about his daily routines, which include a lot of physical activity, a strict diet, and dozens of supplements, all combined in an intervention program he calls Blueprint, the details of which are freely available online[75]. As you probably guessed, his website also offers "biohacking" products such as Blueberry Nut Mix, Cocoa Powder, Nutty Pudding, and supplement mixes that cost more than one hundred dollars per month.

When it comes to consuming supplements, the champion is probably Ray Kurzweil, the futurist, best-selling author, cofounder of Singularity University, and director of engineering at Google. At one point he claimed to take 250 supplements daily in an attempt to "reprogram" his biochemistry—although in 2008 he dialed that number down to 150, and then to 100 by 2015. Kurzweil believes that in the future humans will be able to live forever. To bring this future closer he has supported nonprofit aging-research organizations such as the SENS (Strategies for Engineered Negligible Senescence) Research Foundation. Notably, gerontologist Aubrey de Grey, one of the most famous activists for radical life extension and a cofounder of SENS, does not take any supplements for life extension. Aging gerontology researchers usually take modest numbers of pills if they take any. For example, professor of genetics at Harvard Medical School David Sinclair, one of the most well-known gerontologists and the author of *Lifespan: Why We Age and Why We Don't Have To*, takes only a dozen.

On the one hand, I have great respect for people who value their lives and show remarkable discipline in their efforts to prolong it. I would even go as far as to say that the biohacking community is doing humanity a favor by promoting the idea of radical life extension and self-improvement. I applaud the "Don't Die" slogan that Bryan Johnson is actively spreading. On the other hand, I am skeptical that the efforts of even the most scientifically literate modern biohackers will result in prolonged rather than shortened lifespans. I would also argue that the self-experimentation efforts of biohackers and "self-trackers" have limited scientific value. A single person typically represents a sample size too small to draw meaningful conclusions from, and the millions of dollars biohackers spend would serve much better if used on traditional research. Finally, I fear that the word *biohacking* has become a disguise for anti-scientific alternative medicine and is often used by dishonest or misinformed actors to capitalize on the public's growing interest in life extension. Biohacking might even drive our attention away from what really needs to be done. So, let's try to separate the good ideas from the bad.

There are good reasons why clinical trials tend to involve hundreds of subjects, complex statistical analysis, and precautionary measures such as placebo controls, randomization, and blinding that reduce the biases introduced by the expectations and behaviors of participants and experimenters. Human history is riddled with incorrect conclusions about the effects of medical interventions. Popular treatments of the past include things like the royal touch (based on the idea that French and English monarchs could cure disease by laying on hands), the use of sympathetic powder and weapon salve (remedies for wound healing that were applied not to wounds but to the weapons that inflicted them), the use of steel and brass rods to cure inflammation, and so on. All of this was supported by someone's "personal experience." In his 1842 lectures, American physician Oliver Wendell Holmes Sr. used these examples to demonstrate how unreliable anecdotal evidence can be, and in doing so he helped pave the road toward the science-based medical approaches that would later revolutionize healthcare.

One reason why people fall prey to alternative medicine is related to magical thinking[76], which the anthropologist James George Frazer suggested is a by-product of associative thought. Consider the choice of substances used by Chinese alchemists in their pursuit of the elixir of immortality. They used jade, cinnabar (a source of mercury), and drinkable gold simply because those materials are long-lasting and were believed to transfer their longevity to people who ingested them.

For similar nonscientific reasons, phallus-shaped rhinoceroses' horns are believed by some to have aphrodisiac properties, fueling a demand met by brutal poaching. The same magical thinking probably crossed the mind of geologist and biohacker Anatoli Brouchkov, who injected himself with ancient bacteria isolated from a three-million-year-old sample of permafrost[77,78].

But we are not what we consume. Eating a boiled egg does not make you boiled or an egg. Likewise, consuming rhino-horn powder does not cause an erection, ingested gold does not make you resistant to change, being bitten by a radioactive spider does not turn you into Spider-Man, and eating genetically modified plants does not alter your genes.

It's hard to overestimate the power of magical thinking. Consider the widespread addition of plant stem cells to cosmetic products that promise to return skin's youth. Most consumers have probably heard that stem cells have rejuvenating properties, but stem cells function only when they are genetically identical to the other cells of one's body and can replace them when needed, and plant stem cells can't become human cells. Perhaps they have some mechanism of action that science has yet to discover, but the likely reason for their popularity is the intuitive connection between stem cells and rejuvenation. Similar associations most likely made collagen and vitamin supplements so easy to market.

The idea that walnuts are good for the brain is another excellent example of magical thinking. Just because walnuts look like the cerebral cortex, does not mean they provide cognitive benefits, a claim that has little scientific support[79]. Then we have the more general idea that "natural products" are somehow superior, safer and healthier, despite that some of the deadliest poisons, such as botulinum toxin, ricin, strychnine, and tetrodotoxin, are absolutely organic and natural—not to mention the natural microbes that cause syphilis, tuberculosis, and the bubonic plague.

Perhaps my favorite example of deceptive associations is cyclopamine, a naturally occurring alkaloid and potent teratogen found in the beautiful corn lily *Veratrum californicum*.

When ingested by pregnant animals such as sheep, it causes cranial deformations in their progeny, which will be born not with two eyes but with one, in the middle of the head[80]. The results are terrifying; however, the corn lily is all-natural, non-GMO, organic, and free of artificial pesticides.

Like Oliver Wendell Holmes Sr. in the nineteenth century, I am impressed by the especially absurd field of alternative medicine that revolves around the magical principle of "like cures like." I mean, of course, homeopathy, the pseudoscientific practice in which active ingredients are diluted, often to the extent that no molecules of active substance remain. I call homeopathy the *perpetuum mobile* of medicine, because of its fundamental contradiction with natural science. The most bizarre case of its use I know of involved a young girl who dyed her hair. Her parents were instructed to give her a preparation made from an octopus, because the octopus "also changes its color." Magical thinking is intuitive, which explains why homeopathy is so popular despite the numerous studies concluding that it doesn't work[81–83].

Misleading associations are reinforced by the illusion of causality[84]. This is the drawing of false conclusions about a drug's effectiveness based on observed improvement after its use. The trick is simple: *after* does not mean *because of*. Sometimes it rains after a rain dance. Obviously, that does not mean that synchronized human movement can change the weather. Likewise, sometimes people recover from illnesses after taking placebos, such as sugar pills—just as people sometimes get better without any treatment, or after eating a banana. This is not necessarily because of some psychological placebo effect (which can alleviate pain or anxiety while lacking major health benefits[85]), but because our bodies are able to repair themselves, and we have a sophisticated immune system that is often capable of doing its job. Given time, our health tends to regress toward normal, but we are easily seduced into concluding that our actions made us well, just as the "superstitious" pigeons in the experiments of behaviorist Burrhus Skinner learned to repeat whatever behavior they performed before getting a snack from a food dispenser. The poor birds desperately flapped their wings or banged their heads on their cage walls, while food was provided at regular intervals, with no reference to what the bird was doing[86].

While homeopathy has little to do with biohacking (although Ray Kurzweil's coauthor on *Transcend: Nine Steps to Living Well Forever* and *Fantastic Voyage: Live Long Enough to Live Forever*, Terry Grossman, is a licensed homeopath), traces of similar magical thinking can be seen in how various supplements are selected and promoted while the illusion of causality haunts those who experiment on themselves. The popularity of placebo medicine demonstrates just how common this kind of self-deception is. This style of thinking can lead even intelligent people, such as Ray Kurzweil, to use and promote alkalinizing devices even though no legitimate research supports the health benefits of alkaline water[87]; or to talk about avoiding "electromagnetic pollution" due to the perceived harm caused by computer monitors, hair dryers, electric shavers, and other electrical appliances. This is why it's always worth asking yourself whether you have been convinced by the science behind a certain intervention or by some pleasant or unpleasant association in your mind.

At first, I hinted, but now I will say directly, that modern biohacking has many hallmarks of alternative medicine. To quote Australian comedian and musician Tim Minchin, "By definition . . . alternative medicine has either not been proved to work, or been proved not to work. You know what they call 'alternative medicine' that's been proved to work? Medicine." To slow the aging process, biohackers regularly take supplements and medication for which evidence is very limited. In a good scenario we have some hypothesis as to why a certain molecule should work (say it promotes autophagy—the ability of cells to digest their unnecessary components) and some promising data from animal studies (say twenty percent increased lifespan in mice). That is the evidence level of preclinical trials. And we know that around ninety percent of treatments that show promising results in preclinical tests fail in human studies[88]. Furthermore, not all animal studies are reproducible.

In 2012, an article in a journal called *Biomaterials* reported an extraordinary claim. Oral administration of C60 fullerenes (hollow spherical molecules that consist of sixty carbon atoms each) dissolved in olive oil nearly doubled the lifespan of rats compared with control animals that received only olive oil or water[89]. While the sample size wasn't large, the reported effect was too impressive to be dismissed as a statistical error. There simply was no precedent for a single molecule or combination of molecules nearly doubling a mammalian lifespan. Influenced by the article, myriad websites started selling C60-containing products as dietary supplements. Nine years later, replications of this research appeared and the results were . . . absolutely negative.

A 2021 paper in *Rejuvenation Research* found that mice treated with C60 plus olive oil had lifespans similar to those of mice treated with water. Surprisingly, mice treated with just olive oil, and without the fullerene, actually had their lifespan reduced[90]. The authors of another study published in the same year, in *GeroScience*, failed to observe significant benefits of C60 plus olive oil on lifespan or healthspan when subjects were compared with untreated controls. In fact, the authors found that when exposed to light, C60 in olive oil can form toxic substances that can cause diseases and increase mortality[91].

As these results emerged, one coauthor of the original 2012 publication expressed concern that C60 products were being sold without certified toxicity tests and that, although C60 itself has neither acute nor chronic toxicity, its preparations can be "highly toxic" due to light exposure, impurities, or defective adjuvants. He concluded that clinical trials in humans are necessary and warned that, without them, C60 consumption could cause "a public health issue[92]."

Despite warnings from researchers, C60 fullerenes are readily advertised online: "C60 Fullerene for a Longer Life"; "C60 (Carbon 60) Nobel Prize Winning Antioxidant";

"Award-Winning Molecule That Could Extend Your Life"; "C-60: The Miracle Molecule for Biohacking Pets, Hair Loss, EMF, & Cancer"; "C60 (Fullerene) in Organic Extra Virgin Olive Oil." Here are some of the "key benefits" listed on an "organic" website that sells this stuff: "anti-viral," "anti-aging," "anti-inflammatory," "UV and radiation protection," "non-GMO," "vegan friendly," and "Paleo friendly." In case you wonder where people from the Paleolithic era got their C60 nanoparticles: it can be found in soot, and can even be detected in interstellar space[93] and young planetary nebulae[94]. And yes, it's non-GMO, because it does not contain any genes. But I doubt that's a good reason to take it. Despite scientific concerns, C60 is taken by a number of biohackers, and even David Sinclair has admitted to trying C60 products.

Another example of a controversial biohacking intervention is resveratrol. The story of the controversy begins in 1992, with the so-called French paradox—the epidemiological observation that the people of France had a low incidence of coronary heart disease despite having large amounts of saturated fats[95] in their diet. The authors of the *Lancet* article that reported this observation suspected that it could be explained by habits of moderate wine consumption[96].

Aside from alcohol, red wine contains numerous compounds, such as vitamins, quercetin, anthocyanins, caffeic acid, and resveratrol. In 2003, David Sinclair and his colleagues reported in the journal *Nature* that resveratrol extends the lifespan of yeast by seventy percent via activation of a gene called *Sir2*[97,] a known player in the longevity of these organisms[98]. At roughly the same time, studies on worms[99] and fruit flies[100] also reported longevity benefits from increased *Sir2* activity. In Sinclair's work, the human version of this gene, called *SIRT1* (or *sirtuin 1*) was also affected by resveratrol and increased cell survival. *SIRT1* is an enzyme capable of modifying chromatin (DNA) structure, and regulating the activity of many genes. Functionally, *SIRT1* appears to counteract many hallmarks of aging: it reduces inflammation, promotes the growth and division of mitochondria, and improves insulin sensitivity[101–103]. There was a potential universal pro-longevity gene, and a molecule (resveratrol) that apparently could target it.

In 2004, Sinclair co-founded a biotechnology company called Sirtris Pharmaceuticals that focused on researching the health benefits of resveratrol and its derivatives. Sinclair stated that resveratrol was "as close to a miraculous molecule as you can find," and that "one hundred years from now, people will maybe be taking these molecules on a daily basis to prevent heart disease, stroke, and cancer[104]." In 2006, his lab reported that the compound significantly improved the survival of mice fed a high-calorie diet[105]. The resulting scientific article did not, however, mention whether resveratrol affected the lifespan of mice fed a standard diet.

The finding was also an unlikely explanation for the French paradox: the smallest concentration of resveratrol used in the study was 5.2 mg/kg per day, which for a human is roughly equivalent to the resveratrol-like compounds in a barrel's worth (158 liters) of wine per day. To be honest, this calculation does not account for the differences between the metabolic rates of mice and humans (larger animals tend to require less active ingredient per unit of body weight[106]). But even after applying the commonly used mouse-to-human conversion factor of 12.33[107], we still get an unrealistic 12.8 liters of wine per day. However, large resveratrol doses can be acquired via supplementation, and the benefits of this molecule were further supported by findings that it can increase the median and maximum lifespan of short-lived fish in a dose-dependent manner[108].

In 2008, Sirtris Pharmaceuticals was sold for 720 million dollars to GlaxoSmithKline. This happened just before additional studies questioned whether resveratrol and its derivatives are direct activators of *SIRT1*[109]. The initial findings could be explained by an experimental artifact: the studied compounds affected the interaction between *SIRT1* and its native substrates only when the substrates were artificially tagged with a fluorescent molecule; this is a common visualization technique in molecular biology[110].

The final blow to resveratrol came a few years later. Every year the Interventions Testing Program (ITP) of the National Institute on Aging[111] examines the effects of three to four selected chemical interventions on mouse lifespan in several independent organizations. The ITP is probably the most reliable source of information about the effects of drugs and supplements on murine lifespan. In 2013 the ITP finished its independent evaluation of whether resveratrol prolongs the lifespan of genetically heterogeneous mice, and found no statistically significant effect[112]. This confirmed the results of an earlier study that found longevity-promoting effects for rapamycin (another potential anti-aging drug, which will be discussed in later chapters) in mice, but not resveratrol[113]. A few months after the ITP published its results, GlaxoSmithKline shut Sirtris Pharmaceuticals down.

It was around that moment that the role of resveratrol's purported target, *Sir2/SIRT1*, in animals was also questioned. Independent researchers tried to replicate the effects of increased *Sir2* activity on the lifespan of worms and flies and found that the previously reported effects on aging were not observed when appropriate controls were used and genetic backgrounds of animals were standardized[114]. While mice engineered to lack both copies of the *SIRT1* gene have shorter lives and do not benefit from calorie restriction[115] (a known way to increase murine lifespan), increase of the gene's activity beyond its normal levels was usually not potent enough to affect animal longevity[116,117]. One exception was presented in a study that found a 9% and 16% increased median lifespan in male and female mice, respectively, with increased *SIRT1* activity in their brains[118], but that's another story.

Anyway, it is safe to say that resveratrol has not been proved to be a longevity-boosting intervention in mice on a healthy diet. In some recent studies it also failed to prolong the lifespan of fruit flies[119], and human life-extension data is lacking. Nevertheless, resveratrol supplements are marketed for "healthy aging," sometimes with the usual "does not contain GMO" labels. Many biohackers take resveratrol. To be fair,

Bryan Johnson did not include it in his Blueprint; David Sinclair, however, takes resveratrol, arguing that the molecule is safe and that there is evidence for its beneficial effects on healthspan. There are, indeed, more than 150 trials of resveratrol in various stages of completion, although the drug has a neutral effect on most diseases[120]. Some studies have even reported undesirable effects of high resveratrol doses, including increased levels of low-density lipoprotein (LDL), or so-called "bad" cholesterol[121].

After the "era of resveratrol" a new activator of sirtuins became a center of attention for biohackers. The enzymatic activity of sirtuins depends on the availability of a small molecule called nicotinamide adenine nucleotide (NAD), in its oxidized form (NAD+)[122]. NAD+ is essential to metabolism and found in all living cells. As we age, our levels of NAD+ fall dramatically, presumably due to the increased activity of enzymes that degrade NAD+ and its precursors[123] and to decreased activity of enzymes that synthesize them[124]. Hence, some researchers have proposed that aging can be slowed by inhibiting the former[125] or boosting the latter. A simpler idea is to provide the body with additional NAD+ precursors, such as nicotinamide mononucleotide (NMN) and nicotinamide riboside (NR), both of which have become heavily marketed supplements.

In 2022, the global market for NR reached 91 million dollars, while the market for NMN exceeded 250 million dollars[126]. Unsurprisingly, NR is part of Bryan Johnson's Blueprint; NMN is preferred by David Sinclair. Sinclair even cofounded Metro International Biotech (MIB), a pharmaceutical company that manufactures and researches clinical-grade preparations of NMN. Coinciding with the registration of NMN clinical trials by MIB, was the purging of the molecule from the FDA's list of supplements, which is not supposed to include molecules subject to "substantial clinical investigations." The FDA's action may secure future NMN sales for MIB, although it's unclear whether the FDA's ruling will be enforced[127]. Obviously, the topic is financially very hot, but the most salient question is whether NAD+ boosters can help us live longer.

A 2016 article in *Science* reported that NR supplementation provided a modest increase in mouse lifespan[128], but this finding was not reproduced by two subsequent studies[129], one of which was performed within the framework of the Interventions Testing Program[130]. Various health benefits were reported for NMN in animals[131], including increased insulin sensitivity[132], improved mitochondrial and vascular function[133,134], increased fertility, and improvements in cognitive[132] and physical activity[135]. There is also a widely circulated claim that NMN increases mouse lifespan by more than twenty percent, but unfortunately, this claim is not based on a peer-reviewed research article but on a US patent[136].

So far, no peer-reviewed study has shown that NMN supplementation can increase mammalian lifespan, and there are very few published reports regarding the long-term safety or efficacy of these products in humans[132]. That said, in 2019 a promising article in *Cell Metabolism* reported a ten percent increase in median lifespan and a fifteen percent increase in maximum lifespan for aged female mice treated with extracellular vesicles containing an enzyme that synthesizes NMN[124]. A 2023 study also found that increased activity of genes involved in NAD+ biosynthesis is a feature of long-lived mammalian species and of mice subjected to lifespan–extending interventions[137]. So, the jury is still out. The science is exciting, but once again we see a pattern of products being marketed and sold without proper investigation in a field full of financial conflicts of interest.

Now let's examine the interesting case of metformin, an oral antidiabetic drug that many biohackers use; it is part of Bryan Johnson's Blueprint and is also taken by David Sinclair and Ray Kurzweil. First of all, there is no denying that this is an essential drug for people with type 2 diabetes. Metformin can also delay and reduce the risk of type 2 diabetes[138]. However, whether the drug increases lifespan of healthy subjects remains uncertain[139].

The drug works primarily by decreasing glucose production by liver cells[140]. Interest in metformin for biohacking purposes was sparked because it improves insulin sensitivity (in this sense it mimics calorie restriction), and this interest was fueled by several publications. First a 2013 study in *Nature Communications* reported that metformin supplementation improved the median lifespan of male mice, albeit by a modest 4.5% (and was toxic at higher doses)[141]. Then a 2014 observational study compared 78 thousand people who had diabetes and were treated with metformin against matched controls without diabetes. The study found that the first group had a 15% longer median survival time[142] and introduced the idea that nondiabetics might also benefit from the drug. There was a time when I thought it might be a good idea to try metformin, and I even made my first personal biohacking attempt—which didn't last very long, as I experienced adverse effects in the form of muscle pain.

In 2016 the ITP investigated metformin and failed to replicate any longevity benefits of its use in mice[143]. Meanwhile, the observational study, while highly cited, turned out to have an inbuilt methodological flaw. Metformin is usually the first-choice treatment in patients with type 2 diabetes. As diabetes progresses, patients generally require additional medications to maintain adequate blood glucose levels. Since the study examined people who use metformin as a monotherapy, patients on additional medications would not meet the study's inclusion criteria and would be disqualified. This means that only the healthiest people in whom diabetes did not progress would remain in the metformin group. This is obviously a source of bias.

In 2022 another observational study avoided this issue by comparing twins (matching sex, birth years, and familial factors). Of each selected pair, one twin initiated metformin monotherapy due to developing type 2 diabetes and the other twin didn't have such a diagnosis. In this study the people with diabetes taking metformin had significantly higher mortality than controls[144]. Of course, this does not rule out the possibility that the health benefits of metformin are masked by diabetes, but it does negate the initial argument for its use by nondiabetics.

To be fair, some biohackers, such as Bryan Johnson, are taking metformin together with rapamycin—a combination that has been shown by the ITP to increase the lifespans of both male and female mice[143]; however, human data for such interventions is lacking. There seems to be some evidence that metformin use reduces cancer risk, as shown by several meta-analyses[139,145,146], although this effect was not observed in one of the latest studies[147]. Recent reviews point out that, by modern standards, observational data on metformin is largely unreliable outside of its use in preventing diabetes[148] and in a few other medical cases.

Another problem with preemptive use of metformin is that the drug has a number of potential adverse effects. Some of these are somewhat mild; for example, the increased risk of vitamin B12 deficiency[149] can be circumvented by vitamin supplementation. But metformin also carries a small risk of lactic acidosis, a dangerous buildup of lactic acid in the bloodstream. This condition is rare, with fewer than ten cases per 100 thousand person-years, but when it does occur, it's likely to be fatal[150–152]. The adverse effects of metformin are far outweighed by the overall benefits of reduced mortality in patients with diabetes[153], and there are epidemiological studies that show benefits of metformin for patients with preexisting cardiovascular diseases[154,155], but it is unclear if it's worth taking for healthy subjects. Perhaps more clarity will be reached after the completion of the Targeting Aging with Metformin (TAME) trial[156]. This is an upcoming double-blind randomized placebo-controlled trial designed to test the hypothesis that metformin can delay the onset of aging-related diseases in humans. Regardless of the outcome, this research is especially important because it's the first-ever registered anti-aging study in humans—hopefully the first of many.

The abovementioned examples illustrate some of the many pitfalls of longevity research, and of science in general. Different experts can come to different conclusions regarding well-studied molecules. Research findings are often difficult to reproduce. Errors lurk where certainty should be. When it comes to the evaluation of drugs against accepted illnesses, we can at least rely on gold-standard clinical trials, although the problem of reproducibility remains relevant even in those cases[157,158].

Now imagine how reliable our knowledge is regarding the beneficial and adverse effects of molecules that can be marketed without the support of clinical trials. I'm talking about the industry of herbal and dietary supplements, which have become a fad among biohackers.

The supplement industry is worth billions of dollars, with more than seventy percent of Americans taking some dietary supplements daily. The wide use of supplements comes with severe health consequences. In the United States, approximately twenty percent of hepatotoxicity cases are a result of herbal or dietary supplement use[159]. Also in the United States, more than twenty thousand emergency room visits and more than two thousand hospitalizations annually result from the use of herbs or complementary nutritional products. Most commonly, these are products for weight loss or "increased energy[160]."

While each supplement probably deserves its own case, there are universal problems that affect the vast majority of them. Unlike actual medicine, dietary supplements can be marketed without the support of clinical trials, and thus their side effects are severely underreported[161] and their claimed health benefits are often exaggerated. But that's only a small part of the problem: given the leniency of government regulation, supplements are frequently adulterated with active pharmaceutical ingredients without the consumer's awareness.

In 2007, the FDA created a Tainted Dietary Supplement Database. A recent (2022) analysis of that database indicated that more than a thousand unique products had undisclosed additions of various substances, including drugs that were removed from the market due to adverse cardiovascular or oncogenic effects[162] (the total number was 1,950 as of May 2023[163]). For example, weight-loss supplements can contain phenolphthalein. For over a century phenolphthalein was used as a laxative, until evidence of "multiple carcinogenic effects in experimental model systems[164]" emerged. Some weight-loss and sports-performance supplements even contain dimethylamylamine (DMAA), an FDA-banned derivative of amphetamine[165] that has potential for abuse[166]. There have been multiple reports of people dying or suffering cardiac arrest after consuming dietary supplements with DMAA[167,168].

A more common supplement addition is sibutramine, an appetite suppressant that works as a serotonin–norepinephrine reuptake inhibitor, similarly to some antidepressants. It was shown to increase the risk of myocardial infarction and stroke[169] and was removed from the US and multiple other markets. You won't find this substance as an over-the-counter drug, but you may encounter it in numerous "natural," "traditional," or "herbal" remedies. Likewise, supplements for sexual enhancement often conceal the addition of phosphodiesterase-5 inhibitors such as sildenafil (Viagra). The good news is that this drug works, but perhaps consumers deserve to know what actual medication they are taking.

Aside from containing potentially harmful individual components, many supplements can be viewed as being untested combinations of multiple molecules. Their components can interact or cancel one another out. This issue is deepened for supplements based on botanicals or other raw biological materials, which display natural content variability. Concentrations of components can change from batch to batch depending on climate or weather conditions, unpredicted environmental or anthropogenic factors, or even genetic variations, so a consumer might not know the exact dosages of ingredients he is receiving, sometimes leading to overdose. *Breaking Bad*'s Walter White would not approve of the purity of many such products.

Antioxidant supplements are particularly popular among biohackers. Their use is usually justified by references to the once-popular oxidative stress theory of aging (also known as the free radical theory of aging), which linked our deterioration to the molecular damage caused by reactive oxygen species produced by the cells of our bodies. Unfortunately, most studies of antioxidant supplementations in humans do not support their pro-longevity effects, and some antioxidant supplements even increase mortality[170].

A 2008 Cochrane Review of antioxidant supplements that included data on more than 300 thousand people found that "patients consuming the antioxidants were 1.03 times as likely to die as were the controls." One prominent study tested some of the most popular natural antioxidants, such as curcumin, morin, and quercetin, as well as extracts of blueberry, cinnamon, green tea, black tea, pomegranate, sesame, and French maritime pine bark, and found no evidence of murine life extension[171]. The free radical theory of aging has also been extensively criticized[172] and reworked, despite the important role it played in the early period of aging research. We now know that antioxidants can have both beneficial and detrimental health effects.

Next, we have the widespread idea that nearly everyone needs to take vitamin supplements. Indeed, vitamin deficiencies can lead to various health problems. Vitamin-deficiency symptoms include impaired wound healing for vitamin C; night blindness and impaired immunity for vitamin A; bone softening for vitamin D; anemia for B-group vitamins; bleeding, osteoporosis, and cardiovascular diseases for vitamin K; and so on. Folate (vitamin B9) supplementation is considered especially important for pregnant women, as it reduces the risk of fetal neural-tube defects[173].

We can find reports that up to 31% of the US population is at risk of at least one vitamin deficiency or anemia[174] and that vitamin supplementation is associated with a reduction of that risk. That sounds like an immediate call to action, but notice the words *at risk*. When it comes to actual clinical manifestations of these deficiencies, the picture is usually much less troublesome. For example, a recent study examined the rate of all vitamin deficiencies in the US Army, with a sample size of more than 1.3 million person-years[175]. The total rate of all diagnosed vitamin deficiencies was less than 0.1% per person per year. Half of the diagnosed cases were vitamin D deficiencies. For some vitamins, such as vitamins A and E, even the risk of deficiency was less than 1%, according to the abovementioned general-population report.

Vitamins are a frequent component of biohacking interventions. Bryan Johnson's Blueprint includes vitamins E, C, D3, K1, and K2, and the vitamin B complex. David Sinclair has said that he takes vitamins D3 and K2. Unfortunately, vitamin research is filled with contradictory studies and potential conflicts of interest, and there is much uncertainty about the data.

Meta-analyses of antioxidant vitamin E supplements reveal that they either increase all-cause human mortality[176] or have no effect[177] on lifespan, or on aging disorders such as dementia[178]. Animal studies of such supplements do not show consistent beneficial effects[179], and in some cases the effects on lifespan are extremely negative[180]. A large clinical trial involving more than 29 thousand male smokers randomized to either alpha-tocopherol (the most abundant, biologically active form of vitamin E), beta carotene (a precursor of vitamin A), both supplements, or a placebo found no effect on mortality or lung cancer incidence for alpha-tocopherol. Both outcomes worsened for subjects taking beta carotene[181]. Finally, a Cochrane Review found no evidence that vitamin supplements can reduce lung cancer incidence or mortality in healthy people, despite popular claims[182].

The latest systematic reviews did not find that vitamin D supplements or vitamin K supplements reduce all-cause mortality compared with placebo or no treatment, although some reduction of cancer-related mortality was observed for vitamin D[183]. The VITAL trial holds significant importance as a unique, large-scale, placebo-controlled, randomized study that assessed the health impacts of vitamin D and widely used Omega-3 supplements on a diverse cohort of over 25,000 participants. With a median follow-up period of more than five years, the study concluded that neither supplement had any impact on reducing all-cause mortality[184,185].

In 2022, the US Preventive Services Task Force published a summary of eighty-four studies of vitamin and mineral supplements and found that they provide little or no benefit in preventing cardiovascular diseases, cancer, or death. There was only a small benefit from multivitamin use for cancer prevention, and there was an increased risk of lung cancer from the use of beta-carotene by those who were already at high risk of this disease[186].

A 2024 a study published in *Nature Medicine* raised additional concerns about excessive intake of niacin, also known as nicotinic acid—a form of vitamin B3, a popular NAD+ booster and a supplement that is part of Bryan Johnson's Blueprint. The study reported that one of niacin's terminal metabolites contributes to vascular inflammation and an increased risk of major adverse cardiovascular events such as myocardial infarction and stroke[187].

Since 1941 the United States has mandated that flour, cereal, and rice be fortified with niacin. This was done to combat pellagra, a disease caused by lack of vitamin B3 that can lead to four Ds: diarrhea, dermatitis, dementia, and death. This deadly disease united the poorest US citizens during the Great Depression, Soviet children raised in orphanages, and Gulag prisoners. Given their success, food fortification mandates spread to more than fifty countries, saving countless lives. However, in 1974 the recommended doses of niacin were increased, which, together with increased food availability, led to widespread excessive levels of vitamin B3 in some populations. Thus, it is reasonable to question the idea of adding even more niacin in the form of vitamin B3 NAD+ boosters.

By the way, niacin is not just a food supplement, but a medicine prescribed to decrease levels of "bad" LDL cholesterol. The so-called "niacin paradox" is that while niacin improves lipid biomarkers, it does so without the expected improvements in cardiovascular health[188], according to most studies (except some very old ones[189]). In fact, recent reviews have found that niacin intake can marginally increase all-cause mortality[190], which is consistent with the abovementioned *Nature Medicine* study.

One limitation of most vitamin studies is that individual differences between people are often not taken into account; one person might need more of a certain vitamin, and thus benefit from a supplement, while another might need less, or even suffer from its excess. I believe the most reasonable approach is to test whether you actually have a vitamin or mineral deficiency (or are approaching one), and if you do, take supplements of only the specific vitamins that you need. Even better: consult a doctor while doing so.

A balanced, diverse diet might help prevent the need for vitamin supplementation. Diets are another tool for biohacking interventions, and they are often taken to extremes. Ray Kurzweil advocates for a low-carbohydrate diet; Bryan Johnson is on a vegan diet restricted to fewer than two thousand calories. In the "Feast of famine" chapter we will discuss calorie restriction in detail, but for now, let's just say that while this intervention is highly beneficial for the lifespan of many animal species, epidemiological evidence in humans typically finds mortality to be lowest around the recommended daily intake of 2,200 and 2,700 calories for adult women and men respectively. Some studies have found that calorie restriction beyond that point can increase mortality[191], although this is not entirely settled[192].

A 2018 study published in *Lancet Public Health* followed 15,428 US adults who did not report extreme calorie intake. It found a U-shaped association between the percentage of energy obtained from carbohydrates and mortality; participants in the 50%-to-55% range had the lowest risk. This finding was confirmed by a meta-analysis of multiple cohorts, with a total sample size of more than 430 thousand participants; there was significantly increased mortality for both the less-than-40% and more-than-70% carbohydrate groups[193]. A 2019 study published in the *European Heart Journal* followed 24,845 participants for 6.4 years and found that low-carbohydrate diets were associated with increased overall mortality, as well as increased cardiovascular and cancer mortality[194]. This research is consistent with the 2020–2025 USDA Dietary Guidelines for Americans, which recommend that carbohydrates make up from 45% to 65% of total daily calorie intake.

While studies consistently find that eating more fruits and vegetables and less processed meat is beneficial in reducing all-cause mortality, the same can be said for consuming more fish[195], so pursuing a strictly vegan diet is probably not an optimal decision for lifespan extension (although it's difficult to argue when the choice is ethical). Studies consistently find that the Mediterranean diet—which is high in vegetables, legumes, fruits, nuts, and grains, but also in fish and seafood, with some intake of red wine—is beneficial for multiple health outcomes[196-198]. As for red meat, there is accumulating evidence that its high intake is associated with increased risk of several cancers[199]—although this seems to apply mostly to its processed versions, with weaker evidence[200] for detrimental effects of unprocessed meat and poultry, according to recent research[201].

Overall, according to a study[202] connected with the European Prospective Investigation into Cancer and Nutrition (EPIC), practicing just four behaviors (avoiding smoking, being physically active, consuming alcohol moderately, and eating at least five servings of fruits and vegetables per day) may provide about fourteen extra years of life compared with not practicing any of them.

While some biohackers base their diets on medical research—even if outdated or flawed—others prefer to draw conclusions about the benefits of interventions based on self-experimentation. Popular self-tracker Seth Roberts advocated for a Paleo diet, which involves consuming the foods our Stone Age ancestors presumably ate. Somewhat contradictorily, he also concluded that butter made his brain "work better" and admitted to eating half a stick of butter each day[203]. "I really benefit from this, and perhaps other people would too," said Roberts during a speech at a Quantified Self event. Unfortunately, the self-experimenter died of coronary occlusion at the age of sixty. If we follow the biohacking logic that a sample size of one person is sufficient to make causal inference between intervention and outcome, we may as well conclude that mixing Paleo diets with butter is extremely dangerous. Obviously, that would be most unscientific.

If we really want to find out whether butter improves cognition or increases mortality, we should consider higher-quality data. Butter is very high in saturated fat. Randomized dietary-intervention studies show that, in the short term, a meal high in saturated fat has no effect on cognition for lean people[204], and it can actually decrease cognitive performance in obese individuals. Meanwhile, large-scale meta-analyses show that, in the long term and in older populations, the category of greatest saturated-fat intake is associated with increased risk of cognitive impairment[205]. So, butter probably won't make you smarter on any timescale, and limited evidence from self-observation probably led Roberts to an incorrect conclusion.

As for the health hazards and benefits of butter, in 2020 the Cochrane Collaboration performed a meta-analysis of randomized controlled studies of dietary interventions aimed at reducing saturated-fat intake by the participants. The study found that such interventions led to a reduction in cardiovascular events[206]. More recently, a 2023 report commissioned by the World Health Organization concluded that replacing dietary saturated fats with carbohydrates, plant-sourced monounsaturated fats (such as from olive oil or sunflower oil), or polyunsaturated fats (found in fish, nuts, and seeds) reduces all-cause mortality[207]. All this was based on a sample size of more than 1.5 million individuals. Unsurprisingly, WHO guidelines recommend consuming less than 10%[208] of your calories from saturated fat. I think anyone has the right to eat half a stick of butter a day, but it's problematic when low-quality data such as personal experience is used as a foundation for recommendations that contradict the scientific literature.

Of course, sometimes self-trackers do get the right answers. Emeritus professor of computer science and engineering at the

University of California, San Diego Larry Smarr has been continuously tracking more than seventy biomarkers of his body for many years. In 2005, he noticed that he had dramatically elevated levels of inflammatory marker C–reactive protein. This turned out to be predictive of an inflammatory bowel disease (IBD) that was diagnosed much later. In his speech at a Quantified Self event, Smarr claimed to have self-diagnosed IBD before his doctors could notice that anything was wrong: "Now, my doctor who had done the colonoscopy had said, 'You don't have IBD. I'd notice it. You know I've done your colonoscopy; I've been inside, and you don't have it.' And I said, 'Like, you must be doing these all day long.' He said, 'Yeah, I do. I do dozens of these a day.' I said, 'So, that's why you don't have time to read the scientific literature[203].'"

Eventually Smarr had a successful operation on his colon, and it brought his C-reactive protein levels back to normal. A small detail: the surgery happened only after severe symptoms, including bloating and rectal bleeding, appeared[209]. This means that the same outcome could be expected without any self-tracking (unless the patient ignored blood in his stool). Meanwhile, there is growing consensus that in some cases excessive screening for disease can lead to financial, psychological, and physical harm due to overtreatment, unnecessary restrictions and procedures, and increased risk of false-positive diagnosis[210,211].

Nevertheless, a group of biohackers is particularly obsessed with "self-knowledge through numbers." Some scholars have described this cultural phenomenon as "data fetishism[212]." Others find the Quantified Self movement to be a form of epistemological anarchism that, in contrast with normal scientific procedure, is based on ruleless accumulation of observations[213] while ignoring the fact that numbers alone do not necessarily reveal truth. For example, you could measure the exact number of hairs on your head and notice that that number declines with age. That does not mean that hair transplantation will increase your lifespan. Or consider body temperature as a biomarker of infection. You could use antipyretic medication to reduce the number on the thermometer as soon as it rises, but this would not reduce mortality or risk of serious adverse events[214] in most cases. It would be like trying to increase the battery life of your phone by hacking the display to always show fully charged.

I'm not saying that data isn't important. Biomarkers such as glycated hemoglobin HbA1c can reflect average blood glucose levels over the past two or three months, which, together with fasting glucose levels, can be used for detection of diabetes and prediabetes[215]. High levels of LDL cholesterol may be indicative of excessive saturated-fat intake. Blood pressure should be monitored in older individuals to evaluate risks of stroke and heart disease. Inflammation markers such as C-reactive protein and interleukin-6 can be indicative of inflammation. But each of those examples is based on science, not someone's personal experience.

Researchers have been inventing and improving all sorts of methods of measuring aging. Some of the most sophisticated and widely used methods involve epigenetic clocks (such as PhenoAge[216], GrimAge[217,218], and DamAge[219]). The idea was initially developed by German–American geneticist Steve Horvath. Such clocks measure the accumulation of age-specific markers attached to someone's DNA and predict mortality and diseases (much better than one's chronological age). These tools may be highly informative, and essential to speed up drug discovery and human clinical trials. But even so, the improvement of biomarkers after an intervention does not necessarily mean that lifespan has been prolonged—unless the corresponding study was properly designed.

In polish writer and philosopher Stanisław Lem's collection *The Cyberiad*, there is a story in which two nearly omnipotent "constructors," or "wizard-robots," are accosted by a sophisticated space pirate named Mordon who demands that they give up their most valuable treasure—information. Mordon is obsessed with knowledge, and the constructors offer to fulfill his endless desire for data by creating an entity called the Demon of the Second Kind.

This demon is a reference to Maxwell's demon, a thought experiment proposed by physicist James Clerk Maxwell that hypothetically violates the second law of thermodynamics. The demon controls a massless gate between two chambers filled with gas. The demon opens the gate only for fast molecules coming from one side and slow molecules coming from the other, effectively transferring heat from one chamber to the other without usual limitations. The Demon of The Second Kind performs a similar trick, but with information instead of heat: it analyzes the random movements of air molecules, and when that movement produces data that "kind of makes sense," it records that data, offering Mordon an endless stream of information, all absolutely useless, burying the unfortunate pirate under heaps of paper.

How we acquire data and interpret it often matters more than the data itself. We can't achieve meaningful knowledge about life extension without relying on the scientific method. This is why I don't think most approaches used by contemporary biohackers have moved far beyond those of their unsuccessful historic predecessors. Neither unsystematic self-experimentation nor self-quantification provide an incremental increase of knowledge. But there is an old saying that, when you criticize something, you should offer something in return. It might be easier to name the few things that may work than to go through the countless things that probably won't.

Some of the most popular books, articles, documentaries, and blogs about longevity offer concrete and confident advice: *This incredible diet will change your life*; *These seven supplements will make you live longer*; *The secret to a healthy lifestyle is within reach*; *Just apply these three essential hacks to your body*. Confidence sells, especially when a referral link to buy promoted products is provided, or shares in a company that produces them are owned. The revenue is then used to promote the message, and a self-sustaining cycle is achieved. As a result, we find ourselves in a world where the salesman has an unfair advantage over a skeptical researcher in terms of disseminating information. For example, Andrew Steele's book *Ageless: The New Science of Getting Older Without Getting Old* was criticized for not providing enough health advice, but I felt that the author should be commended for not making things up.

When asked for advice, scientists are usually cautious, being aware of the problem of irreproducible research, of contradictions in some articles, and of the fact that new evidence emerges daily. So, the public turns to the salesman, who has no such concerns. However, I think I have found an acceptable solution to the problem of staying honest while submitting to public demands.

I invented a thought experiment. I imagine I am cornered in an alley and taken hostage by a notorious biohacker armed with a high-tech automatic rifle who threatens to kill me unless I give him a list of supplements or drugs he should take to increase his lifespan. The rules are simple: I will have to take the same pills, I must provide peer-reviewed justification for their use, and the risk of adverse effects may not exceed the benefits for a person without underlying medical conditions. Cheating by suggesting safe but useless placebo-medicine such as homeopathy is not permitted. Offering the standard "diet and exercise" will result in a bullet to my knee, which will hinder my future adventures as a traveling science lecturer. Also, the interventions need to be available for members of the general public, and not be some sci-fi invention.

In case I ever find myself in such a peculiar situation, I have prepared a short list of drugs and supplements to suggest. The first candidate molecule would be acarbose. This drug is similar to metformin in terms of lowering blood glucose levels and the risk of diabetes[138,220], but its adverse effects, which are usually limited to the gut[221], are less dangerous than those of metformin. Unlike metformin, acarbose increased the lifespans of mice of both sexes in three different studies by the Interventions Testing Program[143,222,223].

The second suggestion involves combining two amino acids: glycine and taurine. Glycine provided a small, yet consistent and statistically significant positive effect on mouse lifespan in an ITP study[224], conceptually reproducing previous experiments[225]. Being one of the amino acids that our body uses to construct proteins, glycine is probably safe to use. Its mechanism of action is still unclear, but glycine apparently reduces the levels of another amino acid, methionine, thus mimicking methionine restriction and activating autophagy[226]. Also, glycine is required for the synthesis of collagen. It is a minor amino acid in most other proteins, but collagen is an exception, with about a third of its residue being glycine. Some studies have found that increased glycine concentration promotes the production of collagen by isolated chondrocytes, the cells responsible for cartilage formation[227]. If collagen supplements actually worked, it would probably be via supplying additional glycine. I will cover the importance of collagen and other long-lived proteins in the "Fragile, but not that fragile" chapter.

Taurine is not a protein component, but it is among the most abundant amino acids in the brain, retina, and muscle tissues[228]. A 2023 study published in *Science*[229] found that serum taurine concentrations decrease by more than eighty percent in the elderly, when compared with the young. The same study showed that taurine supplementation can prolong the lifespan of mice and roundworms and improves healthspan in monkeys. Several biomarkers of aging are reduced in the process, including DNA damage, impairment of mitochondrial function, cellular senescence, and shortening of chromosome ends. Taurine's importance can be easily demonstrated: its deficiency is known to cause retinopathies and cardiomyopathy in animals[230]. Similar conditions develop in humans who suffer from genetic mutations that impair taurine transport, which are correctable by taurine supplementation[231]. Popular energy drinks typically contain taurine, but not all of their other components are useful, so the supplement route seems to be a better option.

The fourth supplement is a molecule called spermidine. As the name implies, spermidine was originally isolated from semen, but it is found in many living tissues. It can be produced by the gut's microbiota[232], and it is contained in rather high concentrations in dietary products such as chicken liver, dry soybeans, green peas, corn, shellfish, and blue cheese[233]. Several epidemiological studies[234] have found that people who consume more spermidine-containing foods have reduced cardiovascular and all-cause mortality[235], and some research has even linked dietary spermidine intake with reduced aging-related cognitive decline[236]. In one study it modestly increased the lifespan of male and female mice, presumably via an autophagy-activation mechanism[237]. Unfortunately, the ITP has not yet considered this molecule for evaluation, and a recent study found only health-span, not lifespan, benefits of spermidine consumption in male rats[238]. One of the proposed mechanisms of action for spermidine is the induction of autophagy[239] which will be discussed in the "Feast of famine" and "Whatever it takes" chapters.

The final items on my list would be coffee and tea. Large epidemiological studies on millions of people have revealed that moderate coffee consumption (two to four cups per day) is associated with roughly fifteen percent reduced all-cause mortality compared with no coffee use[240]. This includes a reduction of age-related causes of death such as cardiovascular diseases and cancer, and, as a bonus, a reduction in Parkinson's disease risk [241]. Similar longevity-promoting effects have been observed for tea consumption[242,243], and a recent study found that drinking both coffee and tea is associated with a reduction of up to twenty-two percent for all-cause mortality[242]. One important limitation is that people can be genetically predisposed to drink more coffee, and the underlying genetic variants may have health effects of their own, confounding the observed associations. Genetic variants such as in the CYP1A2 protein can also affect caffeine metabolism. For people with some of these variants, coffee may be toxic, increasing the risk of cardiovascular diseases[244,245]. You can check which variant you have with commercially available genetic tests. In any case, make sure you don't drink your tea or coffee too hot, because the consumption of high-temperature beverages is associated with increased risk of esophageal cancer[209].

Epidemiological studies are not ideal for inferring causation, but when they are taken together with mechanistic explanations (such as that caffeine induces autophagy in various model animals), I would say you are better off drinking coffee and tea than not, unless you have individual adverse reactions, the abovementioned genetic traits, or intolerance for components of these beverages.

Of course, I highly doubt that these interventions are sufficient to defeat aging. People vary in dietary preference, supplement use, and lifestyle choices, but nobody has come close to solving the aging puzzle. What has truly helped increase our lifespan over the last hundred years has been incremental scientific advancement. If aging will ever be defeated, it will not be because of some miracle mushroom from ancient Chinese medicine, but will be thanks to sophisticated gene therapies, modern drugs that regulate the activity of specific genes, cellular reprogramming, cell and tissue replacement, artificial organs and other cutting-edge technologies that we will discuss in upcoming chapters.

My main issue with the biohacking industry is that it offers false grails that take our attention away from finding real solutions. Why should we demand anything that actually works when we can alleviate our anxiety with a natural, non-GMO, eco-friendly antioxidant or a magical cocoa-powder brain elixir? Why should anyone invest in science when bullshit sells so well? Our actions and choices shape the societies we live in. If we really want to live longer, we should not be paying attention to overhyped miracle cures, and we should definitely stop enabling those who sell them. Let's incentivize state-of-the-art scientific research instead. Let's keep the snake-oil salesmen at bay; let's educate ourselves and those around us on the scientific method and biology. And for those who really want to experiment on themselves—why not volunteer for clinical trials.

The false grail lies to the left, a mix of herbs and ancient practices. To the right stands the true grail, embodied in the promise of gene therapy. The journey between them is arduous, tangled with thorns, requiring deep exploration into the secrets of human DNA.

CHAPTER 4
Evolution is not your friend

> And why do I deserve to die?
> I'm dominated by this animal that's locked up inside.
>
> —*Korn ("Evolution")*

S ummary: in this chapter we discuss why we age and why some animals evolved to live much longer than others.
On May 14, 2013, Angelina Jolie shared her story of undergoing a double mastectomy for breast cancer prevention[246]. Two years later she had her ovaries removed as well. The decision was based on genetic testing and a family history of disease. The actress's mother, Marcheline Bertrand, died at the age of fifty-six, after an eight-year battle with ovarian and breast cancer. Angelina's grandmother Lois, cousin Francine,

great-aunt Stella, aunt Debbie, and several other relatives also died from cancer.

The cause behind these tragedies was a mutation in a gene called BRCA1 (breast cancer 1) that can be passed through generations. Before her operations, Jolie had an 87% risk of developing breast cancer and a 50% risk of developing ovarian cancer during her lifetime, according to her doctors[247]. The actress's opinion piece in the *New York Times* brought public attention to the importance of preventing life-threatening diseases. The number of genetic tests undertaken by women surged during the following weeks[248], and rates of contralateral risk-reducing mastectomies doubled[249] after the announcement. Women with high-risk profiles were especially affected. Later this was called "the Angelina Jolie effect."

BRCA1 and BRCA2 (another gene linked to ovarian and breast cancer) are important for several reasons: they are involved in the processes of DNA repair, and in initiating the timely self-destruction of cells whose DNA is beyond repair. Thus, the genes encode proteins that function as tumor suppressors. Individuals carrying malfunctioning BRCA1 or BRCA2 genes have an approximately 70% chance of developing breast cancer[247] by the age of eighty if they are female (the risks for Angelina Jolie might have been higher due to other genetic factors indicated by her family history). The probabilities of other types of cancers are elevated as well among BRCA1/BRCA2 mutation carriers. These include a subgroup of leukemias and lymphomas[250], and prostate cancer for men[251]. The genes are so important that, even with the exclusion of cancer-related mortality, women and men with pathogenic BRCA1/BRCA2 mutations may lose several years from their average life expectancy[252]. If having these pathogenic mutations is so clearly disadvantageous, why hasn't natural selection removed them?

Some genetic disorders have simple explanations for their persistence. Humans have two copies of most of their genes. One copy we inherit from our mother, and one from our father. It's like having a backup of an important file in case something happens to the original. If one gene copy is disrupted by mutations, we can still benefit from the other. For example, carriers of a single malfunctioning copy of the CFTR gene[253] (cystic fibrosis transmembrane conductance regulator) are asymptomatic or almost asymptomatic[254]. But if two carriers meet, their children will have a 25% chance of inheriting two broken genes and developing cystic fibrosis (50% from one parent, multiplied by 50% from the other). This is a life-threatening genetic disorder characterized by frequent lung infections, as well as other, diverse symptoms, mainly caused by a buildup of fluids. Because carriers of a single pathogenic CFTR gene variant remain healthy, natural selection is less effective at eliminating the mutation, allowing it to persist in our population.

This explanation doesn't work for BRCA1 and BRCA2 genes, a single malfunctioning copy of either of which is sufficient for a significant shortening of lifespan. Mutations that have notable consequences even if a normal second gene copy exists are called dominant. Despite being dominant, pathogenic gene variants of BRCA1 and BRCA2 can be found in as many as one person per several hundred[255] tested. So, the question remains: how has evolution allowed these damaging mutations to persist in human populations?

As it turns out, there might be more to these genes than their role in causing cancer. The authors of one study compared the number of children born from carriers of damaged BRCA1/BRCA2 variants and controls[256]. To exclude the biases introduced by modern contraceptives and family planning methods, such as genetic testing, the main sample consisted of people born prior to 1930 who had two or more descendants with the undesirable mutations. People born during the same period, whose progeny had normal genotypes served as the control group. As expected, predicted mutation carriers had higher mortality rates, especially from breast cancer (18% of known death causes versus 3% in controls) and ovarian cancer (27% of death causes versus 1% in controls). However, carriers also had on average two more children (6.2 versus 4.2) and a one-year-shorter average birth interval (2.7 years versus 3.6 years).

Among people born after 1930 the reproductive success of mutation carriers was also higher—but to a smaller extent, presumably due to the increasing role of birth control. Similar but smaller fertility benefits were found for BRCA1/BRCA2 mutation carriers versus noncarriers from BRCA-mutated families in more modern samples[257]. The mechanisms behind this effect remain a mystery. Some genetic variants that are associated with increased breast cancer risk have been linked to greater breast size[258]. Although these findings do not include BRCA1/BRCA2 mutations, they provide an example of how human reproductive physiology, cancer, and genetics can interplay with one another. While this research is preliminary[259], the bottom line is that the negative impacts of BRCA1/BRCA2 mutations on later life are likely offset by some advantage that they give to the youthful.

Another proposed example of such evolutionary trade-offs involves Huntington's disease, which is caused by dominant mutations in the huntingtin gene. This disorder manifests itself in uncontrolled, uncoordinated, and involuntary movements; gradual loss of physical and mental abilities; dementia; mood swings; irritability; and an approximately thirteen-year reduction in life expectancy[260]. Once again, a single affected gene copy is sufficient for a clinical outcome, and an affected person has a 50% chance of passing the disorder to his or her offspring. No effective treatment exists. The severity of this condition was portrayed in the TV series *House, MD*. In one of the show's seasons, Dr. Remy Hadley, also known as "Thirteen," struggles with anxiety, knowing that her mother and brother died from the terrible disease. She even postpones genetic testing, afraid to learn whether she is sentenced to mental decline and preterm death.

The symptoms of Huntington's disease usually kick in between the ages of thirty and forty-five and take time to develop, without affecting the most important parts of the human reproductive period. Several studies have found that carriers of the disease are at least as successful in reproduction as noncarriers, despite being disabled for a portion of their adult life[261,262]. And in some populations[263] they are actually more successful, causing the frequency of the pathogenic gene variant to increase over time (with a predicted doubling in 150 years in Canadian Prairies[264,265]), although most studies show a neutral, rather than positive, effect.

Remarkably, transgenic mice carrying the huntingtin mutation that is pathological in humans have a longer lifespan, and have an increased number of offspring[266] if they are male. In humans the disease is associated with a lower risk of various cancers[267,268]. One hypothetical explanation for this antitumor effect is that the modified huntingtin protein increases the rate of naturally occurring programmed cell death in precancerous cells—the same mechanism that causes the death of nerve cells as Huntington's disease progresses[269,270]. Once again, the clearly detrimental effects of genetic mutation in late life appear to be offset by some benefits in early life.

A situation in which one gene controls several traits and there is a trade-off between the beneficial effects of one trait early in life and the negative effects of another trait later in life is called antagonistic pleiotropy. In 1957, biologist George Williams (celebrated for his gene-centered view of evolution, which was developed later in *The Selfish Gene* by Richard Dawkins[271]) proposed antagonistic pleiotropy as an evolutionary explanation for senescence[272]. The idea is that some genes that contribute to our aging also help our development, reproduction, and early survival.

The antagonistic pleiotropy hypothesis is related to the mutation accumulation theory of aging first described by Nobel Prize–winning British biologist Sir Peter Medawar in 1952. Medawar suggested that the force of natural selection acting upon an organism gradually decreases after the reproductive period is reached[273]. A mutation that somehow prevents maturation most certainly will not be passed to subsequent generations, but a mutation that causes a delayed disorder just might. In support of this proposition, recent research has shown that mammalian genes that are upregulated with age typically accumulate more genetic variation, suggesting less efficient removal of unfavorable genetic mutations by natural selection[274].

In an article called "An Unsolved Problem of Biology," Medawar, to illustrate his concept, proposed a thought experiment that involved a laboratory working with test tubes. These fragile glass objects don't age, but they are mortal in the sense that they can be accidentally broken by scientists. Let's assume that a lab constantly requires a hundred functional test tubes and ten percent of them break each month, mimicking natural predation. Each month the broken test tubes are replaced with new ones. If we look at the age distribution of test tubes under these circumstances, we will notice that no matter how long we run the experiment, the proportion of very old test tubes will be relatively small, despite the fact that they are not aging. If the test tubes could somehow produce progeny, most would come from younger objects.

Now imagine that, due to a manufacturing issue, all test tubes break after a certain period of use. If the test tubes broke within the first months, many replacements would be required. But if the spontaneous breakage occurred only after ten years, we probably wouldn't even acquire enough data to notice it, and there would be no reason to improve manufacturing.

The conditions of this thought experiment are quite different from what we see in the biological world. Biological "test tubes" are genetically variable, and thus have different predicted lifespans. A long-living biological test tube would probably have longer-living offspring, which is not taken into account in Medawar's simple model. Also, a biological test tube that has survived predation for multiple years may have acquired experience or grown to a larger body size or obtained other properties that help it survive, or excel at mating and outcompeting younger subjects. Nevertheless, Medawar makes a clear and important point: whether something breaks early or later on matters. Or, in his own words: "The force of natural selection weakens with increasing age—even in a theoretically immortal population, provided only that it is exposed to real hazards of mortality. If a genetic disaster . . . happens late enough in individual life, its consequences may be completely unimportant."

As geneticist Theodosius Dobzhansky once wrote, "Nothing in biology makes sense except in the light of evolution[275]." A "selfish gene," in Richard Dawkins's terminology, that facilitates an organism's early survival and reproduction at the cost of death and suffering at an older age might be very successful at spreading itself. For a longevity-associated mutation to share that favorable fate, it must also somehow improve fecundity. Now this is something scientists can test by experimenting with artificial evolution.

Evolutionary biologist Michael Rose studied the evolution of longevity in *Drosophila melanogaster* fruit flies. Some flies were brought up under regular conditions, while the eggs of others were collected at progressively postponed timings so that only the oldest females could pass their genes on to subsequent generations. Over many generations this intervention led to delayed fecundity, and an average increase of female and male lifespans by 29% and 15% respectively[276].

This clearly designed experiment shows how external conditions can influence the evolution of aging and how genes can influence lifespan. But natural evolution doesn't always favor the fast, the strong, the smart, or the long-living. This is why questions like *Why aren't all monkeys evolving into humans?* make no sense. Evolution does not strive toward an ultimate goal of one perfect species. The value of different adaptations depends on the conditions in which an organism evolves. What is perfect for one situation can be detrimental in another, leading to the diversity of life as we know it.

It is not surprising that animals have a wide distribution of lifespans, just as they vary in size, shape, and behavior. On one side of the lifespan spectrum we have the potentially immortal hydra and the *Turritopsis dohrnii* jellyfish, which can indefinitely switch between its sexually reproductive medusa form and its highly regenerative polyp form[277]; sponges capable of living for hundreds or even thousands of years[278]; species of clams living more than 400 years; 300-year-old deep-sea tube worms[279]; and Greenland sharks, with a lifespan of more than 250 years [280]. The oldest living specimen of the Greenland shark was determined to be 392 ± 120 years old, via radiocarbon dating of its eye lens, making it the longest-lived vertebrate to date[281].

In contrast, the roundworm *Caenorhabditis elegans* lives only 2 to 3 weeks, while the female mayfly *Dolania Americana* emerges from its aquatic larval stage to live for only several minutes in its adult form[282], just long enough to mate and deposit eggs in the river from which it came. Vertebrates can be short-lived too, with coral reef pygmy gobies holding the record by having just a 3.5-week maximum adult lifespan[283], less than the 2 to 3 months of life of *Drosophila melanogaster*[284].

When we look at mammals, we have the bowhead whale living more than 200 years and the forest shrew living for only 2. But we also have the pygmy sperm whale, which lives up to 17 years, and the naked mole-rat, which can survive for more than 39 years (we haven't observed this species long enough to know its maximum lifespan). This distribution of lifespan might appear random, but certain laws can be deduced. Larger animals tend to live longer—although some animals, including humans, exceed weight-predicted expectations. Flying animals live longer than with their counterparts of similar weight; so do animals that live in trees or underground[285]. Nocturnal and poisonous amphibians are more long-lived than their relatives that lack chemical defense[286]; but chemical offense, such as that of venomous snakes, does not provide such benefits[287]. Additionally, animals that live in groups tend to evolve longer lifespans[288]. Longevity-associated traits have one thing in common: they reduce extrinsic (externally caused) mortality by giving animals an easier time evading predation.

Consider a field mouse that by chance acquires some hypothetical mutation that allows it to live for ten or twenty years in good health. At first glance this seems like an extremely beneficial mutation, which natural selection should favor. But if the mouse has, let's say, a ninety percent probability per year of being eaten by a predator, the potential of the mutation is unlikely to be utilized. The mouse will simply not survive long enough to enjoy the rewards of the mutation and produce more offspring. For longer lifespans to evolve, extrinsic mortality must be relatively low.

This hypothesis has been confirmed by the experimental evolution of fruit flies. *Drosophila* populations placed in conditions of higher extrinsic mortality (when scientists act as deadly predators, removing large parts of the populations from the gene pool) evolve higher intrinsic mortality in comparison with similar populations experiencing lower extrinsic mortality (when scientists are more reserved[289]).

The picture, however, isn't always this simple. Interactions between predation intensity and the evolution of lifespan can be mediated by other factors, such as mating costs and the availability of food. The shift from body repair to reproduction when predation levels are high is typical for the "classic" situation, when mating costs are low and food is easily accessible, but outcomes may be different in other conditions[290]. Experimental evolution on roundworms has revealed that while greater random extrinsic mortality does lead to the evolution of reduced lifespan, if it selectively affects only less-fit worms the evolution of an increase in lifespan is possible[291]. Also, increased early-life reproduction is not always coupled with decreased lifespan in these animals[292]. Sometimes you have to pay the price, but sometimes you don't.

Evolutionary trade-offs aren't always between genes that favor the young and disfavor the old. The genetic variants responsible for sickle cell anemia and thalassemia are widespread in Africa, India, and other parts of the world that have something in common: malaria[293]. While it takes two mutated gene copies to acquire one of the just-mentioned heritable blood disorders, a single "bad" copy provides resistance against the dangerous infection[294]. So, the increased risk of one malady becomes tradable for a reduced risk of another, but only when the malaria parasite is present to guide evolution.

Another example of an evolutionary trade-off is provided by Laron syndrome. In this case a mutation in the growth hormone receptor leads to dwarfism and obesity, but carriers of two mutated gene copies enjoy an increased resistance to cancer due to enhanced programmed cell death and recycling of damaged cellular components[295].

Perhaps the most thrilling tale of pleiotropy comes from the world of insects. The small parasitoid wasp *Cotesia glomerata* lays its eggs in caterpillars. To protect the developing larvae, the wasp uses a biological weapon: viral particles that suppress the host's immune system. Unfortunately for the parasitoid, these viral particles can also influence the interaction between the caterpillar and the plant that it eats, causing an increased production of volatile molecules that attract hyperparasitoids, or parasitoids of parasitoids that arrive on the scene to lay their own eggs in the progeny of the original wasp[296]. Now, that's karma.

The antagonistic pleiotropy hypothesis partially answers the question of why many life-extending traits that can be introduced in the lab, with artificial selection or genetic engineering, do not evolve by themselves in natural conditions.

For example, mutations in a gene called *daf-2* can double the lifespan of *Caenorhabditis elegans* worms while also maintaining their youthful behavior[297]. *Daf-2* mutants remain physically active at an age when normal worms stop moving. Aside from their "young looks," longevity, and physical activity, the *daf-2* mutants are also more resistant to bacterial pathogens[298]. However, their fitness is reduced[299]: normal worms outreproduce their long-lived counterparts in a shared environment with resource competition.

There are other examples of one genetic change both increasing the worm's lifespan and decreasing its ability to reproduce. A longevity-promoting protein encoded by the *daf-16* gene is known for its ability to enter a cell's nucleus, where it binds DNA and stimulates the activity of genes that improve the worm's resistance to DNA damage, oxidative stress, and other negative factors. But an artificial increase of *daf-16* activity leads to slower reproduction[300].

Stress-resistant long-lived *age-1* mutants thrive and outcompete normal worms under certain harsh environmental conditions due to their thermotolerance[301,302]. Under normal conditions they have higher hatching success in the later parts of their reproductive period, but this is ultimately offset by their reduced fecundity in early life[303].

Taken together, these examples illustrate evolutionary reasons why immortality, agelessness, or simply the capacity for a long lifespan are not universally widespread in the animal world. Some people might regard this as confirmation that death is natural and we shouldn't mess with the "intended" order of things. But this is a flawed conclusion. Evolution does not have intentions; it is not some wise, divine being whose decisions we should respect, and it is certainly not our friend. Evolution does not have a plan or serve a greater purpose. It is merely an uncaring parent who knows not what the future will or should be.

If evolution has no plan, there is no way we can mess it up. The attribution of intention to natural selection arises commonly because human beings suffer from generally childish tendencies toward teleological thinking[304] (for example, *The rain exists to water the plants*; *Mountains are here so we can go hiking*; or *Flowers grow for our delight*), the projection of our way of thinking about other human minds onto mindless objects and processes. Ironically, this common flaw of reasoning is a by-product of our imperfect evolution.

While nature does not have intentions, we do. Let's assume that the pathogenic BRCA1/BRCA2 or huntingtin gene mutations increase the average number of offspring and provide some form of evolutionary advantage to their carriers. Given the choice, would you rather have that advantage or would you prefer a longer and healthier life? Would you be cheered up by studies suggesting a positive reproductive side to your suffering from a life-threatening condition?

CHAPTER 4

Unlike natural phenomena, people have goals, which don't always align with the best ways of spreading our genes. We are a product of evolution, but we don't owe it anything. In fact, we have been altering this history of life by using technology and medicine that has saved the lives of millions who would otherwise have perished. Friedrich Nietzsche said, "God is dead." Perhaps evolution should be next in line, replaced by the more humane alternative of self-improvement termed "autoevolution" in the works of Stanisław Lem. This is because evolution works by trial and error, which means the deaths of organisms that are "unfit." That is the price that our ancestors had to pay during the billions of years it took to get us from the state of unicellular organisms to where we are now as a biological species. Perhaps the time has come for humanity to take responsibility for how we change and how long we live.

The so-called evolutionary tree of life is, in truth, a tree of death— its branches nurtured by the demise of countless living beings, driving the engine of natural selection. Yet, four individuals resolve to seek a different path forward.

CHAPTER 5
Death after sex

> That's the price you pay.
> Leave behind your heart and cast away.
>
> —*Imagine Dragons ("Natural")*

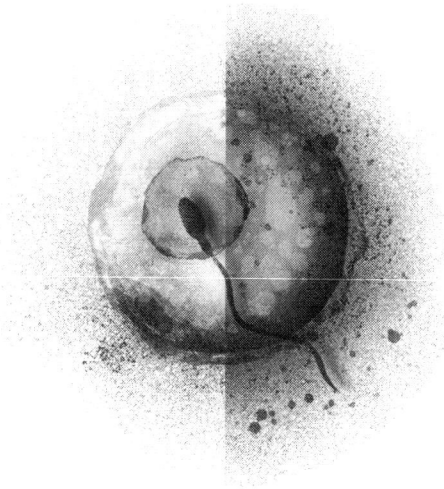

S ummary: in this chapter we further discuss the conflict between survival and reproduction, the driving force behind the longevity of our species, and whether aging is caused by an inbuilt adaptive "death switch".

A hectocotylus is a special tentacle of male octopuses and some other cephalopods that is analogous to a penis. Members of some species use it "the old-fashioned way," but others self-amputate this appendage and present it to the female. In the strangest scenarios the tentacle assumes the role of a swimming homing missile that searches for someone to copulate with. Octopuses are exceptional in other ways too. Some are smart enough to hide and ride the seafloor inside an empty coconut, proving their ability to use tools. This is why I have chosen to stop eating them. They are also adept at solving mazes, disguising themselves, and even mimicking other animals. Even in death they surprise us.

After laying its eggs, the female bumblebee octopus, also known as the Caribbean two-spot octopus (*Octopus hummelincki*), stops feeding and develops a propensity for self-injury[305,306]. She may then self-cannibalize by biting off the tips of her tentacles, beating herself against solid surfaces, and even tearing her own skin. She also exhibits strange, excessive grooming behavior with the simultaneous use of all available tentacles. In 1977 psychologist Jerome Wodinsky tested whether this rapid mental and physical decline was controlled by a pair of optic glands situated between the octopus's brain and optic lobes. These organs produce various hormones and are somewhat analogous to our pituitary gland, which produces signal molecules that distantly regulate growth, metabolism, and reproduction[307]. The hypothesis turned out to be correct: surgical removal of the optic glands canceled female octopuses' biological self-destruction. The animals resumed eating, continued to grow, and survived up to 277 days (175 days on average[308]), far exceeding the usual 40-day survival period without the operation.

Soon researchers described several molecular pathways that are related to the post-reproductive programmed death of the octopus[305]. For example, the transition from feeding to fasting and the terminal stages of self-destruction is accompanied by increased production of a protein that is similar to one produced by some tumors in mammals[309] and even in insects[310,311]. In humans the production of this protein leads to muscle wasting and cachexia, a problem affecting more than half of cancer patients[312]. Thus, there is a certain similarity between the "death switch" of an octopus and how some people die.

Another pathway involved in the death of female bumblebee octopuses is increased steroid hormone production in their optic glands. Steroids play an important role in the death of another animal with a built-in self-destruction mechanism, the Pacific salmon[313]. These fish are known for their migration patterns. They hatch in a river, spend several years feeding in the ocean, and return to their freshwater place of birth to reproduce. During their return journey, they experience progressive organ degeneration, including in their nervous system[314], as their adrenal glands expand and plasma levels of the steroid hormone cortisol dramatically increase. High cortisol levels reduce the activity of the immune system, leading to infections. The Pacific salmon may begin to rot while still alive; hence, it is sometimes referred to as a "zombie fish." But unlike brain-hungry fictional zombies, salmon do not survive for long: their death happens soon after spawning.

Humans share the ability to produce cortisol. We know it as a stress hormone produced by the adrenal glands in response to signals from the pituitary gland. Cortisol triggers a release of sugar from the liver, providing fast energy to help us deal with stressful conditions. Chronically high levels of cortisol can be detrimental to health. Excessive cortisol-like medication, as well as tumors in the adrenal or pituitary glands, can lead to extremely elevated cortisol levels, causing Cushing's syndrome, which is characterized by bone and muscle weakness, fragile skin, fatigue, abdominal obesity, and increased mortality rates[315].

As in the case of the octopus, the programmed death of salmon can be prevented with surgery—in their case by removing their gonads[316]. This reduces the production of sex steroids that drive elevated cortisol levels[317]. In some salmon species death can also be delayed by captivity and avoidance of reproduction, while forced feeding can further sustain the fishes' lives and even allows them to regain color, vigor, and weight[318].

Another example of programmed self-annihilation is the "blue wave of death[319]," or death fluorescence, found in the roundworm *Caenorhabditis elegans*. When a worm is about to die, a bright burst of fluorescence propagates from the frontal parts of its intestine toward its rear end. Death normally occurs within two hours of the start of this process. Death fluorescence can be triggered spontaneously with old age, but also by external causes, such as heat, freeze-thaw, changes in acidity, and some infections. This fluorescence is not the cause of death, but merely an indicator of an underlying cascade that involves the self-destruction of intestinal cells via a process called necrosis.

Unlike the programmed cell death called apoptosis—which protects animals from cancer by removing malfunctioning cells—necrosis, triggered by different molecular mechanisms than apoptosis, leads to an uncontrolled release of cell-death products into extracellular space and is rarely beneficial to an organism. In the case of death fluorescence, necrosis of intestinal cells triggers an influx of calcium ions into neighboring cells, causing further necrotic events. This "death switch" can be turned off by deactivating various genes involved in this signal transmission. One such gene is called *inx-16*. The innexin proteins it encodes form channels called gap junctions between cells and allow calcium to transfer from one cell to another. Mutated worms lacking such channels are long-lived and can survive conditions that would otherwise trigger the blue wave of death.

Similar damaging mechanisms can be found in other animals. In mice a channel called connexin-36 is implicated in the spread of cerebral ischemia—acute brain injury caused by the impairment of blood flow to the brain. Deficiency of connexin channels is neuroprotective in mice. It leads to better functional outcomes and decreases brain damage after ischemia[320] (reduced blood and oxygen flow). The protein levels of some connexin channels have also been linked to the progression of traumatic brain injury in humans[321]. Normally gap-junction proteins play an important role in communicating electrical signals between neurons, coordinating heart muscle cells to synchronize contractions, and passing chemical signals and metabolites between various types of cells, but during pathological processes they may propagate injury between coupled cells, leading to severe organ damage and even death in some cases[322,323].

These three examples show that the death of some animals can be triggered by specific "death switches" and that certain components of these switches are present in mammals. This and the fact that members of different species have distinct and dramatically varying lifespan characteristics has led some researchers to consider whether the aging of most organisms could be genetically and "purposefully" programmed[324], perhaps even adaptive. Gerontologist Vladimir Skulachev proposed the term *phenoptosis* (similar to *apoptosis*[325]). If aging is indeed controlled by some heritable program, we might benefit from finding ways to turn it off. But before we start removing pituitary glands and gonads or changing our genes, let's recall an important piece of biological wisdom: humans are not octopuses.

Drawing similarities between humans and other animals requires careful consideration. Let me give you a vivid example. In a species of the cannibalistic orb-weaving spiders of the genus *Argiope* the male spider experiences programmed death after copulation. This is not the typical scenario of a female spider killing a male as he tries to escape after sex. The male *Argiope aurantia* dies without any participation from the female, right after it inserts its second sperm-containing pedipalp appendage into the female's genital opening. Death occurs via rapid cessation of heartbeat[326].

Researchers studying this species know that the female is not at fault, not only because biting does not take place prior to the male's death but also because of anecdotal observations of a male spider mistakenly attempting to copulate with the carcass of a mealworm beetle and dying in the same manner. It appears that death is triggered by the increase in hemolymph (spider blood) pressure necessary to inflate the pedipalp appendage. The evolutionary advantage of having such a death switch might not seem obvious, but there is one explanation I personally find amusing:

"Males fight over access to the moulting female, and try to dislodge any male that has a palp inserted. Because the palps of dead males are fixed in the inflated state and are therefore harder to remove, dead males act as whole-body mating plugs, often preventing other males from copulating[326]".

According to a recent study published in *JAMA Cardiology*, "Sudden cardiac death [in humans] may occur in various circumstances, including physical exercise and sexual intercourse," and "0.2% of natural deaths that underwent autopsy were associated with sexual activity and predominantly involved middle-aged men[327]." I believe it's safe to assume that death after sex does not serve a whole-body-mating-plug function in humans. This role is fulfilled by a social phenomenon called jealousy and the activity of hormones, such as oxytocin and vasopressin that influence pair bonding[328]. Hence the abovementioned caution for translating scientific findings from one species to another.

To understand whether humans have anything resembling the "kill switches" of other animals, we should understand how programmed organismal death evolves and where it can be found in nature. Is it an exception or a rule? Is it in any way related to conventional aging?

First of all, it should be noted that this phenomenon is not limited to animals. Plants have special molecules to trigger programmed senescence—such as the gaseous hormone ethylene[329], which we use to accelerate fruit ripening, as in the case of the banana from your local supermarket[330].

A number of plants are monocarpic: they reproduce only once and are destined to die after flowering. Examples include bamboo, yucca, and agave. The relevant fundamental biological property of such plants is the massive translocation of resources from roots, stems, and leaves to their reproductive organs. The devotion of virtually all available resources to producing progeny makes death inevitable for many of these organisms[331]. This is known as the "big bang" reproductive strategy.

This reproductive strategy has both positive and negative properties. The risks include possible total reproductive failure if the single act of reproduction meets unfavorable environmental conditions. This can be partially circumvented by adaptations such as using dormant seeds, and offspring's use of the nutrients and the convenient growth sites provided by the deaths of adult plants. The advantage is that the relative reproductive output of monocarpic plants can be much higher than that of non-monocarpic plants. Such features can evolve gradually when mortality rates between first and second reproduction attempts are high. If a second attempt is less likely to become successful, the first attempt becomes more valuable, so more resources are dedicated to it, further lowering the success of subsequent reproductions. The "big bang" reproductive strategy thus becomes an adaptive evolutionary conclusion.

The case of the Pacific salmon is similar. Its reproduction has many costs, including lengthy migrations and difficult transitions between the ocean and freshwater[332]. The likelihood of repeating the difficult task of spawning isn't great, so it makes evolutionary sense for a salmon to invest all its resources in a single reproductive attempt, even at the cost of the salmon's life. This ultimate sacrifice of the parent is not in vain: the additional resources increase egg number, egg size, and offspring survival. The "selfish" genes that control such behavior and physiology will be passed on to subsequent generations.

Similar trade-offs occur in some marsupials, such as *Antechinus stuartii*—a small, shrew-like animal[333]. In the wild, females of this species breed with multiple males, resulting in sperm competition[334]. In such conditions, evolution favors increased relative testis size and sperm production in males, ensuring greater fertilization success and more offspring. In this species the time window for mating is small, and surviving for a second reproductive season is unlikely. As a result, males end up having sex until death. In captivity, these animals are able to copulate for up to fifteen hours nonstop[335]. Eventually the deaths of males typically occur during mating season, as their cortisol levels rise and their immune systems collapse. If there is a proper way to go, this is probably it. If only this pleasurable ending weren't darkened by the various infections, parasites, and ulcerations that a dying male has to tolerate while pursuing his evolutionary objective.

The evolutionary cause for the "death switch" in octopuses is less obvious. One hypothesis is that self-destruction prevents cannibalization of offspring by their mother[305]. It is known that octopuses can feed on their own kind and that females often kill males after reproduction, just like many female spiders and praying mantises. Perhaps the complete cessation of feeding behavior was simpler to evolve than the development of the means of discriminating between one's own offspring and a tasty snack. The deaths of larger, older octopuses may also lead to increased food availability and reduced competition for the young.

As for the evolutionary cause of the roundworm *Caenorhabditis elegans*'s necrotic cascade, scientists are still divided. These worms come in two sexes: males and hermaphrodites, with males being typically more long-lived. As we discussed in the previous chapter, increased intrinsic mortality may evolve under circumstances of increased extrinsic mortality. Excessive extrinsic mortality for hermaphrodites often comes from unfavorable conditions, such as the absence of food, which leads to internal hatching of larvae, which then eat their parent[336]. This reduces the evolutionary importance of a longer lifespan. In harsh environments, a programmed death mechanism might actually improve the survival of the young by supplying them with nourishment. If that is the case, perhaps *Caenorhabditis elegans* deserves the Mother of the Year award. Or Hermaphrodite of the Year, to be more accurate.

Another hypothesis is antagonistic pleiotropy: mutations that prevent the blue wave of death could be detrimental for early-life reproduction. Although worms deficient in innexin-16 live longer and have better survival, they grow slowly and have a starved appearance. Other necrosis-protected mutants have been found to benefit from increased resistance to death caused by temperature or infections, but that doesn't necessarily translate to reproductive success. In fact, it doesn't always lead to increased average lifespan, as some mutants turn out to be short-lived[319].

It is easy to notice that known cases of programmed death evolved under circumstances that are not applicable to humans. Although we allocate a lot of resources to reproduction, the amount isn't usually critical, and we have many chances to mate. Our young do not typically cannibalize us, and we do not tend to eat our young; in fact, we usually take care of them and invest in their well-being for many years after their birth.

The investments that humans make in their immature offspring might be a key to explaining a rather unique human property: we don't die after having all the reproductive sex we can get. Or, in scientific terms: we have greatly extended postmenopausal longevity[337]. If evolution is all about spreading genes, then there should be almost no post-reproductive periods in the life cycles of normal species. Even in chimpanzees, survival rates and reproductive ability are reduced at comparable paces with age, yet we are predisposed to maintain relatively low mortality rates even after ceasing reproduction entirely[338].

The phenomenon of menopause in women can be partially explained by the risk of mortality due to pregnancy and childbirth. Upon having a certain number of offspring, it might be beneficial to dedicate more resources to assisting the survival of already-existing children. This effect can be taken a step further, as grandparents may contribute to the survival of their grandchildren. Observations of still-existing hunter-gatherer tribes, such as the Hadza of northern Tanzania, reveal that the efforts of grandmother gatherers provide substantial nutritional benefits for the children of their daughters[339]. Food sharing might be especially important for humans because of our large brains, which require a lot of energy. This raises the question of whether our post-reproductive periods are truly post-reproductive or, as biologist George Williams wrote, "no one is post-reproductive until his youngest child is self-sufficient[272]."

Aged individuals are important carriers of skills and knowledge that can be passed down to subsequent generations.

Many of us are lucky to know this from personal experience. I have a vivid memory of spending time with my grandparents at the Kozinka River, near Kiev. My grandfather taught me mathematics, French, and chess, and my grandmother, with her background in medicine, once saved me from a nasty tick. As human culture evolved, having caring and experienced relatives became ever more beneficial, giving an evolutionary advantage to carriers of longevity-promoting genes. This is a realistic explanation of why we live much longer than other primate species.

Recently an interesting, although speculative, addition to the idea of grandparental importance in the evolution of human longevity was published in an article called "The active grandparent hypothesis: Physical activity and the evolution of extended human healthspans and lifespans[340]." The authors tried to explain why lifelong physical activity is so strongly linked to age-related mortality and morbidity. They argue that in the past grandparents would have been most helpful for their grandchildren if they remained active as hunters or gatherers. Thus, natural selection could favor healthy post-reproductive longevity in people who maintain high levels of physical activity at an older age. The more active the grandfather, the more valuable he is to the survival of his grandchildren, and the stronger the selective pressure for his survival. In any case, it is likely that we owe our relatively long lives to the caring efforts of ancestral generations of grandmas and grandpas.

So, is aging programmed? Nobody denies that "death programs" exist in some species, although none have been found in humans so far. Also, known examples of "kill switches" bear little resemblance to "regular" aging, which proceeds normally even if these mechanisms are somehow disabled. Neither the salmon nor the octopuses or roundworms become ageless or immortal once their "death programs" are turned off.

The true controversy is in whether aging organisms can have a selective advantage over their hypothetical ageless counterparts. Perhaps aging is truly universally beneficial for gene spreading, and thus it is selected for? Perhaps it helps prevent overpopulation, frees up resources for the young, accelerates the turnover of generations, promotes faster adaptation in a population, and removes individuals with great burdens of transmissible diseases, and thus increases the chances of group survival? This isn't an entirely crazy thought. Some bacteria have abortive infection systems that facilitate altruistic cell death in response to viral infection and protect the entire bacterial population[341]. Could aging play a similar role for animals?

The idea of an adaptive genetic program of death and aging arising through natural selection was first presented by biologist August Weismann, in 1891. He argued that eliminating older, worn-out members of a population could benefit their remaining kin:

It is of no importance to the species whether the individual lives longer or shorter, but it is of importance that the individual should be enabled to do its work toward the maintenance of the species. This work is reproduction, or the formation of a sufficient number of new individuals to compensate the species for those which die. As soon as the individual has performed its share in this work of compensation, it ceases to be of any value to the species, it has fulfilled its duty and it may die[342].

Harsh words! While group selection may be important in certain cases, especially in clonal populations (such as bacteria), which are genetically homogeneous, there are several strong arguments against the adaptive theory of aging. First of all, it seems unlikely that an aging mechanism would be adaptive for nearly all species on the planet despite the numerous variations in ecological conditions and genetic backgrounds. Yet agelessness is an exception reserved for very few organisms. Second, even the most vital genetic mechanisms can be broken by mutation, but we don't see any spontaneously appearing immortalized individuals.

Let's use a metaphor in which a simple multicellular organism is a species and its cells are individual members of that species. From that organism's perspective, various genetic mechanisms that promote programmed death of damaged cells can indeed be beneficial for survival. However, no matter how rigorously these mechanisms operate, sooner or later selfish mutant cells that avoid self-destruction will come into existence and prevail because of their inherent evolutionary advantage (unfortunately, causing cancer). If a genetic mechanism were all that prevented organisms from becoming ageless, it would surely fail from time to time, with spectacular results.

This does not match our observations. Let's take humanity as an example. Our population has now exceeded eight billion. The human genome consists of around three billion nucleotides distributed among twenty-three chromosomes. Current public databases cover more than one billion verified single-nucleotide genetic variations called polymorphisms[343]. Meanwhile, every child has around fifty novel mutations that were not present in either of his or her parents[344]. This means that for almost any nucleotide position in the human genome there should be some living person with a mutation in it—unless that mutation is lethal. Yet, despite all this genetic diversity, we don't see people who do not age. Perhaps some rare combination of mutations could do the trick, but so far, no evidence of this has been seen.

Let's pause for a second, because there is a peculiar apparent exception. In 2009, endocrinologist Richard Walker and his colleagues published the case of a sixteen-year-old girl named Brooke Greenberg[345] a girl who had stopped growing. Despite being sixteen years old, the girl looked like an infant, with the biometry of an eleven-month-old. Unfortunately, her educational performance was poor. Tests of her intellectual development were similar to those of someone between one and eight months old. Her lack of growth did not translate into lack of aging or into immortality: she had shorter than usual telomeres (the ends of chromosomes; they shorten with age in most tissues), and she died at the age of twenty. Several other patients with this mysterious syndrome were later identified and had their epigenetic clocks (which measure age-related reversible DNA modifications) compared with healthy controls.

No differences in this measure of biological aging were observed[346].

There is also a rare disorder called pituitary dwarfism, which is characterized by reduced production of growth hormone. Affected people in their mid-thirties can look like teenagers. Unlike the most common and recognizable kind of dwarfism, called achondroplasia and caused by a different biological mechanism, this rarer kind of dwarfism preserves the regular proportionality of the human body while delaying and diminishing growth. As in the case of Brooke Greenberg, this does not prevent other aspects of aging, or death.

In 1932, marine biologist George Bidder proposed his hypothesis of the cessation of somatic growth aging, which suggested that the decline of our health could be partially a by-product of regulatory mechanisms that are necessary for maturation but unfortunately continue their work after their job is done[347,348]. While the provided examples of people continuing to experience symptoms of aging despite growth reduction seem to refute this hypothesis, we should consider that the underlying health conditions might have multiple effects, both beneficial and detrimental for lifespan. The bottom line is that while growth and maturation can be slowed by some not-fully-understood mechanisms, we have yet to find any examples of aging being completely turned off in a representative of a species that ages normally.

An alternative to the adaptive aging hypothesis is the idea that aging is detrimental or neutral to fitness but difficult and costly to avoid. This is called the disposable soma (or disposable body) theory of aging and was initially formulated by biologist Thomas Kirkwood[349]. According to this theory, there is a biological trade-off between maintenance, reproduction, and growth. At some point excessive body repair becomes detrimental to fitness because an organism has limited resources, and their allocation to extended lifespan is accompanied by fewer investments in reproduction and growth. Likewise, disproportionately elevated growth and reproduction can result in diminished body maintenance and a shorter lifespan.

Unfortunately for us, from the evolutionary standpoint reproduction and growth are essential but the body is merely an instrument for the effective spreading of genes, just as a pine cone exists only to protect the seeds it contains from cold weather and hungry animals.

In to the disposable soma theory, our bodies, including our brains—with all their complex cognitive functionalities, behavioral patterns, and emotions, everything that makes us human—are reduced to disposable vessels for their first-class passengers: our reproductive cells. From this viewpoint, the quest for radical life extension can be seen as a revolution of the oppressed hardworking cellular majority against the privileged one percent (actually a much smaller proportion) that are not as concerned with the fate of the vessel because they have lifeboats, a rocket to Mars, or some other metaphorical escape vehicle. But it's the vessel we want to preserve, because that vessel is us.

One remarkable illustration for the disposable soma theory in its modern version is longevity distribution in ants. Worker ants of the species *Temnothorax nylanderi* live an average of 250 days and rarely beyond three years, while queens can live for decades. This difference is explained by evolutionary theories of aging: natural selection works in favor of increasing life expectancy only when it promotes reproduction. Worker ants do not reproduce, so long-term preservation of their bodies is not critical for vertical gene transfer. They are disposable.

The situation for the worker changes dramatically once when we introduce a new player: *Anomotaenia brevis*, a parasitic species of tapeworm, armed with hooks and suckers, for which the ant serves as an intermediate host. While the fate of an individual worker ant may not be vital for the reproductive success of its colony, it is crucial for the success of the larval worm and the spread of its own genes. After all, the larva needs that ant to survive until it is eaten by the parasite's definitive host, the woodpecker. To achieve the goal of extending the ant's lifespan, the larva changes the activity of hundreds of ant genes[350,351] until its mortality rate becomes almost identical to that of its queen. An infected ant also receives more care from other ants, likely due to elevated production of chemical attractants.

In a similar example, the lifespan of the large flour beetle *Tenebrio molitor* (especially the female's) is extended by the larva of a rat tapeworm[352]. The exact mechanism is not known, but it may be related to redistribution of resources in the host organism. Infected individuals have reduced fertility and spend less effort on reproduction. The selfish genes of the parasite, unlike the genes of the beetle, are not "interested" in the reproduction of the host, but are "interested" in preserving the host's body to increase the likelihood of its being eaten during the worm's lifetime. Ironically, the beetle's body (soma) is, in a sense, less disposable from the point of view of the parasite than that of its own germ line. Yes, that's the kind of relationship the hardworking body and the reproductive "elites" have.

In *The Extended Phenotype*, evolutionary biologist Richard Dawkins argues that the idea of a phenotype should not be limited to the biological processes that take place inside an organism but can also include the influence of genes on the environment, and even on other animals. Ant and termite mounds, beaver dams, spiderwebs, and bird nests are all examples of the extended phenotype. The extended phenotype also includes the influence of parasites on their hosts—such as the hairworm manipulating a cricket to jump into water, allowing the parasite to swim in search of a breeding partner. Dawkins writes that animal behavior tends to maximize the survival of the genes responsible for that behavior, regardless of whether those genes are in the body of the animal that performs the behavior. This is exactly what we see in the case of longevity-promoting parasites.

To illustrate this concept further, let me give you a final example of a parasite protecting its host. The worm *Pomphorhynchus laevis* uses the small amphipod crustacean *Gammarus pulex* as one of its hosts and a fish as another. In the early stages of its development the larva cannot infect the fish, so it is interested in saving the life of the crustacean and forces it to hide. When the larva reaches the necessary age, this behavior is reversed and the host's risk of being eaten increases[353]. Once again, the tendency toward longer or shorter lifespan is dependent on the needs of the selfish genes that are in control of the situation. Unfortunately, even the almighty parasites have so far failed, or refused, to grant agelessness or immortality to their hosts.

Archimedes once said, "Give me a lever long enough and a fulcrum on which to place it and I shall move the world." There is an appeal for biologists to say: *Give me a program of aging important enough and the genetic editing tools to disable it and I shall make you ageless.* Unfortunately, this doesn't seem to be achievable.

Known death switches don't work as we might hope. Turning them off can prolong the lives of octopuses, salmon, and roundworms, but it doesn't stop aging entirely. Furthermore, the mechanisms by which death switches operate differ from those of regular aging. We also don't need death switches to explain the exponential increase in mortality rates—this is easily accounted for by damage accumulation, as described by simple mathematical models of aging[354]. It appears, after all, that "aging is a product of evolutionary neglect, not evolutionary intent[355]".

This implies that there are no shortcuts. Dealing with aging will require addressing the various types of molecular, cellular, tissue, and organ damage that accumulate over time. It is unlikely that any single treatment will be good enough, so different causes of aging will need to be addressed simultaneously. While we shouldn't count on a single longevity gene or immortality pill, that doesn't mean we can't succeed in our quest against death. After all, the genetic programs stored within our cells hold all the necessary instructions for creating each and every organ of our body in its youngest and most functional form. Surely maintaining these structures shouldn't be more difficult than creating them from scratch. Or maybe it is?

Our bodies are like the Titanic, bound for an inevitable collision. Our germline cells are the privileged few who make it onto lifeboats. Yet, we are more than just these few survivors—we are the entire ship, every cell woven into the vessel of our being.

CHAPTER 6
Feast of famine

> You never should settle for
> the lifetime that is handed to you
> Feast or famine!
>
> —*Black Friday the musical*

S ummary: in this chapter we discuss calorie restriction, the molecular pathways that link nutrition with aging, autophagy and the rate-of-living theory.

In Neil Gaiman's TV series *Good Omens*, based on his and Terry Pratchett's novel of the same name, we get to see a modern version of the Four Horsemen of the Apocalypse. Pestilence has been substituted with Pollution, and Famine has become a thin businessman, a best-selling author on foodless dieting who enjoys spreading the art of fine cuisine with extremely small portions at hefty prices. "I have never seen a roomful of rich people so hungry before," Famine notes delightedly while observing the guests of a highly rated restaurant. He would be pleased with the popularity of fasting-related longevity diets.

The public's excessive fascination with fasting and methods of undertaking it has led to some awkward situations. In 2020, Yoshinori Ohsumi, a Nobel Prize laureate for his research on autophagy—the process of self-devouring through which cells get rid of unnecessary or dysfunctional components—was invited to give a lecture in Moscow. The organizers attempted to attract attendees by describing him as an "author of a unique intermittent fasting method" and titled his lecture "Autophagy and the Peculiarities of Longevity-Promoting Fasting." A large gathering of people attended, but many were puzzled: instead of practical health advice, they received a rather complex lecture about "protein-protein interactions," "P13 kinase complexes," and "ubiquitin-like conjugation"—concepts that might be too difficult for non-biologists to understand.

During the event's Q and A session attendees probed the presenter on diets and fasting intervals. To these people Ohsumi apologized for not having answers. "I am a researcher who worked, basically, with yeast. . . . I have not deeply studied how autophagy affects longevity." Eventually one attendee said, "I think many people gathered here because you were described as an author of a unique intermittent-fasting method and they would like to hear from you that you have not authored such a method, that there are no official sources where one could read about it, and that this does not comply with the scientific worldview." Ohsumi replied, "I have never said that fasting helps autophagy; I have never made this claim. Apparently, something went wrong. Maybe I will change my opinion when I realize that fasting is very healthy, but this was probably a misunderstanding."

It's remarkable that although Nobel Prize–winning biologists are unsure about the health effects of fasting you can find dozens of best-selling authors claiming precise knowledge on the matter. Indeed, there is little doubt that overeating leads to excessive weight gain, which can be a serious health problem[356]. A body mass index (defined as one's weight in kilograms divided by the square of one's height in meters) greater than 35 is associated with an approximately 30% increase in mortality rate[357]. But is there evidence that eating less can promote the lifespan of people without excessive weight?

Calorie restriction (CR), or undernutrition without malnutrition, has, since 1935, been known to increase the lifespan of certain mammals. This phenomenon gained increasing attention in 1986 when scientists revealed how dramatic the effect can be. They reported a comparison of several groups of mice. The mice in one group ate as much as they wanted, while others received roughly 75%, 45%, or 35% of the normal calorie amount[358]. Remarkably, the most restricted animals lived up to 65% longer than the unrestricted group! Some mice lived as long as four and a half years, which is much more than the two-to-three-year lifespan of these animals. Note that the current record for mouse lifespan is 1,819 days (roughly 5 years), achieved by a mutant dwarf mouse lacking the growth hormone receptor[359].

Since then, similar findings regarding CR have been obtained for species ranging from fruit flies to dogs and fish. Mixed findings have been reported for rhesus monkeys, with some studies finding beneficial effects[360] of CR on longevity, and others not so much[361], probably due to differences in study design[362]. Life extension via calorie restriction has also been observed in lemurs[363]. However, caloric restriction has not extended lifespan, and has even decreased it, for some other species, including butterflies, houseflies, certain strains of yeast[364], and some lines of mice. In fact it takes the loss of just one gene to reverse the effects of CR on murine lifespan[365,366]! These mutant mice live longer when well-fed, but suffer metabolic deterioration and death when subjected to CR. So, although CR appears to be one of the most universal lifespan–extending interventions, there are numerous exceptions. The effects of CR vary from species to species and are dependent on sex and genotype, and the question remains: does it works in humans?

For good reason it is emphasized that calorie restriction must pursued be without malnutrition. The infamous Minnesota Starvation Experiment[367], the findings of which were published in 1950, was designed to mimic the conditions of warfare and explored the effects of 40% calorie restriction in humans. In addition to roughly 25% weight loss, the volunteers experienced numerous adverse health conditions, including painful lower-limb edema[368], chronic weakness, depression, loss of sex drive, and severe emotional distress. It is unlikely that such a study could be repeated today, for ethical reasons. However, a less severe 15%-to-25% calorie restriction without malnutrition was explored in recent human studies and found to be safe[369,370].

A number of human studies reported that caloric restriction can lead to positive changes in certain biomarkers of aging[371,372]. However, there are difficulties with some of the most important ones. A 2023 study published in *Nature Aging* investigated the changes in the biological age of 220 adults who were randomized into either a 25% calorie restriction or an "eat as much as you want" control diet for two years. Biological age was measured via epigenetic clocks that look at aging-related changes in DNA methylation patterns. One of these clocks, called GrimAge, has been particularly effective in predicting human lifespan[218], healthspan, and mortality[217], usually outperforming other such clocks[373]. Unfortunately, there was no difference in predicted biological age between the experimental groups and the control groups according to this clock[374]. Another epigenetic clock, called DunedinPACE, did show some slowing of aging, but the effect sizes were small.

This is in line with the mixed results we see from human epidemiological studies investigating the relationship between human calorie intake and mortality[192]. For example, a study of more than 28 thousand Swedes found the lowest all-cause mortality in people approximately meeting national recommendations for energy intake[191], which is around 2,200 and 2,700 kilocalories per day for adult women and men respectively. Another study divided 7,700 Japanese participants into five groups based on their energy intake. After adjustments for smoking status, alcohol consumption, and dietary preferences, there was increased all-cause mortality in the group with the highest energy intake, but only for men[375]. There were no significant differences in mortality rates between the remaining groups. A similar study for older Spaniards found an association between a surplus of energy intake and higher mortality; however, there was no positive effect for calorie consumption below the energy requirements estimated based on participants' age, sex, weight, and levels of physical activity[376]. I think the best way to summarize these findings is to say that it's probably a good idea not to overeat, but it's not necessarily beneficial to fast unless your true goal is to lose weight.

Even in mice, for which the benefits of calorie restriction are well established, the picture turned out to be more complex than initially thought. A 2024 study published in Nature replicated the beneficial effects of caloric restriction on mouse mortality. However, it also revealed that genetic factors played a more significant role in determining lifespan than dietary restriction[377]. Also, mice kept on a restricted diet are hungry most of the time, and they tend to eat all available daily calories at once and have a long fasting interval afterward. In 2022, a study found that shorter fasting intervals reduce the longevity-promoting effects of calorie restriction[378]. The authors discovered that it's best when animals are fed during the time of the day–night cycle when they are normally active. This is called circadian, or twenty-four-hour, alignment of feeding.

Circadian alignment of eating turned out to be a factor in human obesity. Several studies have shown that losing weight is easier with a high-calorie breakfast and low-calorie dinner than vice versa[379], even if total calorie intake is the same. Meanwhile, multiple studies have found that skipping breakfast is associated with increased risk of cardiovascular disease and all-cause mortality[380]. Likewise, moving dinner calories to breakfast was associated with decreased mortality among US citizens with diabetes[381], although a similar study did not find excessive dinner calories to be harmful in the general US population[192].

Unfortunately, epidemiological studies have limitations—for example, groups that differ in dietary patterns or energy intake might have unaccounted differences in occupation, lifestyle, habits, or even genetics that may introduce bias—so we can't say that the matter is finally settled. Either way, it appears that the beneficial effects of caloric restriction, even when they are present, involve fine-tuned genetic mechanisms that respond to food intake. If we understand these mechanisms, we might be able to target them with specific interventions. To understand what such interventions might be, let's review the complicated interactions between nutrition, metabolism, and aging.

Just as fire involves heat-producing chemical reactions between fuels and oxidants, cellular respiration involves the oxidation of biological fuels that we consume. During this process the chemical energy of various nutrients is used to create ATP, the main molecular currency that our cells use for energy-consuming biological processes.

Oxygen wasn't always a noteworthy component of Earth's atmosphere. It became prominent more than two billion years ago, due to the activity of photosynthesizing cyanobacteria[382], and this wasn't a happy occasion for the planet's inhabitants. The so-called "great oxidation event" was an ecological catastrophe of unprecedented scale and triggered mass extinction[383].

Our ancient unicellular ancestors had to evolve antioxidant enzymes, such as catalase, that accelerate the decomposition of dangerous forms of oxygen, such as hydrogen peroxide (H_2O_2). These enzymes are now found in almost all oxygen-exposed organisms ranging from bacteria and plants to animals. But our unicellular ancestors also learned to benefit from oxygen by allowing respiration-capable bacteria to live inside them. We now know the heirs of these bacteria as mitochondria, our cells' primary energy generators. This was a great evolutionary leap, but the dangers of oxygen have not disappeared completely.

Amounts of body-mass-normalized consumption of O2, or production of CO2, and of heat released from a body are widely used metrics of resting metabolism rate. In 1908, German physiologist Max Rubner proposed the rate-of-living theory of aging, which stated that animals with slower metabolism may be more long-lived than animals with faster metabolism. To quote ancient Chinese philosopher Lao-tzu, "The flame that burns twice as bright burns half as long." Modern mechanistic versions of Rubner's theory assume that higher metabolic rates lead to increased production of free radicals and other oxidants that cause accumulating damage to DNA, proteins, and cellular structures such as mitochondria, accelerating organismal deterioration. Indeed, mutation-induced mild inhibition of respiration has been shown to prolong the lifespan of roundworms[384]—although, surprisingly, this has been associated not with decreased but with increased production of reactive oxygen species.

One important mediator of metabolic rate is temperature. Temperature is a measure of the average kinetic energy—the energy of motion—per molecule in a system. At higher temperatures the frequency of interactions between molecules, including those that are biochemically damaging, is increased. The greater the energy put into a system, the larger the number of molecules that are excited, and the greater amount of random activity that is generated. Humans are warm-blooded animals that keep a constant internal temperature of about 36.6°C, or 98.6°F. Even small variations in temperature can cause severe complications, because so many processes within the human body are fine-tuned to work under precise thermal conditions. The same is true for most other mammals, and birds. But genetic interventions can alter this regulation. Transgenic mice with a mutation that leads to a small—0.3°C to 0.5°C—drop in body temperature have been found to live longer (12% lifespan increase in males, and 20% in females[385]), which is consistent with the rate-of-living theory.

In animals lacking endothermy, such as insects and reptiles, metabolic rates vary within a range of body temperatures that are usually compatible with life. A decrease of room temperature, and hence body temperature, from 28°C to 20°C provides a threefold increase in *Drosophila* lifespan. This coincides with decreased metabolic rates measured via the heat flow from their bodies[386]. Similar effects of temperature on longevity have been found for fish[387], aphids[388], yeast[389], and *Caenorhabditis elegans* roundworms[390,391].

For humans it was long assumed that metabolism slows with age, leading to age-related weight gain even when normal eating behavior and physical activity are maintained. However, recent studies have shown that weight-adjusted energy expenditure is usually stable at ages twenty to sixty and is reduced only in the oldest people, partially because certain health conditions make it more difficult to execute some vital functions. For example, chronic obstructive pulmonary disease leads to increased resting-energy expenditure[392], due to the difficulty it causes with breathing. Elderly people with favorable profiles of senescence who have no physical or cognitive impairments, chronic conditions, or blood-profile alterations have lower resting metabolic rates than their equally aged but less healthy counterparts. So higher metabolic rates can be a marker of metabolic inefficiency and problematic aging[393].

On the surface there seems to be a lot of theoretical and empirical support for the rate-of-living theory. However, there are also numerous problems with it. Bats can be extremely long-lived despite high metabolic rates. Likewise, marsupials, despite having lower average metabolic rates, are short-lived compared with placental mammals[394]. Birds tend to have higher body temperatures and basal metabolic rates than size-equivalent mammals, but they usually live longer[395]. These differences are better understood as consequences of evolution under reduced extrinsic mortality that leads to greater lifespan. A recent study examined the metabolic rates for more than eight hundred land vertebrates, including birds, mammals, reptiles, and amphibians, and found no association between basal rate of living and longevity when animal body mass and other confounding variables were accounted for[396]. Other studies also confirmed that across tetrapod species the presence of defensive measures against external threats (such as a turtle's shell or a bird's ability to fly) is a much stronger predictor of longevity than metabolism rate or body temperature[397].

There are other data points that don't fit the rate-of-living theory. For example, mice and hamsters living in high ambient temperatures experience decreased metabolic rates[398] but have reduced lifespan. The increase of *Caenorhabditis elegans* lifespan at lower temperatures is at least partially explained by responses from longevity-associated genes[391] and not by the temperatures themselves. This issue is highlighted by the case of *Drosophila*, for which short-term exposure to cold can lead to an increase in lifespan without the necessity of constant rate-of-living changes. Once again, we observe that adaptive genetic programs that respond to environmental factors play a more crucial role in regard to lifespan modification than nutrient intake or basal metabolic rates.

This makes a lot of sense. Given the importance of food availability for growth, maintenance, and reproduction, it is not surprising that organisms have evolved complex regulatory pathways and nutrient-sensing signals that affect cellular metabolism. One product of such evolution is insulin, a hormone that is produced by our pancreatic islets in response to increasing blood glucose levels. Insulin promotes the absorption of glucose by fat, liver, and muscle cells, as well as the conversion of this sugar into a storage molecule called glycogen. Decreased insulin activity due either to the reduction of its production or to loss of insulin sensitivity by cells leads to diabetes, a medical condition involving high blood glucose levels, with significantly increased prevalence among the elderly[399]. It causes muscle loss, disability, frailty, and increased mortality.

Many genes involved in nutrient-sensing pathways have turned out to be connected with aging and have become targets for anti-aging therapies. In 1993, gerontologist Cynthia Kenyon and her colleagues discovered mutations in the *Caenorhabditis elegans* gene *daf-2* that doubled the organism's lifespan[297]. Furthermore, they allowed the worms to retain normal levels of physical activity up to an older age. It turned out that *daf-2* is a receptor of an insulin-like growth factor that, just like insulin, is released after feeding. The cells of mutant worms lacking this receptor behave as if they are lacking food even when food is sufficiently provided, and this seems to promote lifespan. The question is, how?

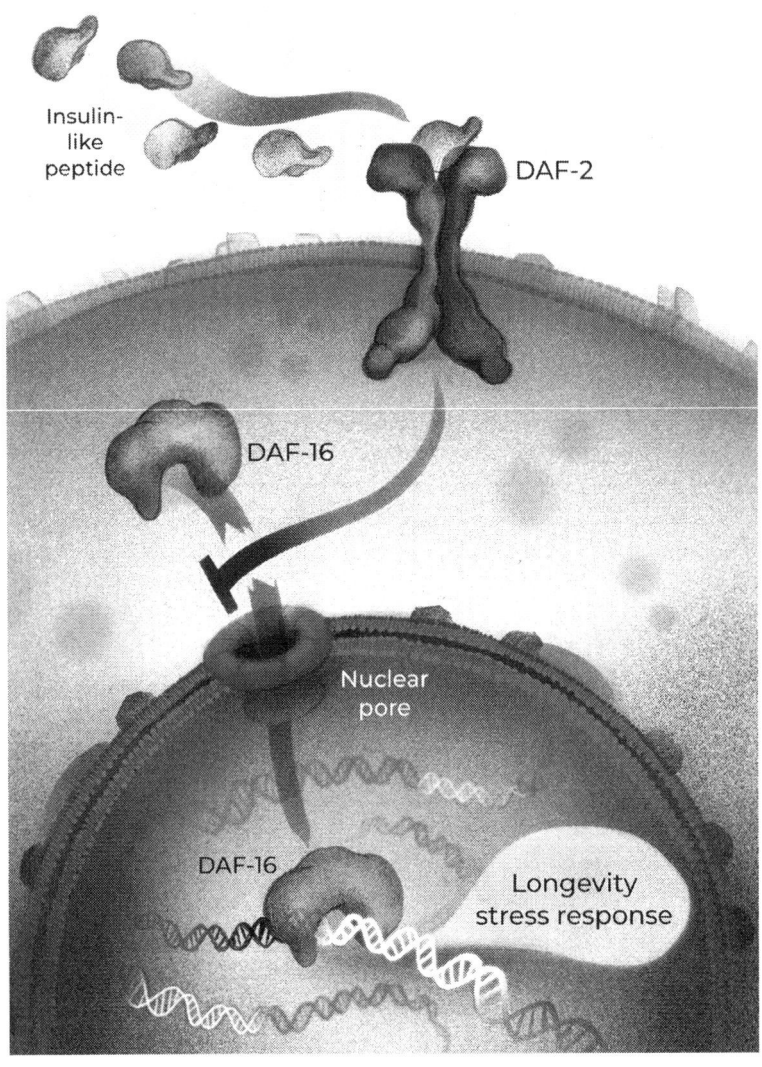

It turns out that the life-extending property of *daf-2* mutants relies on another gene, called *daf-16*. When insulin signaling is reduced either because food is lacking or because a *daf-2* receptor is malfunctioning, a *daf-16*–encoded protein moves from a cell's cytoplasm into the nucleus, where it binds to DNA and activates various maintenance and self-preservation genes. These genes may improve DNA repair, reduce oxidative damage, protect proteins from misfolding and heat shock, or improve the clearance of damaged cellular components.

Similar regulatory mechanisms are found in mice. The worm's *daf-2* is analogous to the insulin-like growth factor receptor (IGF1R) in mammals. Mice lacking both copies of the corresponding gene do not survive; however, male and female mutants with one gene copy removed live thirty-three percent longer than average, while displaying normal appearance, physical activity, and even fertility in lab settings. Like *daf-2*–mutant worms, they are also more resistant to induced oxidative damage[400]. Likewise, the mice's functional analogue of the *daf-16* gene, called *FoxO3*, is required for calorie restriction to increase longevity in these mammals[401]. This gene and its relatives play a very important role in aging in animals ranging from worms to humans. It even holds the secret to the hydra's immortality.

The Lernaean Hydra of Greek and Roman mythology was supreme at regeneration; when one of its many heads was cut off, two new ones would grow in its place. However, it had a weakness: Hercules claimed his victory with the help of Iolaus, who used a flaming torch to cauterize the stumps of the serpent's severed heads. The real-life hydra is a smaller and much more delicate freshwater animal, but it is also capable of growing new "heads" from parts of its polyp body. This property is used for asexual reproduction when growth conditions are good and food is plentiful. Perhaps the ability to produce progeny from parts of its body coevolved with its ability not to age. Of course hydras aren't really immortal; these animals can die from disease and extrinsic events and can even develop spontaneous tumors that resemble cancer[402]. However, their fertility and survival rates usually[403] do not decline with age[404].

Like the Lernaean Hydra, the real-life hydra also has a weakness. Its version of *daf-16*, called *FoxO*, plays a critical role in the animal's remarkable rejuvenation ability. When the gene is tuned down with the help of genetic-engineering techniques, the hydra suffers increased cell specialization and reduced growth and self-renewal[405]. Also, the innate immune system of the modified animal weakens and produces fewer antimicrobial peptides. Conversely, artificially enhanced activity of the *FoxO* gene increases stem cell proliferation, further highlighting its role in the hydra's extraordinary capacity to tackle aging.

FoxO, *FoxO3*, and *daf-16* can be called "master genes," because they encode proteins capable of regulating the activities of other genes by binding to regulatory DNA sequences. Such regulatory proteins are called transcription factors. If the genome is a cookbook, then transcription factors are removable colored sticky page tags that highlight recipes that need to be produced. Along with roundworms' *daf-16* and murine *FoxO3*, the hydra's *FoxO* belongs to an ancient *Fox* family of genes that evolved in unicellular eukaryotes long before the group we know as animals came into existence[406]. This aging-related gene family is extremely old.

The first gene of this family was discovered in *Drosophila*. It was called *forkhead*, because mutations in it modified the fly's embryogenesis, leading to a characteristic spiked-head appearance—a "forkhead[407]". It's a long-established tradition for geneticists to be playful in naming genes according to the morphological features of mutations that involve them. When legs start growing on a fly's head you name the responsible gene *antennapedia*. If an embryo gets covered with spiky protrusions, you call the gene *hedgehog*. You then call its mammalian counterparts *Indian hedgehog*, *Desert hedgehog*, and *Sonic hedgehog*, with a potential inhibitor named *Robotnikinin*. Some of these names have unintentionally stirred controversy.

In humans, *Sonic hedgehog* is linked to a severe disorder called holoprosencephaly, which causes brain, skull, and facial defects, such as having only one eye, in the middle of the forehead. Thus, the gene's name might not be well received by a patient's family members.

Since their initial discovery, hundreds of *Fox* genes have been found across the animal family tree. Some organisms have only one *Fox* gene, while others, such as humans, have many, due to long histories of gene duplication. While *Fox* genes differ by their genetic sequences, they are highly conserved, allowing scientists to identify their relatedness between species despite hundreds of millions of years of evolution. It also appears that they have retained a degree of similarity in their functions and in the regulatory pathways they participate in.

It is thus not surprising that elevated levels of *Fox* genes have been observed in long-lived mammals, such as the bat *Myotis brandtii*. These bats are exceptional creatures, capable of living for more than forty years despite their very small size[408]. Bats have some of the longest size-proportionate lifespans, and their cells are highly resistant to oxidative stress and heat shock[409], which helps them survive high temperatures and increased levels of metabolic activity during flight. An important part of this protection is granted by special heat shock proteins (HSPs), which help other proteins obtain and maintain their three-dimensional structures despite temperature changes and other molecular hazards. HSPs are among the many protective proteins upregulated by *Fox* genes in response to stressful conditions[410].

It is convenient that to achieve the longevity-promoting effects of an organism's *Fox* genes a genetic engineer does not need to increase their activity in all of the organism's cells. An artificial boost of *daf-16* in roundworm gut cells leads to its increased activity in other tissues as well[411], a phenomenon known as *FoxO*-to-*FoxO* signaling. This indicates that affected cells produce signals that spread to other parts of an animal's body. Likewise, it is sufficient to genetically engineer the head fat body tissues in *Drosophila* for more *dFoxO* (*Drosophila*

FoxO) gene activity to prolong the lifespan of both male and female flies and provide them with resistance to a poisonous substance called paraquat, which induces oxidative stress[412,413].

The practical consequences of these features are immense because they increase the viability of gene therapy approaches in enhancing adult animal lifespan. Gene therapy is the delivery of one or several genes into the cells of an organism. Usually this is done by injecting viral envelopes in which harmful genetic material has been substituted with functional copies of beneficial genes, just like in some COVID-19 vaccines. Adult multicellular animals typically have so many cells that only a small subset will be affected by the viral particles, and the others will remain unchanged. This is why it's so exciting that changing the genes in a relatively small number of cells is sometimes sufficient to increase the lifespan of an entire organism[414].

Like the *Fox* genes of worms, mice, and fruit flies, human *Fox* genes are also linked to longevity, cellular maintenance, and nutrient sensing. For example, human cells with reduced *FoxO1* activity are more susceptible to oxidative damage than others[415,416], while increased *FoxO3* activity improves resistance to oxidative stress. One of the most convincing pieces of evidence for the role of *FoxO*-related genes in human aging comes from genetic comparisons between people of average longevity and people of exceptional longevity[417].

A study of Japanese men with an average age of 78 years found that certain genetic variants of the *FoxO3a* gene were strongly associated with increased lifespan and better physical function and self-rated health, as well as reduced age-related diseases such as cancer, coronary heart disease, and diabetes[418]. The first author of this work was Dr. Bradley Willcox, a leading investigator of this gene's role in longevity. The findings were replicated in samples of centenarians (people who are 100 or more years old) and nonagenarians (people who are between 90 and 100 years old) from other populations, such as Germans and Han Chinese, and extended to women[419,420]. It appears that a healthy body helps maintain a healthy mind;

there are *FoxO3* gene variants that are strongly associated with human intelligence[421,422]. The *FoxO3* longevity variant seems to exert its effects through multiple molecular mechanisms, including modulation of inflammation signaling and protection against the shortening of chromosome ends[423].

Other evidence for the importance of *Fox* genes comes from studies of nonhuman primates, our shorter-lived relatives. For example, *FoxO3a* is downregulated in the blood vessels of monkeys as they age. Deactivation of *FoxO3a* in human vascular endothelial cells leads to major defects that are similar to those observed in aged monkey arteries[424]. Also, it appears that the remarkably enhanced lifespan of humans already comes with higher activity levels of some *Fox* genes than in chimpanzees[425]. Perhaps evolution has already optimized our nutrient-signaling pathways for promoting longer lives. After all, human lifespan has been increasing over the past several million years, implying that natural selection has favored anti-aging genetic adaptations.

This could explain why the beneficial effects of calorie restriction are not as evident for our species as for a number of other mammals. This also brings hope: if evolution has increased the lifespan of our species through these mechanisms, perhaps we can develop this success further. Unfortunately, I haven't found a single study that attempted to increase mammalian lifespan by enhancing the activity of *Fox* genes via gene therapies. I think it's definitely worth investigating.

We have discussed one important molecular mechanism that links nutrition to lifespan, but there are others. Let's return to Yoshinori Ohsumi and the secrets of autophagy—the intracellular digestion of misfolded or aggregated proteins, malfunctioning or superfluous organelles, remnants of infectious particles, and other deleterious cellular components[426-428]. During this process, undesired materials are captured in a double-membrane vesicle called an autophagosome, which then merges with another membraned structure, called a lysosome, which contains digestive enzymes[428]. Eventually, "junk" gets recycled and repurposed.

Before venturing forth, we must acknowledge that autophagy isn't always a good thing. As the saying goes, one man's trash is another man's treasure. Autophagosomes are not one hundred percent selective and can accidentally capture useful cellular components. Also, "junk" isn't always bad to have, as it may be a form of nutrient storage that is usable during periods of starvation. Indeed, nutrient deficiency increases autophagy in various species[429]. Of course, this latter function is more relevant to organisms that are at high risk of encountering periods of starvation, which most modern humans avoid. From this view, it could be that our autophagy isn't activated often enough—hence the popular idea that fasting may be beneficial for one's health.

When I think about the molecular mechanisms that prevent cells from discarding their own junk, mechanisms shaped by the evolution of our ancestors in the harsh conditions they had to survive in, I remember a story from my favorite TV series, *South Park*. In the episode "Insheeption," school psychologist Mr. Mackey suffers from compulsive hoarding, a mental disorder characterized by having difficulty discarding even the least valuable items, such as a dirty piece of paper or an old half-eaten sandwich. A team of armed agents parodying the movie *Inception*, along with the most prominent dream walker of all time, Freddy Krueger, is sent into Mr. Mackey's dreams within dreams. They find out that the cause of his hoarding lies in his past, when he was sexually assaulted by Woodsy Owl, an environmentalist mascot whose tagline is *Give a hoot! Don't pollute*. Afraid of the owl's punishment, Mr. Mackey doesn't throw away any trash.

Normally, autophagy is suppressed by a protein called TOR, the Woodsy Owl of our cell, the molecular god of our junk. The role of TOR in autophagy was discovered by Takeshi Noda and Yoshinori Ohsumi in 1998. They found that a drug called rapamycin is capable of inducing autophagy in yeast cells, even if they live in a nutrient-rich medium that normally suppresses the process[430]. Then they found that this drug works by binding to and deactivating TOR, which stands for *target of rapamycin*. Today we know that TOR is evolutionarily conserved in species ranging from yeast to humans and that rapamycin triggers mammalian autophagy in more or less the same manner.

Normal physiological TOR activation is dependent on nutrient, amino acid, and growth factor sensing[431]. Certain diets have been found to activate TOR in animals and promote aging, while those that inhibit TOR do the opposite[432]. Increased concentrations of amino acids—specifically methionine[433] and branched-chain amino acids such as leucine, isoleucine, and valine, which are abundant in meat, dairy, and legumes—can activate TOR, and thus repress autophagy. Methionine restriction has been found to increase lifespan by forty-two percent in rats[434], basically mimicking the longevity-enhancing effects of calorie restriction. Smaller effects have been observed in normal and quickly aging (progeroid) mice[435,436]. This has led many researchers to believe that perhaps rapamycin, a potent TOR inhibitor, can help slow some aging processes.

In 2009, the journal *Nature* published the first study to show that rapamycin can extend mammalian lifespan[437]. The study involved rats that were already six hundred days old, roughly equivalent to sixty human years. Some scientists joked that this was because rapamycin is so expensive to produce that the authors wanted to save money by using animals at death's door. That isn't entirely true, because the study involved an additional population of younger animals. But there was a problem with the original study: the control mice were accidentally fed a different diet than those receiving rapamycin, and they had lower mortality rates prior to treatment, confounding the results[438].

However, the longevity- and healthspan-extending properties of rapamycin in rodents have been reproduced in numerous subsequent studies[113,439-442], including studies by the Interventions Testing Program, typically with 10%-to-25% increased lifespan, with greater effects at higher doses and in females.

Rapamycin is probably the most widely studied anti-aging drug for rodents. For a human, that effect would translate to eight to twenty additional years of life. Studies on aging rodents have also revealed that, aside from enhancing autophagy, rapamycin can reduce oxidative stress and cell death. Its use is associated with decreased presence of neurodegeneration markers in the brain[443,444] and counteracts age-related hearing loss[445]. Unfortunately, the use of this drug on healthy human subjects would likely have several downsides.

Rapamycin was first isolated from a species of bacteria found on Easter Island (Rapa Nui in the native language) and initially used as an antifungal agent. It was later found to have additional properties, including tumor suppression[446], and it became famous as an immune-system suppressant that was used, in combination with other medications[447], to prevent organ rejection after transplantation. Immunosuppression is not the best side effect for an anti-aging medicine to have. Another effect of rapamycin is the promotion of insulin resistance and impaired glucose tolerance[448] —also to be avoided.

Paradoxically, a 2023 meta-analysis revealed that, despite being an immunosuppressant, rapamycin does not impair but actually improves mice's resilience to pathogens (while dietary restriction slightly hinders it[449]). There are several possible explanations for this. Perhaps rapamycin reduces some overzealous immune-system reactions that can damage organ tissues. Maybe the drug improves immune function in the long run by slowing down aging. Indeed, one placebo-controlled human trial of an analogue of rapamycin showed a surprising reduction of infection rates in the elderly[450]. Another hypothesis is that rapamycin has antimicrobial properties.

It is possible that the adverse effects of rapamycin aren't as critical as they were deemed in the past, and that rapamycin will still become an anti-aging drug for humans—perhaps in smaller doses or in combination with other drugs. It was discovered that intermittent rapamycin use in mice can provide some benefits with fewer side effects when compared with chronic treatment[451,452]. Additional animal studies have paired rapamycin with the antidiabetic drugs metformin and acarbose, to compensate for increased risk of developing insulin resistance, and found additive beneficial effects on lifespan[143,453].

Unfortunately, the results of clinical trials of rapamycin as a treatment for human age-related disorders are lackluster and currently insufficient to draw meaningful conclusions. Some studies show beneficial effects of topical rapamycin on skin aging, but sample sizes are very small[454]. There are ongoing randomized double-blind clinical trials to test whether rapamycin can reduce certain clinical manifestations of aging (such as visceral fat accumulation, loss of bone density, liver and renal function), also with relatively small sample numbers[455], and we have yet to find out their results.

It is annoying how little we know about the anti-aging properties of rapamycin and similar molecules in practice, and this shows how financially starved the field of aging research is. Some researchers such as Dr. Brad Stanfield have even turned to crowdfunding[456].

Even if rapamycin does not turn out to be the true Grail that we seek, the possibility that mTOR inhibition is beneficial for longevity has facilitated the exploration of other drugs with similar effects. For example, alpha-ketoglutarate is a popular, though controversial, supplement taken by many athletes[457] and some biohackers. It is also a common metabolite in cells, so it's probably safe. In 2014, a prominent study showed that supplementation with alpha-ketoglutarate increases the lifespan of *Caenorhabiditis elegans* by approximately fifty percent[458]. It also induced autophagy and delayed age-related decline in rapid coordinated movement and other hallmarks of senescence. The authors found that mutant worms with impaired *Fox* gene activity had a smaller, but substantial, lifespan increase under alpha-ketoglutarate treatment, indicating that some benefits were due to *Fox* gene activation, though not all of them. On the other hand, alpha-ketoglutarate failed to further increase the lifespan of worms that had had their version of TOR deactivated via a gene-silencing technique called RNA interference. This suggested that alpha-ketoglutarate acts via TOR inhibition.

Several years later, alpha-ketoglutarate was shown to increase the lifespan of *Drosophila* flies, though only modestly (up to 8.5%). Flies receiving the supplement had reduced TOR activity and increased autophagy compared with controls[459], supporting the abovementioned conclusion. A study on mice also reproduced the life-extending and frailty-reducing properties of alpha-ketoglutarate (also to a much smaller extent than in worms), but surprisingly did not find a reduction in mTOR activity in tissue samples[460]. The lifespan extension from inception of treatment was 8% to 20% in females; and positive, although statistically insignificant, in males.

Alternatively, alpha-ketoglutarate's pro-longevity action could be related to inhibition of an enzyme called ATP synthase[458], the main cellular energy generator—which, as its name implies, synthesizes ATP. As I've already mentioned, ATP is the "energy currency" of the cell. The energy stored in its chemical bonds is used for processes ranging from chemical synthesis to muscle contraction and the transmission of neural impulses. *ATP* stands for *adenosine triphosphate*, because it consists of adenine, which is one of the nucleobases used in RNA and DNA; a sugar (ribose); and a triphosphate group. ATP can lose a phosphate, creating ADP (adenosine diphosphate), while simultaneously providing energy for other coupled processes, fueling various cellular biochemistries and machinery.

ATP synthase reverses this process by creating ATP from ADP and inorganic phosphate. In a sense, ATP synthase acts like a turbine, but instead of using the flow of water, it uses the flow of protons (H+) between the outer and inner membranes of mitochondria. The proton gradient required to maintain this flow is produced by a network of mitochondrial proteins involved in respiration. Respiration creates a proton gradient, and that proton gradient is used to generate ATP via ATP synthase.

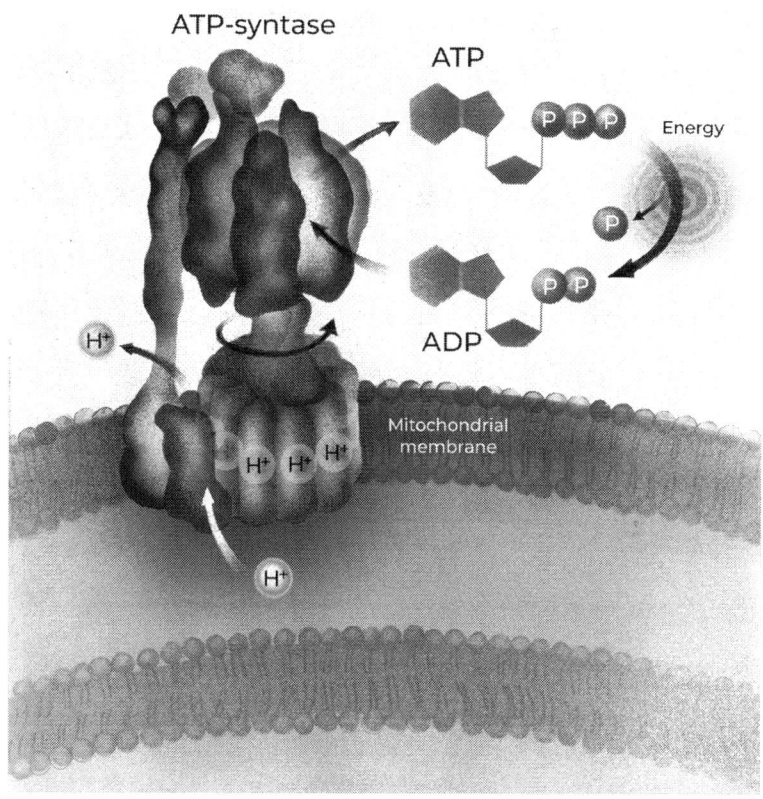

Targeted inhibition of *Caenorhabditis elegans* ATP synthase reduces oxygen consumption and metabolic rates and provides up to a sixty percent increase in the worm's lifespan[451]—at the cost of delayed development, small adult body size, and sterility. So perhaps this is a case where the rate-of-living theory works. Dr. Alexander Tyshkovskiy and his colleagues studied the biological effects of seventeen longevity interventions in mice and found that lifespan extension is indeed often associated with the activity of genes involved in cellular respiration and the conversion of energy resulting from nutrient oxidation into the form of ATP[461].

Aside from rapamycin and alpha-ketoglutarate, other molecules are actively being studied for potential autophagy and additional longevity-enhancing effects. Databases such as geroprotector.org and DrugAge[462] are dedicated for exploring pro-longevity compounds in various models. Among the molecules that increase the lifespan of *Caenorhabditis elegans*[463], caffeine is a widely available mTOR inhibitor[464–466]. Both natural and decaffeinated coffee trigger autophagy in mice, indicating that the popular beverage has other molecules other than caffeine that stimulate this process[467].

As you have probably noticed, most drugs studied for life extension are common ingredients in food or in drugs that are used for other purposes—such as the immunosuppressant rapamycin, the questionable[468] sports supplement alpha-ketoglutarate, and the antidiabetic drug metformin[469]. We find ourselves in this awkward situation for several reasons.

One problem is that aging is not generally considered a pathology. Only recently did the World Health Organization include "aging associated decline in intrinsic capacity" in its International Classification of Diseases (ICD-11). Initially the term *old age* was used, but it was abandoned due to protests and controversy. As an article in *The Lancet Healthy Longevity* stated, "Old age is not a disease, but ageism is[470]." While I agree that the newer terminology is preferable, the recognition of biological aging as a health problem is vital to create a framework of streamlined clinical trials of anti-aging therapies.

Another problem is the insufficient funding of anti-aging research. I am concerned that so many anti-aging therapies that have achieved promising results in one model organism have not been replicated in other organisms and have never been subjected to human clinical trials. It took six years to test whether a strong pro-longevity effect of alpha-ketoglutarate found in roundworms is reproducible in a model mammal. We haven't developed any gene therapy based on increased *Fox* gene activity, and we still haven't tested whether rapamycin can improve human lifespan, more than fifty years after its discovery. That's how slowly we are advancing in our understanding of aging.

While these issues persist, we will remain uncertain about the efficacy of many potential longevity-promoting interventions, and Yoshinori Ohsumi and others will have no choice but to continue expressing doubts about the health benefits of fasting and autophagy.

Let's summarize what we have learned from this chapter. There is evidence both in favor of and against the rate-of-living hypothesis, indicating that the aging process does not boil down to a single factor. While metabolism and respiration are linked to certain types of damage that accumulate with age, this has less to do with rates of living and more to do with the genes and molecular pathways that react to the intake of nutrients and regulate processes such as cellular maintenance and autophagy. Regulatory *Fox* genes, such as *FoxO3a*, are among the important longevity-promoting actors, while mTOR appears to play a grim role. While calorie restriction works in many species, including mice and some primates, epidemiological data does not currently support the premise that it has large longevity benefits in humans—beyond promoting healthy body mass index and preventing overeating. Evolution has already provided us with the potential for living relatively long lives, but we might do better if we understood the mechanisms of aging in detail and learned how to target them specifically. This is something we will address in the next chapters.

Can we escape the destructive fire ignited by our own metabolism? To win this race, we'll need revolutionary medicine—one that can rewrite the very regulations of our genes.

CHAPTER 7
The immortal cell

> When I grow up,
> I will be smart enough to answer
> All the questions that you need to know
> The answers to before you're grown up.
>
> —*Matilda the Musical*

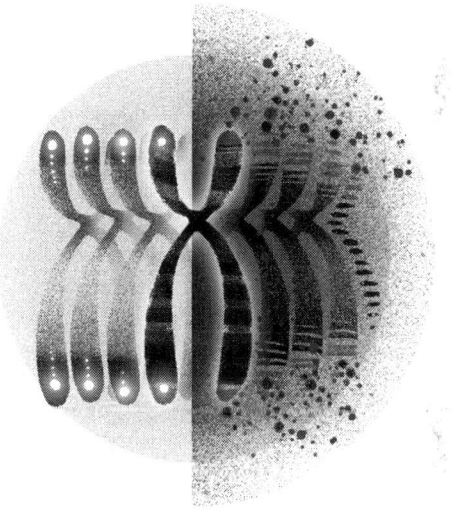

Summary: in this chapter we discuss the secrets of the immortal germ line, animal cloning, and how the biological age of our cells can be reset.

Every birth is a miracle. This cliché makes a surprising amount of sense in the context of longevity research: two cells from older adults fuse and form a completely new organism, whose age clock is reset. This process can be repeated indefinitely, as cells have been dividing and propagating genetic information for billions of years—since the origin of life. This

139

gives us hope that by learning the germ line's secret of immortality we will find ways to reverse the aging process. What do we know about the miracle of rejuvenation? Well, first of all, it does not always require sexual reproduction, fertilization, or an embryo.

King's lomatia is a shrub that grows up to eight meters tall, with red flowers that yield neither fruit nor seeds. This plant species has three sets of chromosomes and is thus sterile. Only vegetative reproduction can occur: a fallen branch grows new roots, stems, and leaves, but is genetically identical to its parent. Biologically speaking, it is a clone. The entire population of King's lomatia consists of just several hundred clones growing in a small region of southwestern Tasmania. This may be the oldest living individual plant, cloning itself over and over for 43 thousand years based on the fossil record[471].

Another of the oldest living organisms on Earth is King Clone, a twenty-meter *Larrea tridentata* creosote bush ring with yellow flowers in the Mojave Desert. It represents a single clonal colony of one original plant, and its estimated age is 11,700 years[472]. These are extreme examples, but many plants clone themselves for hundreds or thousands of years[473], even though their tissues age and die.

A similar phenomenon can be observed in some species of fungus. In Michigan a 2,500-year-old honey mushroom (*Armillaria gallica*) clonal individual was identified covering 75 hectares of forest floor[474] and weighing approximately 400 tons. These organisms are characterized by remarkably low mutation rates[475], which is helpful for long-term survival.

Among multicellular organisms, plants and fungi are not unique in their vegetative reproduction. Certain species of starfish, such as *Coscinasterias tenuispina*, are capable of establishing large clonal populations that reproduce through fission, a splitting of the body into two new organisms. They also have remarkable regenerative ability; a whole starfish can grow from just a small body fragment. The regenerating parts of the starfish appear younger and have lengthier telomeres—sequences of DNA on the ends of chromosomes; they typically shorten in aged tissues. The adherence to vegetative, nonsexual reproduction in starfish is also associated with increased telomere length[476].

As we will see, telomeres are an important piece of the puzzle of aging. In 1961, gerontologist Leonard Hayflick and geneticist Paul Moorhead published an article called "The serial cultivation of human diploid cell strains[477]" in which they showed that human cells cannot divide indefinitely. In a 1965 article called "The limited in vitro lifetime of human diploid cell strains," Hayflick details that cultivated cells can undergo no more than 40 to 60 divisions[478], depending on their type. Hayflick hypothesized that this limit (now called the Hayflick limit) is related to cellular and organismal aging.

Fifty cellular divisions are enough to make 1,126 trillion cells out of one. That's about 30 humans' worth of cells, given that a single human body contains, on average, 37.2 trillion cells[479]. Approximately 0.25% of our cells are replaced each day[480], so it's also about 30 years' worth of cells. Just one more cellular division would double that and bring us closer to the average human lifespan. Yet another cellular division would double that again, up to the maximum observed human lifespan, about 120 years. Note that these calculations are only rough estimates, but this analysis shows that our fate relies on the number of times our cells can divide. This is why it's important to understand where the Hayflick limit comes from, why some organisms are unaffected by it, and why it's reset in our offspring.

In 1973, biologist Alexey Olovnikov noticed that the mechanism by which DNA is copied contains a certain inbuilt flaw[481] that he called "the heel of Achilles of the DNA double helix." When DNA needs to be copied, the double helix is separated into two single strands. Each strand serves as a template for creating a new double helix. The enzyme that copies DNA is called DNA polymerase, and it requires a short preexisting stretch of nucleotides called a primer to start its work. In living cells, the primer typically consists of RNA. It temporarily binds to the single-stranded DNA, forming a short DNA/RNA double-helical structure that is then elongated into a new DNA double helix by addition of adenine (A), thymine (T), guanine (G), and cytosine (C).

RNA primers in the middle of a replicating DNA molecule are removed and replaced with DNA, but this cannot be done at the ends of chromosomes, because DNA polymerase specializes in copying DNA in only one direction (from the so-called 5' end toward the 3' end). This issue does not arise in bacterial or mitochondrial chromosomes, because they are circular and thus have no ends. For linear chromosomes, such as those in the nuclei of plant, animal, and fungal cells, this process results in chromosome shortening with each division. Hence, there is a limit to how many times a cell can divide before losing essential regions of its DNA.

Olovnikov predicted that special mechanisms must exist to counter this shortening, at least in some cells (such as the immortal germ line), and that finding the responsible gene could be vital in the quest to delay cellular aging.

The mechanism predicted by Olovnikov was discovered in 1984 by biochemists Carol Greider and Elizabeth Blackburn[482]. An enzyme called telomerase, first identified in a unicellular organism called *Tetrahymena*, had the ability to add six-letter repeatable DNA telomere strands to the ends of chromosomes[483]. The enzyme turned out to be a ribonucleoprotein, which is a protein that incorporates an RNA molecule into its structure. The RNA molecule of the *Tetrahymena* telomerase contained the nucleotide sequence used as a template for synthesizing telomere DNA. In 2009, Blackburn and Greider were awarded the Nobel Prize in Physiology or Medicine.

Human telomerase is slightly different from that of *Tetrahymena*, but it also adds six-letter repetitions of DNA sequences to the ends of chromosomes[484]. While telomerase is not commonly used by most types of specialized body cells, in line with the Hayflick limit, the enzyme is active in germ line cells and early-stage embryos[485,486]. An increase in telomerase activity is one of many things that need to happen for the miracle of rejuvenation to occur.

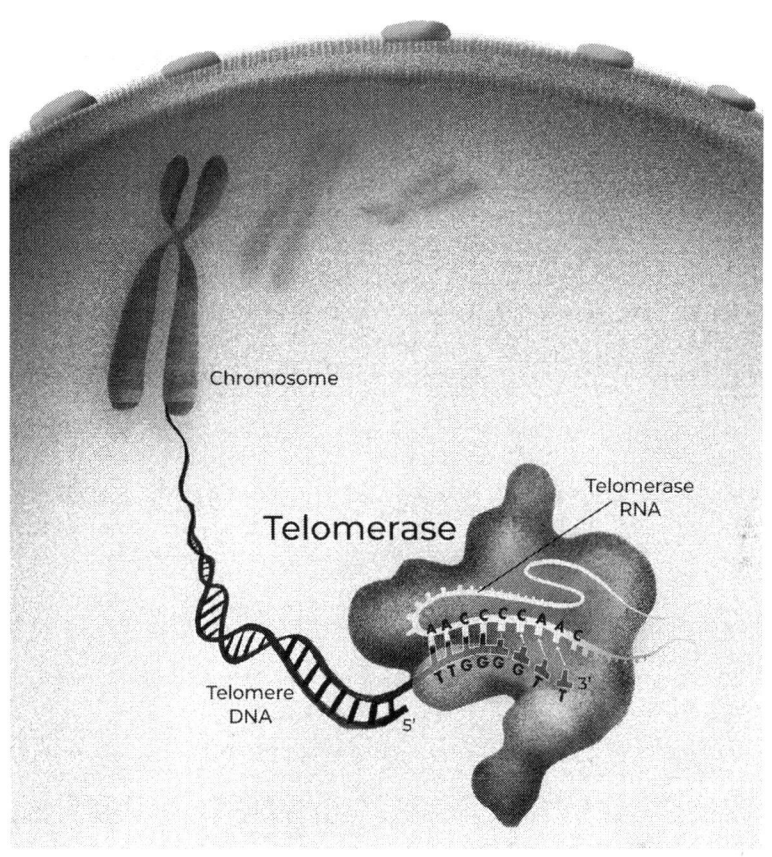

For some time, the Hayflick limit was an important theoretical concern in artificial animal cloning, such as in the case of Dolly the sheep. Scientists had already succeeded in transferring DNA-containing nuclei from one embryo to another before Dolly's creation—for example, this technique was used in the Soviet Union in 1987 to obtain cloned mice (including Masha the mouse[487])—but Dolly's DNA came from a nucleus taken from an adult animal's cell, which already had shortened telomeres. That nucleus was then placed in an enucleated egg cell (from which the original nucleus was removed in advance), and the resulting artificial embryo was then implanted in a surrogate sheep mother. It was unclear whether the clone would inherit the biological and telomere age of the original, older animal or the age timer would be reset, as happens during normal embryonic development.

Some people, including a few biologists, found it hard to believe that the procedure was possible, resulting in claims that Dolly was fake and that her body was burned as part of a conspiracy to hide this fact. In reality, the stuffed body of the sheep can be seen in the National Museum of Scotland[488], and several independent DNA analyses have confirmed that she was indeed a clone[489,490]. Dolly died at the age of six and a half year, and many people assumed that her early death was related to accelerated aging. However, Dolly was euthanized after developing progressive lung disease, a fairly common form of lung cancer that is caused by a retrovirus and has nothing to do with cloning.

A sample size of one animal is not sufficient to evaluate the lifespan of clones. Fortunately, Dolly was not the only sheep clone. An analysis of 13 sheep, including 4 obtained from the same cell line that gave rise to Dolly, did not find any evidence of premature aging[491]. Many clones lived for more than 9 years. And while there were inconsistent reports about slightly shorter telomere lengths of cloned sheep when compared to controls, telomere length remained normal in cloned mice and cows[492]. Scientists even succeeded in cloning mice for up to 25 generations, creating more than 500 viable offspring from a single animal[493]. Clones of many other animal species have also been created, including genetically modified macaques[494], which means human cloning is within our reach. So far, cloning research has yielded no detectable aging-related abnormalities, and molecular data confirms that telomere length is reset during early embryonic development in mammals, and that this length is species specific, regardless of the initial telomere length of a donor nucleus[495].

This is how it works in healthy animals. However, genetic mutations that lead to short-telomere syndromes can cause premature aging. The manifestations of these diseases tend to occur at an early age, with increased severity in subsequent generations, because offspring inherit not only the genetic mutation that impairs normal telomere maintenance but also the reduced telomere lengths from parental organisms[496]. Short telomere syndromes can occur not only due to malfunctioning of the telomerase enzyme itself, but also due to mutations in associated proteins.

For example, congenital dyskeratosis is caused by mutations in a protein called dyskerin[497], which stabilizes the RNA component of the telomerase complex[498]. Affected individuals have less telomerase activity and shorter telomeres, resulting in bone marrow failure, abnormal skin pigmentation, premature graying of the hair, and other issues. They are also at a roughly seventy-fold increased risk of head and neck cancers and acute myeloid leukemia, and at a roughly five-hundred-fold increased risk of myelodysplastic syndromes—conditions under which not enough healthy blood cells are produced[499]. These issues are partially explained by the severely diminished replicative potential of sufferers' cells and by increased DNA-damage signaling, which leads to cellular senescence. Additional evidence for the importance of telomeres in aging comes from interspecies comparisons; telomere shortening is negatively correlated with mammalian maximum lifespan[500].

Telomere research inspired a series of experiments aimed at increasing lifespan through enhancing telomerase activity in animals. The primary concern was that telomerase activity is known to rise in tumor cells, so telomere shortening could be one of many mechanisms of cancer prevention, and the reason why most cells don't produce this enzyme. There is some data showing that genetic variants associated with longer telomeres in humans are associated with higher risk of several types of cancers[501], but this evidence does not prove causation.

On the other hand, cells of genetically modified mice that entirely lacked telomerase experienced frequent chromosomal abnormalities, including end-to-end fusions, and were quite capable of forming tumors as shown by leading telomere researcher Maria Blasco and collegues[502]. Despite this contradictory data it seemed logical to counteract the possible negative consequences of increased telomerase activity by tumor suppression. Some early experiments involved crossing genetically modified mice that had one or three tumor-resistance genes with partners possessing enhanced telomerase activity. As a result, 9% and 26% median lifespan extension was observed in the offspring compared with the original genetically modified controls, with only the tumor suppressors in place[503]. A combination of three cancer-resistance genes and increased telomerase (TERT) activity yielded a remarkable 40% increase in median lifespan in comparison with mice that have just one cancer-resistance gene and that have normal longevity. More recently, scientists generated mice with extremely long telomeres without changing any of their genes. These mice were significantly leaner than control mice, were less susceptible to cancer, and experienced median and maximum lifespan extensions of 12.75% and 8.4% respectively[504].

The abovementioned modifications were present from the beginning of the animals' lives, and what we really want is a way to help aged individuals. In 2012, the results of the first experimental telomerase-based gene therapy of aging was reported in normal adult mice[505]. The authors of the study used safe genetically engineered adenoviruses to deliver the telomerase TERT gene into tissues of animals aged one or two years. This resulted in 24% and 13% median lifespan extension respectively. Most importantly, the researchers did not observe increased incidence of cancer. This could be because the viral delivery systems (or simply vectors) target cells that are less proliferative and less susceptible to cancer—because the role of telomerase in cancer has been overestimated—or because of yet unknown reasons.

This finding was replicated in a more recent study, published in 2022. The authors examined telomerase gene therapy delivered by a genetically modified cytomegalovirus (CMV vector) that was either injected or administered intranasally. Mice treated with the Telomerase reverse transcriptase gene lived, on average, 41.4% longer than controls, irrespective of delivery method[506]. This result was reached even though treatment started at 18 months, a rather old age for mice. The maximum lifespan achieved was 3.4 years, which is impressive, considering that half of the mice of this strain normally die by the age of 2.3 years[507]. The authors also claim that they did not observe an increased risk of cancer in treated animals.

Such therapy has certain limitations[508], since the CMV vector used is capable of replication. This might have unexpected consequences if translated into humans. Imagine a cure for aging that can be transmitted through saliva, blood, tears, or even semen! This might be a good thing if the treatment works and we want to save more lives, and it would be a great idea for science fiction (imagine a fictional scientist spreading the gift of immortality by kissing every stranger she meets), but I can't begin to imagine the public reaction and the ethical debates that would arise. There is also the question of safety: cytomegalovirus infections are globally widespread in humans, and associated with an increased risk of cardiovascular diseases[506] and immune system decline[509].

A few words should be said about some of the authors of this study. George Church is a professor of genetics at Harvard Medical School who has been involved in some of the most impressive life science projects, ranging from altering the fundamental genetic codes of organisms[24] to editing elephant DNA in an attempt to bring back the woolly mammoth[510], and inventing original methods for DNA sequencing and synthesis[511]. It's as if his last name demands that he play God. Colossal, a company he co-founded, made headlines when it created genetically modified white wolves with mutations inspired by ancient DNA from the extinct dire wolf, as well as "woolly mice" with curly hair resembling that of a mammoth.

Another author is Liz Parrish, a controversial figure who is not a scientist by training, but is the CEO of the American biotechnology company BioViva, with a focus on aging treatments. She became well-known as the first human to receive telomerase gene therapy with the adenoviral vector, which was performed via her personal initiative at her own risk. This was widely regarded as a publicity stunt rather than a scientific experiment, so I was pleasantly surprised to see her involved with academic research on the topic. Conflicts of interest among some authors, along with the fact that the study results seem a bit too good to be true, make me suspicious—but I haven't found any evidence that anything is wrong with the research.

So, the good news is that we might be reaching a point where something can be done about the telomeres in our cells. However, clinical trials on human volunteers are warranted before any action can be recommended. Such therapy may be very inexpensive, because many people need it and the necessary materials are easy to produce. The technological and manufacturing processes of anti-aging gene therapy could be similar to those involved in the creation of vector-based COVID-19 vaccines, which were made available for hundreds of millions of people on short notice, and distributed free of charge in many countries. This inspires hope that someday the same could be true for anti-aging gene therapies. The bad news is that chromosome-end shortening is not the only cause of aging, and telomerase elongation is not sufficient to sustain the miracle of germ line immortality by itself.

There was another theoretical problem that could have prevented the successful cloning of sheep and other animals. In the case of Dolly, a nucleus was taken from a highly specialized mammary gland cell. Cellular specialization is accompanied by epigenetic changes, including the addition of biochemical markers, such as -CH3 methyl groups (methylation), to certain regions of DNA, and modification of histones, the molecular spools around which DNA winds itself. These epigenetic changes affect gene activity (expression[512]), and the quantity and types of RNA and proteins being made, affecting the morphology and functionality of cells, if the genome is a cookbook, the epigenetic modifications are erasable commentary on the recipes within it.

The abovementioned epigenetic changes take place in the cell nucleus. Thus, it was not obvious that an egg cell would transform into an embryo upon receiving the genetic contents of a specialized cell. One could as easily have predicted that we would end up with a mammary gland cell instead. But that didn't happen. This means that somehow the epigenetic markers are reset when cloning is successful. This is especially important given that epigenetic alterations are a hallmark not only of cell specialization, but also of aging[513,514].

Researchers have observed that massive epigenetic reprogramming takes place in embryos resulting from somatic (body) cell nuclear transfer, as evidenced by reduced methylation rates[515]. The same process occurs during normal embryonic development in mammals; a substantial part of the genome is demethylated and then, after some time, remethylated in a cell- or tissue-specific manner[516]. Indeed, studies show that embryogenesis involves a natural rejuvenating event during which biological age (measured by epigenetic clocks) appears to decrease[517]. All this makes biological sense because embryonic development requires complex, precise, and coordinated action of many genes. Any severe epigenetic alterations from the norm could be unpredictably deleterious, so their prevention should be favored by natural selection.

In fact, normally reproducing mammals experience two waves of global genomic demethylation and remethylation. Primordial germ cells, the ancestors of both male and female gametes, have their methylation profiles erased before they specialize into sperm or oocyte cells. Then, during specialization, sex-specific methylation profiles are acquired[518]. Sperm cells initially have very high methylation levels, but the paternal genome undergoes active demethylation before the embryo even starts copying its DNA. Oocytes have relatively low methylation levels to begin with, and maternal genomes undergo passive demethylation as they are copied in the absence of DNA methylation enzymes during early development. Novel global methylation patterns are established upon implantation of the embryo.

This process is necessary to ensure the totipotency of embryonic cells—their ability to divide and produce all the differentiated cell types of the future organism[519]. Totipotent cells have the highest differentiation potential, and they can form any embryonic structure, as well as extraembryonic structures such as the placenta. Stem cells that are formed in later developmental stages have narrower spectra of differentiation[520]. For example, hematopoietic stem cells are multipotent, capable of developing into several types of blood cells. A myeloid stem cell is considered oligopotent; it can divide into some white blood cells, but not into red blood cells. Cellular differentiation is influenced by both external factors, such as chemical and mechanical interactions with other cells, and internal factors, such as the epigenetic profiles they inherit from previous divisions.

In 2007, a group led by Shinya Yamanaka published one of the most influential life science papers of the twenty-first century[521]; it has since been cited more than twenty thousand times, and it earned him a Nobel Prize in Physiology or Medicine five years after its publication. The authors found a way to turn specialized adult mammalian cells into nonspecialized induced pluripotent stem (iPS) cells that highly resemble the cells of an embryo. This requires the activation of only four genes, or Yamanaka factors—a set of transcription factors that regulate the activity of other genes. The iPS cells have increased telomerase activity, their epigenetic profiles are similar to those of normal pluripotent cells with epigenetic age close to zero[522], and they can differentiate into all kinds of specialized cells.

The most unusual recent demonstration of the unrivaled potential of this biotechnology has been the creation of baby mice with the genetic material of two male parents[523]. To achieve this, researchers created iPS cells from the cells of adult male mice with normal XY sex chromosomes. The sex of the cells was converted to female by doubling the X chromosomes and discarding the male Y chromosomes. The resulting cells were differentiated into oocytes that were fertilized by the sperm of the second male mouse, yielding viable offspring of both sexes. If I was a believer, I would say that God probably used a similar technique to create Eve from the rib of Adam. But I'm not.

The main practical application of iPS cells is in the generation of transplantable tissues and organs (a topic for a separate chapter, "The twenty-first–century cure"), but for now we will focus on the idea that the induction of Yamanaka factors can rejuvenate adult animals. The information theory of aging states that epigenetic disturbances are an important cause of our deterioration, and that this type of damage can be reversed if the "epigenetic settings" of our cells are rolled back to "factory default"[524]. This is where Yamanaka factors come into play. This idea, although brilliant at its core, turned out to be more difficult in practice than in theory as we quickly learned that cellular reprogramming can be a double-edged sword.

Early studies involving genetically modified mice with a premature aging syndrome (progeria) found that continuous induction of Yamanaka factors actually increases mortality, presumably due to the loss of important specialized cells due to dedifferentiation. Furthermore, one Yamanaka factor happens to be an especially potent oncogene that is highly active in the majority of human cancers[525]; this is not surprising given that active proliferation is something that cancer cells and stem cells have in common. Fortunately, the use of only three genes out of the initial four is sufficient for cellular reprogramming[526].

It also turned out that many of the adverse effects of epigenetic reprogramming can be avoided if the activity of Yamanaka factors is induced cyclically for just two days a week. This was achievable in murine experiments thanks to sophisticated genetic engineering. Test mice can be genetically designed in such a way that the activity of Yamanaka factors in their cells is strictly regulated by an outside chemical such as doxycycline. Such mice need to drink water with doxycycline for any effect to occur. This "lock and key" principle allowed researchers to ensure that cellular reprogramming was merely partial and not complete. Even though cellular reprogramming was partial, this intervention reduced DNA damage, cellular senescence, and the number of defects in the nuclear envelope; and, most importantly, it extended the animals' median lifespan by roughly thirty percent[526].

Mice with progeria are a popular model in gerontology because their abbreviated lifespan makes them easier to study than their longer-lived counterparts. Unfortunately, their aging may not be entirely representative of that of normal animals. However, the tissue-rejuvenating effects[527] of partial cellular reprogramming have been replicated in regular mice, improving various biomarkers of aging[528], reversing vision loss, and improving nerve- and muscle-cell regeneration[529]. One of the most impressive recent works in this field was performed by a team of scientists that included Steve Horvath, George Church, Vadim Gladyshev, and David Sinclair. The researchers tested a gene therapy with three Yamanaka factors and achieved a significant restoration of visual acuity in a mouse model of glaucoma, an eye condition related to optic nerve damage. We know that optic nerve tissues are difficult to regenerate, yet the therapy promoted neural regeneration after injury[530].

The fact that partial cellular reprogramming can be achieved via gene therapy is excellent news. Such an intervention can be applied in adult organisms and does not require any genetic manipulation prior to birth, and some studies are truly inspiring. In one study, old mice (roughly equivalent to seventy-seven-year-old humans) were treated with adenoviral vectors carrying a gene system for doxycycline-induced Yamanaka factors. Cyclic doxycycline exposure doubled their remaining lifespan compared with control animals. This sounds impressive, but the actual median lifespan of the mice increased by a modest six percent[531]. After all, the animals didn't have much remaining lifespan to begin with. Perhaps if the therapy had started earlier the overall results would have been better.

I wouldn't be surprised if such tools became applicable to humans in the near future, although much research is still required. So far, nuclear reprogramming has been shown to reset the epigenetic clocks of normally aged human cells[532] cultivated outside of the human body, but we are far from conducting similar experiments on living people. This is in part because we fear the potential adverse effects of such gene therapies. One promising idea is the creation of small molecules or chemical cocktails that mimic the biological effects of Yamanaka factors without the use of gene therapies. Several candidate molecular combinations have already been tested and shown to restore youthful gene activity profiles in aged cells without compromising their cellular identity[533].

While epigenetic reset is still far from being a practical solution for human life extension, iPS cells will save and prolong lives in the near future by helping us create human tissues and organs[534]. In the meantime, a research initiative called Altos Labs received three billion dollars in funding from investors including entrepreneur Yuri Milner to develop healthspan-increasing interventions based on cellular reprogramming[535]. While not much is known about the exact research that will be performed at Altos Labs, the team includes such prominent scientists as Shinya Yamanaka. More likely than not, some of this research will pave the way to increased human lifespan and healthspan.

We have discussed two important components of the germ line rejuvenation miracle: telomere lengthening and cellular reprogramming. However, the picture is hardly complete. There is still the issue of "junk" that tends to accumulate in aging cells. Multiple divisions of embryonic cells during a new organism's rapid growth are one way to reduce the harmful effects of undesired biological components, through the dilution of those components among trillions of new cells. But this doesn't work as well with replicating junk, such as mutated and malfunctioning mitochondria. Furthermore, it follows from the disposable soma theory that natural selection should favor improved maintenance of the germ line. If that is the case, then by studying the additional layers of protection that our reproductive and embryonic cells are privileged to benefit from, we might learn how to translate them to other cells of our body—for cellular equality! And indeed, dedicated mechanisms of "junk removal" have been linked specifically to germ line maintenance.

Just like other cells, the unfertilized oocytes of *Caenorhabditis elegans* tend to accumulate all sorts of junk. Their interaction with sperm increases the activity of an enzyme known as lysosomal V-ATPase. This is essentially a pump that resides in the membranes of lysosomes, the structures that provide autophagosomes with digestive enzymes. Once activated, lysosomal V-ATPases start utilizing energy stored in the form of ATP to pump protons inside lysosomes, acidifying their contents, essentially activating them[536]—just as the your stomach acids activate digestive enzymes. This induces autophagy, and serves as an additional mechanism of embryo rejuvenation. Conversely, deactivation of lysosomal V-ATPases accelerates age-dependent protein aggregation in normal body cells[537]. Perhaps drugs, or therapies, that stimulate the activity of these proton pumps will promote autophagy. Hopefully interactions with sperm will not be necessary for this to work.

When it comes to aggregated "protein junk," autophagy is not our only ally. Another important component of our garbage disposal system is presented by proteasomes—large cylindrical complexes that are located in the cytoplasm and nucleus and can shred proteins into small chains of utilizable amino acids. Unwanted proteins are tagged for degradation by small proteins called ubiquitins, which leads to addition of more ubiquitins. As the ubiquitin chain grows, it binds the proteasome complex that mercilessly shreds marked targets[538]. The proteasomes themselves can be recycled by autophagy when they are damaged or unneeded[539]. Research shows that the activity of proteasomes is increased in embryonic cells and induced pluripotent stem cells, hinting that they could be another important mechanism of cellular rejuvenation[540,540].

Malfunctioning mitochondria are a more concerning kind of junk that oocytes need to get rid of. A normal cell usually contains thousands of mitochondria[541]. These organelles have their own, circular, DNA, and a number of functional genes that can acquire mutations, including damaging ones, but somehow functional mitochondria have been passed down from cell to cell for more than a billion years. For us the mitochondria of oocytes are especially important, because they are normally transferred to children. While severe mitochondrial disorders are rare, hundreds of pathogenic mutations in these organelles are associated with medical conditions, such as diabetes, cancer, infertility, myopathies, and neurodegenerative diseases.

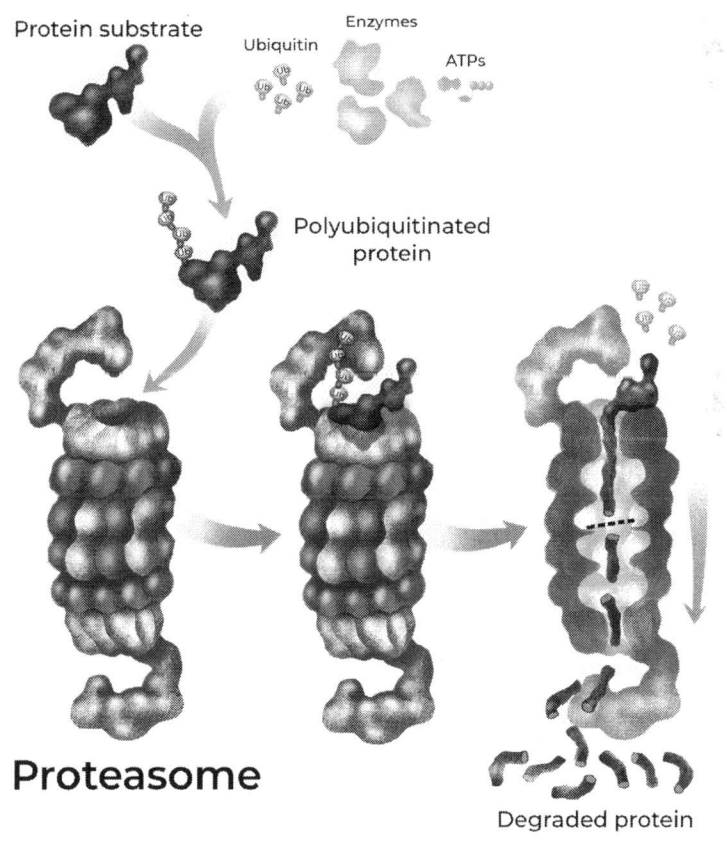

One recently developed approach that prevents these issues in children is called mitochondrial replacement therapy[542]. If you have ever heard about children from three biological parents, this is how they occur. During mitochondrial replacement therapy, a woman (A) with a mitochondrial disorder and a man (B) create an embryo via in vitro fertilization. The nucleus from that embryo is then transferred into an enucleated egg cell from a woman (C) with healthy mitochondria. This technique is similar to animal cloning, except the nucleus is taken from a non-adult cell. The embryo is then implanted, typically, in the primary biological mother (A). As a result, a child is born having the nuclear DNA of two parents (A and B) and the mitochondrial DNA of a third parent (C). Since mitochondria have only thirty-seven genes and the nuclear genome has more than twenty thousand, it might be more accurate to say that the child has 2.002 parents, but that is beside the point.

This is a working interventional solution to the problem of heritable mitochondrial diseases, but how does nature normally handle it? We know that it does so successfully. Experiments on mice show that severe mutations introduced into mitochondria are typically eliminated from the germ line within several generations[543]. As it turns out, this miracle is partially due to the ability of female bodies to harness the power of natural selection, life's secret weapon against the chaotic pull of entropy.

It happens that the female body produces an excessively large number of oogonia, precursors of oocytes. The excess is number of cells will never be used for reproduction, and only a small fraction reach maturity, while the majority die via programmed cell death, to which oogonia are especially susceptible. Damaged, malfunctioning mitochondria are potent inductors of programmed cell death, leading to the survival of the fittest. Only the germ line cells that have the healthiest mitochondria will survive and produce progeny. Since mitochondria are inherited almost[544] exclusively through female germ lines, each embryo is likely to receive the best organelles.

As we age the mitochondria of our body's cells also acquire detrimental mutations, contributing to the aging process[545]. Perhaps a solution can be found in improving the clearance of cells with damaged mitochondria while simultaneously providing healthy cellular replacements by restoring the pool of active stem cells through epigenetic reprogramming and telomere elongation. The goal is to ensure that, as in the case of oogonia, the best cells of our body receive an evolutionary advantage. Call it cellular eugenics at the scale of a single organism.

Alternatively, we might benefit from mitophagy, which is autophagy that is selective for dysfunctional forms of mitochondria[546]. The molecular mechanism behind mitophagy was recently uncovered, and it is truly beautiful. The selectivity of mitochondria degradation is achieved thanks to a protein called PINK1, a mitochondrial damage sensor. A functional, breathing mitochondrion has internal negative electric potential, and its membrane is hyperpolarized; this condition is necessary to transfer PINK1 from the outer mitochondrial membrane to the inner membrane for processing. When a mitochondrion's electric potential disappears due to the organelle's poor condition, PINK1 accumulates on its outer membrane and recruits a protein, called parkin, which essentially tags the organelle for destruction[547]. You might wonder why the protein is named parkin. Mutations in both parkin and PINK1 proteins are associated with early-onset familial Parkinson's disease, emphasizing the importance of mitochondrial quality control in preventing age-related neurodegeneration. Perhaps the search for Parkinson's disease treatments based on the activation of mitophagy will not only benefit Parkinson's patients, but will also contribute to our set of anti-aging tools[548].

This leaves us with one final component of the rejuvenation miracle: we need a solution for mutations in nuclear DNA. When it comes to ensuring the health of the subsequent generations and the removal of unwanted mutations, nature's solution isn't a kind one; it's the premature death of affected individuals due to their reduced fitness. This barbarism has obviously outlived itself and should not be tolerated further.

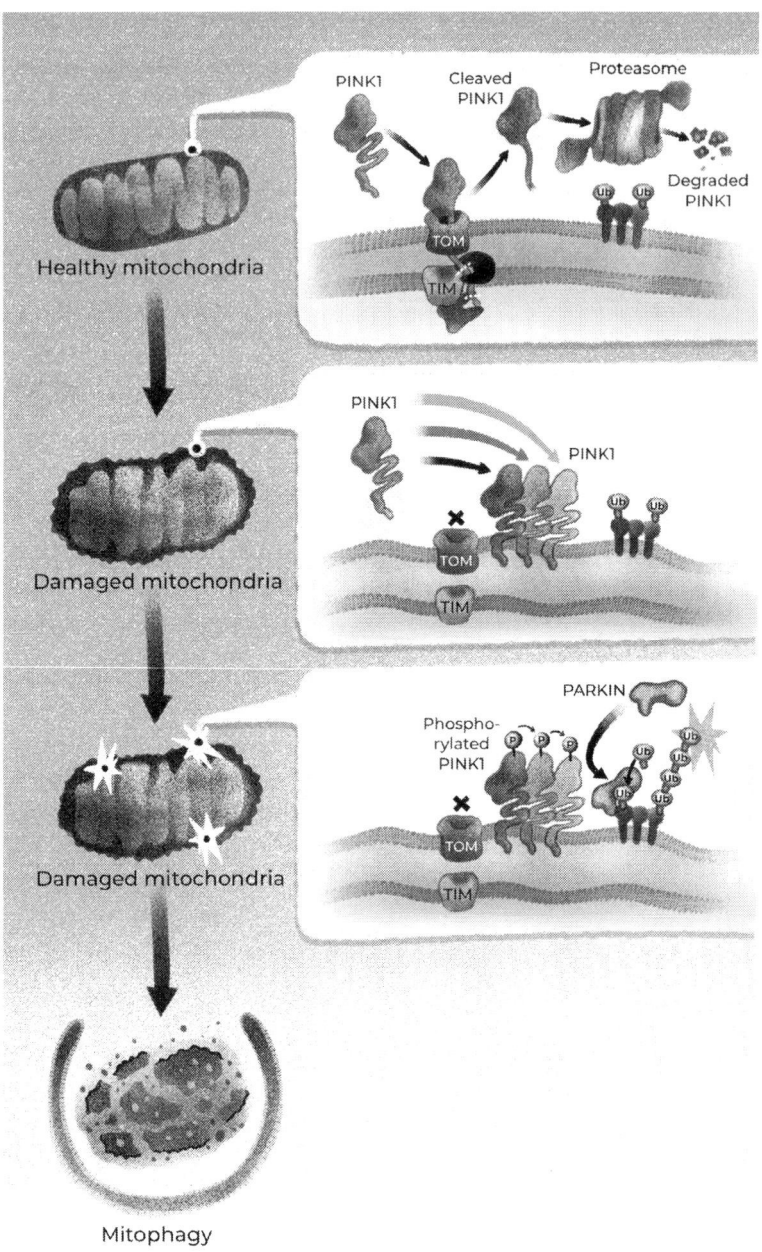

Methods such as preimplantation genetic testing and genetic engineering will help us eradicate most inherited diseases in the future, while gene therapy can help those already affected. But in the context of aging, we should be concerned not with germ line mutations that affect only future generations, but with those of our adult bodies, the somatic ones.

Once a mutation has occurred in a cell of the body, it is likely to remain, and even to spread when the mutated cell divides. This is especially problematic in the long run, because an accumulation of mutations can lead to cancer. Evolution has provided normal, unmutated cells with several advantages within our bodies: damaged cells can become nondividing, or senescent (the topic of our next chapter), or they can undergo apoptosis or be removed forcibly by the immune system (the topic of the chapter after the next). But these mechanisms are far from perfect, as evidenced by the great frequency of cancer among aged people.

These imperfections occur partly because for the purpose of evolution our bodies are disposable carriers of genetic information. Somatic mutations are not passed down to the next generation, so they are less relevant to natural selection than germ line mutations, and repairing somatic mutations is metabolically costly. Unsurprisingly, the privileged mammalian germ line is characterized by a much lower mutation rate than somatic cells[549-551]. This implies that there might be differences between the DNA repair mechanisms of the germ line and those of somatic cells, and that perhaps we can learn how to unleash the full potential of DNA repair in the latter.

A protein called ATM regulates a cell's reaction to DNA breaks by activating DNA repair, programmed cell death, and cell cycle arrest. In *Caenorhabditis elegans*, only germ line cells activate ATM in response to DNA damage[552]; somatic cells do not. The probable evolutionary reason for this is that these worms lack regenerative potential and have a comparatively small number of cells. By essentially ignoring DNA damage, they keep their cells intact. After all, damaged cells are better than no cells, and cancer isn't much of an issue for an organism whose lifespan is only several weeks.

Mutations in human ATM genes lead to Louis–Bar syndrome, a rare genetic disorder characterized by brain impairments, poor coordination, narrowing of blood vessels, increased risk of infection, reduced DNA repair, and a predisposition to cancer[553]. ATM activity is high in most human cell types[554], but in some people it declines with age[555]. This leads to the idea that boosting the activity of ATM (or other genes with similar functions) via gene therapy might be useful for slowing aging. Boosting the activity of maintenance *Fox* genes (discussed in the previous chapter) might also help reduce DNA damage, and we will discuss additional options later in this book.

One animal that is extremely good at repairing its genes is the tardigrade *Hypsibius exemplaris*. It can survive doses of radiation that are a thousand times greater than the lethal dose for humans. As it turns out, the tardigrade does this by sensing damage caused by radiation and rapidly increasing the activity of multiple DNA repair genes[556]. This might be another example of the untapped potential of animal cells. The genes necessary for DNA repair may already be in place, though not sufficiently activated in all cell types, particularly in the disposable soma. Systemic activation of these dormant mechanisms may be useful to slow down aging.

In conclusion, it seems that the miracle of immortality relies on several features of germ line and embryonic cells: epigenetic reprogramming, telomere elongation, enhanced autophagy, selective mitophagy, and control of DNA damage through repair mechanisms and apoptosis. Telomerase activity, mitophagy, and autophagy can be induced, but we still need to find work-arounds for the adverse effects of epigenetic reprogramming. Figuring out how to improve DNA repair seems to be among the most difficult tasks we need to accomplish, although we now know of several genes that promote this important process. An alternative to improving DNA repair and maintenance in our cells would be to increase the clearance of severely mutated cells by strengthening the existing mechanisms that inhibit cellular division, promote programmed cell death, and improve immune system vigilance. Germ line cells are more sensitive to DNA damage than somatic cells and more easily undergo apoptosis[557], and perhaps these qualities can be transferred to other cells of our body.

Seven individuals tap into the eternal power of the germline, seeking to escape the certainty of death. Meanwhile, others exploit this same power to bring new lives into the world, fated to die.

CHAPTER 8
Programmed cell aging

> Things get damaged; things get broken.
> —*Depeche Mode ("Precious")*

Summary: in this chapter we discuss the concept of cellular aging, its connection with inflammation and aging-related disorders, and whether the removal of senescent cells can prolong an organism's lifespan and healthspan.

As we age, our bodies tend to accumulate cells with "zombie-like" properties. These senescent cells are not dead, but they have lost one essential property of living cells—the ability to replicate. The other reason they might be called zombie cells is that they can spread their properties of senescence to neighboring cells, similarly to the spread of a zombie infection. Sometimes this involves the production of pro-inflammatory molecules or virus-like particles[558]. Despite the negative health impacts of some zombie cells, and their contribution to aging, they have a number of redeeming properties. The same can be said about the "necromantic" molecular processes that create them.

Bad things happen to even the best of cells. Irreparable DNA damage, telomere shortening, chromatin-organization alterations, dysregulation of gene activity, and other factors can lead to a permanent halt of cellular proliferation[559]. Intuitively, one might think that cells stop dividing due to accumulated malfunctions in their molecular machinery. However, in most cases specific genetic programs and regulatory proteins are actively involved in cell-cycle arrest.

For example, the ATM protein can sense DNA damage and produce an "error signal" that stimulates genetic repair. Eventually a damaged cell either repairs the damage, so that the "error signal" goes away, or "decides" that further replication is not warranted. It either dies, or becomes "undead" via cellular senescence[560].

This feature is beneficial for the organism, because it prevents cells that are damaged beyond repair from becoming cancerous. This is one of multiple mechanisms that help the healthy and undamaged cells of our body to maintain a sort of evolutionary advantage over the malfunctioning ones, at least for some time. The alternative to senescence is programmed cell death (apoptosis), and the choice between these two options often varies depending on the severity of a cell's damage and on the type of cell[561]. Both solutions appear to have evolved as antitumor mechanisms, and they compete with and sometimes substitute each other. Apoptosis removes a cell completely, while senescence allows it to remain intact, preserving some of its functions.

Several types of cellular senescence have been described based on underlying causes of cell-cycle arrest. Replicative senescence, described by Leonard Hayflick and Paul Moorhead, occurs when, after a certain number of divisions, cells stop proliferating due to telomere shortening. Among mammals, replicative senescence is more pronounced in larger species. Massive animals such as elephants or whales have more cells and must rely more heavily on this tumor-suppressing mechanism to prevent cancer[562].

DNA damage caused by ultraviolet light[563], ionizing radiation[564], or reactive oxygen species[565] can cause stress-induced premature senescence, which is somewhat intertwined with replicative senescence. For example, oxidative stress can accelerate telomere shortening[566,567].

Mutations that activate genes that promote cancer progression can cause so-called oncogene-induced senescence, a fail-safe program that restricts the proliferation of abnormal cells[568]. A common example of this is the formation of human melanocytic nevi. These pigmented precancerous lesions result from mutation-driven proliferation of melanin-producing skin cells. Oncogene-induced cellular senescence halts this proliferation and prevents melanoma[569].

Cellular senescence also plays a role during mammalian early development[570,571], by preventing undesirable cell expansion. It also helps with the elimination of transitory embryonic structures. This kind of developmentally programmed senescence is not induced by DNA damage, and the resultant senescent cells are mostly cleared by white blood cells called macrophages, leading to tissue remodeling[572]. This example highlights the difference between a senescent cell and a cell that is simply chronologically old or damaged.

As we age, the number of senescent cells in our body increases, and some of these cells can have deleterious effects. Cells that undergo senescence due to DNA damage typically acquire a senescence-associated secretory phenotype (SASP). They secrete growth factors and signal molecules, such as cytokines, that affect surrounding tissues and often cause inflammation. One example of a small pro-inflammatory signal protein is interleukin-6, which is known as a mediator of fever. It also plays a role in autoimmune diseases[573], and its increased production is associated with poor clinical outcomes in certain cancers. As cells halt proliferation, their secretion of interleukin-6 can increase by a factor of forty in some cell types[574].

Inflammation can be either a good thing or a bad thing. On one hand, it is an important process in antimicrobial protection and tissue repair and recovery[575]. On the other, it can cause degenerative joint disease (osteoarthritis), plaque buildup on artery walls (atherosclerosis), and a number of other conditions that are common in the elderly. Senescent cells can increase the risk of age-related chronic systemic inflammation which in turn increases the risk of cancer, dementia, depression, and chronic kidney disease. It is also associated with frailty[576], and it is a powerful predictor of all-cause mortality[577]. Not only can cellular senescence cause these problems, but it can also be driven by inflammation, creating a deadly positive feedback loop.

Numerous experimental studies have revealed just how damaging senescent cells can be to an organism. For example, the transplantation of senescent cells from older to younger mammals can cause tissue damage, increase cancer risk, and reduce lifespan. Let me describe some of these studies, so that you can get the idea.

In one study, scientists collected either normal or precancerous epithelial cells that had one of several mutations that are typically required for cancer, meaning that the cells were just a few steps away from causing a disease. Then the scientists mixed these cells with either senescent connective tissue cells called fibroblasts or their presenescent counterparts still capable of dividing. They found that senescent cells stimulated much greater growth of precancerous cell lines compared with presenescent cells.

Furthermore, injection of precancerous epithelial cells with the addition of senescent fibroblasts into mice markedly increased tumor incidence, size, and growth rate compared with control conditions, under which presenescent fibroblasts or no fibroblasts were used instead of senescent ones. This showed that, while cancer requires changes in a cell's DNA, surrounding factors can influence precancerous cells' elimination rate and their ability to proliferate[578].

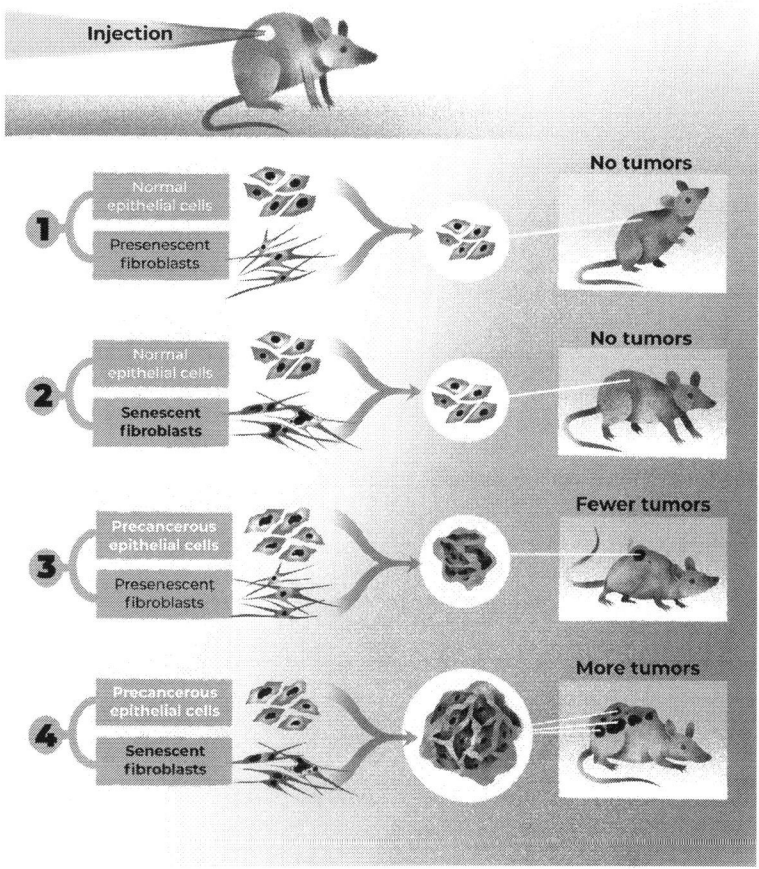

In another experiment, pig scattered-tubular cells (STCs), which are normally involved in renal recovery, were subjected to senescence by irradiation. Control and senescent STCs were then transplanted into the kidneys of young mice. Unlike their normal counterparts, transferred senescent STCs induced renal inflammation, damage, and fibrosis[579].

Yet another study involved the transplantation of either senescent or normal adipose-derived stem cells into mice[580]. Compared with control conditions, about one additional senescent cell per ten thousand cells of a mouse's body was sufficient to cause a 5.2-fold increased risk of death during the following year of observation. Affected mice experienced impairments in walking speed and hanging endurance, which are important indicators of aging. The effect was more pronounced in older mice and increased with the number of senescent cells that were transferred.

These and other experiments suggest that senescent cells can be detrimental to an animal's longevity, and that lifespan extension could be achieved by reducing their number in that animal's body. This aligns with the common view that cellular senescence is an example of antagonistic pleiotropy, a mechanism that has mostly positive effects in an organism's early years and mostly negative effects later on. Indeed, animals lacking a gene encoding an important senescence-inducing protein called p16Ink4a die young because of high cancer incidence[581].

Before we examine potential avenues to rid ourselves of senescent cells, it is important to note that not all senescent cells are the same. They don't always secrete pro-inflammatory molecules[582], and sometimes they may even provide benefits to an organism. For example, some of them can stimulate the regeneration of tissues such as the skin[583] and muscles[584], and they are normally removed by the immune system after their job is done. The senescence of certain hepatic cells is important for the prevention of excessive liver fibrosis and cirrhosis[585]. And senescence of pancreatic beta cells can enhance insulin secretion[586].

In the cnidarian *Hydractinia symbiolongicarpus* (a somewhat distant relative of the immortal hydra) amputation injury induces senescence in a small number of cells, which can persist a few hours before being expelled. Chemical signals from these cells can reprogram their neighbors into a more proliferative state, which drives the animal's whole-body regeneration[587].

Even naked mole-rats maintain developmentally programmed cellular senescence, as well as senescence induced by oncogenes and irradiation[588], and this does not prevent them from having an extremely long lifespan. However, a 2023 article reported that the fibroblasts of naked mole-rats, unlike those of mice, undergo delayed and progressive programmed cell death following DNA damage-induced cellular senescence[589,590]. If we are to learn a lesson from these animals, perhaps it is that we should keep the mechanisms of cellular senescence intact while finding ways to selectively remove the most damaging types of "zombie cells" after their "transformation."

Scientists from the Mayo Clinic College of Medicine and from the University of Groningen developed a clever approach to do this[591]. Genes have adjacent regulatory DNA regions called promoters. Regulatory proteins called transcription factors (such as the products of *Fox* genes discussed in previous chapters) bind to these regions and modify gene activity. Different transcription factors recognize and bind to different promoters based on the unique DNA sequences of the latter. And different conditions lead to the activation of different transcription factors. This is how a cell "knows" which genes to activate in a given situation.

A significant proportion of senescent cells are characterized by increased activity of a tumor-suppressor gene called *p16Ink4a* (mentioned earlier), which leads to cell-cycle arrest and upregulation of the senescence-induced secretory phenotype that causes inflammation[592,593]. As you can guess, *p16Ink4a* has a unique adjacent regulatory-promoter DNA sequence. If we copy that DNA sequence into another gene, that other gene will be activated under the same conditions as *p16Ink4a*. That's smart genetic engineering.

Now let's attach *p16Ink4a*'s promoter to a gene that causes programmed cell death and put this new kill switch into a cell. As soon as the cell decides to become senescent, it will simply destroy itself instead. But let's add an extra layer of control, and make it so the switch works only in the presence of a synthetic agent. Let's call this molecule AP20187. Now transgenic mice born with this new controlling element embedded in their DNA can be given AP20187 at any time to selectively induce the death of their senescent cells while keeping normal cells intact.

CHAPTER 8

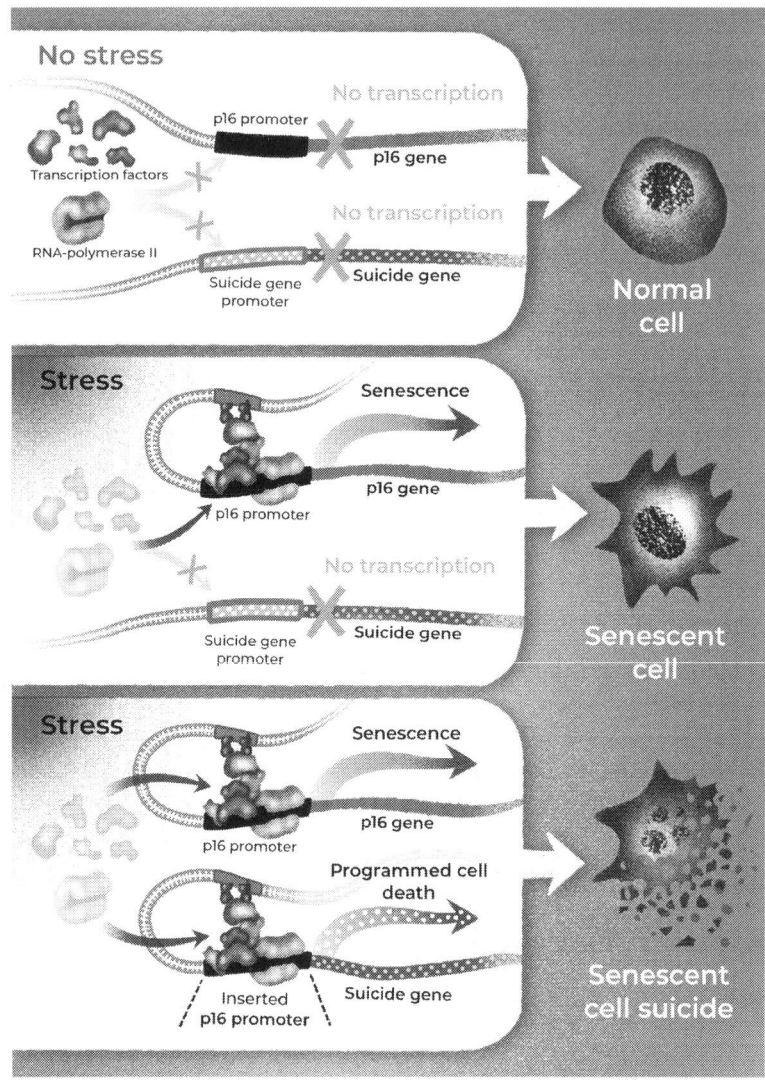

All this was initially tested on a short-lived line of mice with progeria. The administration of AP20187 caused the clearance of senescent cells and delayed the acquisition of age-related dysfunctions in the studied animals. The study's authors then inserted the same genetic switch into normal mice. Injections of AP20187 twice a week increased the animals' median lifespan by around twenty-five percent compared with vehicle-treated controls[594], but unfortunately, there was hardly any change in maximum lifespan. Mice that received the drug had a healthier appearance in old age and experienced less age-dependent reduction in their exploratory behavior and spontaneous activity.

Obviously, this was a proof of principle, but it would be difficult to introduce a similar genetic switch into human adults, given that we have so many cells. This led to a search for alternative approaches and the discovery of senolytic drugs, which can selectively remove senescent cells without genetic engineering. Some senolytics have been shown to increase the lifespan of model animals while reducing age-related dysfunction in the tissues of the heart, lungs, brain, and other organs.

One combination of senolytics was used in a study we discussed earlier, which involved the transplantation of adipose-derived stem cells into mice. The authors could track the transplanted normal and senescent stem cells because they came from genetically modified mice donors and carried a gene for luciferase, an enzyme that produces bioluminescence[580]. They found that the combination of two senolytics, called dasatinib and quercetin, could selectively remove the transferred senescent luminescent cells but not normal luminescent cells. They then treated normal 24–27-month-old mice (equivalent to 75–90-year-old humans) with either these two drugs or a vehicle. Treated mice had a 36% higher median post-treatment lifespan.

In another study, the clearance of senescent cells in mice was performed either with the senolytics dasatinib and quercetin or via genetic engineering and the AP20187 activator. Both methods improved mouse lung function[595]. Meanwhile, dasatinib and quercetin were used to treat diabetic kidney disease in a human clinical trial, leading to a reduction of senescent-cell burden in some tissues, although in a sample of only nine humans[596]. These preliminary advances provide hope that some aspects of human aging will be treated by senolytics in the near future—although larger and more diverse human trials are warranted before we can be entirely sure of the safety and efficacy of these drugs[597].

In the meantime, there is growing evidence that another senolytic, called navitoclax, may help combat age-related increases of cardiovascular diseases. As we age, our hearts accumulate cardiomyocytes with senescence-associated secretory phenotypes that promote fibrosis and hypertrophy, which is a risk factor for sudden death[598]. Navitoclax can trigger selective apoptosis of senescent heart-muscle cells in mice and attenuates these age-related problems[599]. In another study, the drug also reduced myocardial senescence in mouse hearts, as well as the damage caused to this vital organ by ischemia-reperfusion injury[600] (the death and dysfunction of cells following restoration of blood flow). This senolytic treatment also dramatically increased mouse survival rates following induced myocardial infarction[601]. Remarkably, twenty-three-month-old mice treated with the drug had survival rates similar to those of three-month-old mice also subjected to infarction and twice as great as those of untreated geriatric animals.

Cellular senescence also affects the mammalian brain. The brain consists of neurons that form connections and transfer electrical signals between one another, as well as numerous supporting cells, including star-shaped astrocytes, which provide neurons with nutrients, encircle endothelial cells of blood vessels to maintain the protective blood–brain barrier, provide structural support, and serve other functions. The brain also contains neuronal stem cells that can self-renew and turn into the abovementioned and other cell types (this will be discussed in more detail in the "ageless brain" chapter). When the brain becomes damaged (let's say, via a stroke), neural stem cells can become activated and produce more neurons[602]. This activation is mediated by the brain's main immune cells, called microglia, which scavenge the central nervous system for infectious agents and remove damaged neurons and unnecessary connections between them[603], sculpting our neural circuits.

Proper function of microglia can protect the brain from various disorders, including Alzheimer's disease, but these cells also have the power to damage neurons and contribute to neurodegeneration[604].

While neurons do not divide, and thus do not experience telomere shortening and replicative senescence, they can still suffer from DNA damage[605] and undergo age-related physiological alterations that affect their functions. For example, senescent-like neurons have decreased mitochondrial volume[606], experience increased production of reactive oxygen species, and increasingly secrete interleukin-6, contributing to inflammation. But it's not just neurons that suffer; senescent astrocytes in the brain also produce pro-inflammatory cytokines[607]. Even neuronal stem cells undergo types of senescence that involve the activation of pro-inflammatory genes[608].

We know that this is a serious problem for the aging brain. Artificially increased production of interleukin-6 in the central nervous systems of mice causes neurological disorders[609], and cerebrospinal fluid concentrations of interleukin-6 are typically increased in patients with Alzheimer's or Parkinson's disease[610]. This further highlights the link between cellular senescence, inflammation, and the aging brain[611].

But there is hope. Several studies have reported preliminary findings that treatment with a combination of the senolytics dasatinib and quercetin[612], as well as other methods of senescent-cell removal[597,613,613,614], can reduce brain inflammation and improve cognition in several mouse models. Most studies confirm the senolytic effect by evaluating the loss of senescent non-neuron support cells after treatment. But shouldn't we be concerned about the possible removal of aged neurons[615], the cells that are most difficult to replace?

Indeed, some senolytics, such as navitoclax, have been shown to clear senescent-like neurons in a mouse model of chemically induced peripheral neuropathy[616]. One might think that losing irreplaceable neurons is a bad thing, but the effect was beneficial, leading to an alleviation of pathology. Another study reported that the senolytic dasatinib plus quercetin could remove neurons containing harmful protein aggregates in a mouse model of Alzheimer's disease[617]. While some neurons were lost, the treatment prevented cortical brain atrophy in the long term. So perhaps losing some damaged neurons today is a better option than losing healthy ones due to systemic inflammatory processes tomorrow. Only future studies will allow us to estimate the risk-to-benefit ratios for such interventions for humans.

Another good thing about senolytics is that, in theory, they should work in a hit-and-run manner, without the necessity of constant use. Once accumulated senescent cells are removed, it takes time for new ones to form. However, the potential harm of senolytics should be carefully considered. For example, dasatinib was originally developed to treat chronic myeloid leukemia[618], and when used in therapeutic doses designed for this purpose, it causes side effects that range from fever to coughing up blood[619]. After all, the intended effect is to trigger the deaths of multiple cells. Fortunately, the dosage of proposed anti-aging regiments of dasatinib and the frequency of use are substantially more tolerable[620] and should be safer for healthy individuals.

The reported side effects of quercetin are typically more mild and infrequent than those of dasatinib[620]. Quercetin is a flavonoid found in many fruits and vegetables, although in subtherapeutic concentrations. Capers have one of the greatest concentrations of quercetin, but you would have to eat about six hundred grams of capers per day to achieve the intake of quercetin used in preliminary human trials[620]. Quercetin works by inhibiting certain proteins that inhibit apoptosis and thus prevent the self-elimination of senescent cells. When this inhibition is lifted, the cells become free to kill themselves. The mechanism of apoptosis prevention targeted by quercetin is typically shared by both cancerous and senescent cells, and for this reason, a number of anticancer drugs are currently being repurposed for the treatment of aging. For example, navitoclax, which we mentioned earlier, is an emerging anticancer drug with a mechanism of action[621] similar to quercetin's.

Fighting aging and cancer at the same time sounds like an idea worth investigating.

Unfortunately, some studies of the beneficial effects of senolytics have been difficult to reproduce. One very popular senolytic among biohackers is fisetin. Its popularity is probably due to its "organic" nature: it is an antioxidant present in low concentrations in a number of fruits and vegetables, including grapes, apples, onions, and cucumbers. Fisetin is structurally similar to quercetin, and as a supplement it gained especially wide attention after a study showed that it improved the median and maximum lifespan of aged mice by about ten percent[622]. However, the Interventions Testing Program subsequently reported no effect of fisetin on murine lifespan[623]. Such reproducibility problems may come from due to a combination of two factors: the preference of scientists and academic journals' editors to publish positive findings, and the possibility of misinterpreting the lower-than-average animal lifespan in the control group as the beneficial effect of a drug or intervention.

Research on senescent cell research is an evolving field, with new discoveries and corrections made every year. Scientists from the group led by Andrei Gudkov have found that some senescent cells are actually a subclass of macrophages that can eat other cells[624,625] as part of the innate immune response. While it didn't surprise me that some suspected "zombie cells" also turned out to be ruthless cannibals, Gudkov's team made an important point, that some effects of senolytics might be due to the elimination of macrophages. These effects can be beneficial and detrimental at the same time, given that macrophages are implicated both in tissue repair and in age-related diseases such as atherosclerosis and osteoarthritis. Researchers also found that some cells with high p16ink4a activity are capable of proliferation and are free of many of the detrimental properties of senescent cells[626]. This means that the way we identify and target senescent cells might be far from perfect. However, the scientists still confirmed that *p16ink-k4a* activation is associated with reduced proliferation and the senescence-associated secretory phenotype, and that the number of cells with *p16ink4a* activity increases with aging and inflammation[592].

In conclusion, I would say that the accumulation of senescent cells and the rise of systemic inflammation contribute significantly to aging. The good news is that this is a problem that we can potentially solve. Modern senolytics have a comprehensible mechanism of action, and their use has sound theoretical and empirical support. There are a number of ongoing human clinical trials of senolytic drugs aimed at reducing frailty, inflammation, chronic kidney disease, Alzheimer's disease[612], osteoarthritis[627], and other consequences of aging. Hopefully we will soon know more about their effects, and perhaps we will invent better and more selective senolytics in the future. If the clinical trials are successful in humans, we just might get a few extra years to live, increasing our chances of surviving until other aging problems are solved. Gerontologist Aubrey de Grey calls this condition "longevity escape velocity": when the speed of our biological aging becomes less than the pace at which we create new anti-aging interventions, thanks to technological and medical advancements.

Perhaps it would be especially beneficial to pair senolytic solutions with cellular-rejuvenation techniques, so that dead old cells are better replaced by new ones. As I mentioned before, the goal is not to eliminate cellular senescence but to harness it as a means to identify, remove, and replace the most-damaged cells in our body. Cellular senescence itself provides an advantage to our most-normal and most-functional cells, because they avoid it, and it is one of our several inbuilt cancer-preventing solutions. In the next chapter we will discuss what other anticancer mechanisms are at our disposal.

The flames of inflammation rage, and cellular senescence spreads like wildfire. Eight individuals stand as guardians, quelling the blaze by removing and replacing the body's smoldering components.

CHAPTER 9
The rise and fall of living utopias

> Oceans rise; empires fall.
> We have seen each other through it all.
> And when push comes to shove,
> I will send a fully armed battalion to remind you of my love!
>
> —King George III ("You'll Be Back", Hamilton)

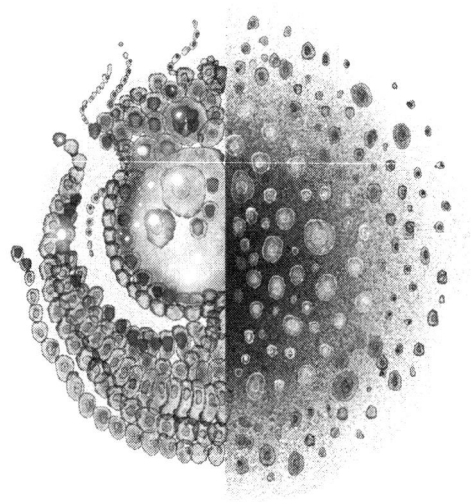

Summary: in this chapter we discuss the interconnections between programmed cell death, cancer, and the immune system; why cancer is an evolutionary process; and how knowledge of this can be used against it.

Pseudomonas fluorescens is a rod-shaped bacterium capable of evolving cooperation when placed in a static liquid broth. As members of this species quickly use up all available oxygen it becomes increasingly beneficial for them to stay near the surface. A simple mutation that leads to an overproduction of cellulosic polymer allows the bacteria to join together and form a surface biofilm or bacterial mat with better access to air. The benefits of this cooperation outweigh its metabolic costs for individual organisms, and thus utopia is achieved.

But the good times don't last very long. Since the cellulosic polymer is a shared resource that anyone can use, "cheaters" without the mutation gain an advantage. They do not contribute to the production of polymers, yet they stick to the colony and reap the collective benefits. The shortsightedness of Darwinian evolution allows these defectors to multiply efficiently until there are just too many of them for the colony to sustain. As a result, such bacterial mats collapse within several days[628].

The above-described chain of events not only is related to biological processes, such as cancer progression and the origin and maintenance of multicellularity, but is also reminiscent of a concept introduced by British economist William Forster Lloyd known as the tragedy of commons[629].

Imagine a pasture that anyone can use freely. A herdsman is rationally seeking a way to maximize his profits. He notices that the benefit of adding an animal to his herd outweighs his individual losses due to overgrazing—costs shared by the entire community. So, he increases his herd size. The problem is that other herdsmen are likely to notice the same thing. As a result, overgrazing costs pile up and each herder ends up less successful than he would have if all actors had cooperated and set a sensible limit to herd size.

Social implications of this and similar concepts are visible in such examples as excessive military spending. In a utopian setting of mutual cooperation, a population's military costs would approach zero and all resources would be allocated for the population's benefit: education, science, proper nutrition, health care, entertainment, and various services. However, in reality a country that spends little on its military can be subdued by countries with more aggressive policies. As one country spends more, other countries are forced to do the same, causing a positive feedback loop. Everyone loses as a result. Likewise, a regime that relies heavily on corruption may be beneficial for its elites, up to the point where the whole system crumbles under the inefficiency introduced by said corruption.

Just as prosperous human enterprises require mutually beneficial agreements between individuals, a multicellular organism requires cooperation between its cells. This includes division of labor, as well as mechanisms for communication, collective action, and ways of keeping potential cheats in check. The cheats in this analogy are cancerous cells, which are very efficient in their own reproduction and resource accumulation, but at the cost of harming their collective environment — one's body.

Thus, it is not surprising that animals evolve inbuilt genetic programs that enable cell-cycle arrest and apoptosis, as well as DNA-repair mechanisms that prevent cheats from appearing in the first place. Likewise, elements of the immune system enforce collective order by eliminating defectors. Unfortunately, over time the most "successful" cells develop mutations that allow them to evade these measures, eventually leading to the demise of the animal those cells constitute. In a sense, metastatic spread of cancer is similar to the collapse of the bacterial mat.

Normally, this process ends with the death of an organism. However, nature, in its horrifying brilliance, has come up with several unique exceptions — when cancers outlive their hosts and find new habitats to exploit and destroy. These examples represent remarkable cases of unicellular parasites evolving from multicellular animals. Devil facial tumor disease is transmitted between Tasmanian devils, carnivorous marsupials through biting. In fact, there are at least two genetically distinct transmissible cancers in these animals[630], each continuing to exist for many generations, long past the death of its initial host. Likewise, the canine transmissible venereal tumor first emerged approximately eleven thousand years ago[631], and since then it has spread to dogs all around the world. Some transmissible cancers found in bivalve mollusks move through water and are even capable of infecting more than one species[632].

It is fortunate that no such diseases are known in humans. The closest thing we have to a cancer with independence is the immortal HeLa cell line, which has been cultivated in laboratories since 1951 and is used for biomedical research. This cell line was derived from the cancerous tissue of a woman named Henrietta Lacks[633]. The woman died, but her cells will live on for ages. The life of a unicellular entity is quite different from the experience of a cell that is exists as part of a larger organism. Unexpected adaptations tend to give an organism an advantage, and cause evolution to start moving in a new direction. Even in the relatively short time HeLa cells have existed, they have undergone substantial genomic evolution, including catastrophic chromosome shattering and rearrangement[634]. This has made some scientists question whether HeLa cells can still be considered a suitable model for human biology. Evolutionary biologist Leigh Van Valen even suggested that HeLa cells should be considered a new microbial species of mammalian origin and proposed the name *Helacyton gartleri*.

That said, there has been at least one case of interspecies cancer transmission that affected a human patient. A man with HIV consciously decided not to take antiviral medication. Seven years after diagnosis, he was admitted to a hospital with fever, cough, fatigue, and weight loss. Biopsies revealed atypical cells of nonhuman origin in his lungs and lymph nodes. Eventually it was determined that the patient was infected with a tapeworm called *Hymenolepis nana*. The poor parasite suffered from cancer, and its malignant cells spread to its immunocompromised human host, eventually killing him[635].

The key to "immortality" and unfortunate "success" in cancer cells lies in the process of pathological Darwinian evolution going on right inside our bodies. As cells multiply, they acquire mutations in their DNA. Cells that are "lucky" enough to acquire mutations that help them survive have an easier time reproducing themselves. A single beneficial mutation is followed by another and then another, causing ever-increasing fitness. Some mutations that suspend DNA repair make the accumulation of other mutations more likely, boosting the speed of this deadly process.

I can't stress enough how important it is to understand cancer as an evolutionary process, if we want to treat such diseases efficiently. Consider the problem of relapse after chemotherapy treatment. Chemotherapy leads to the death of cancer cells, but cells that survive are often chemoresistant[636], meaning that the same treatment will be much less effective in the future. Cancer can evolve drug resistance just as bacteria do when they are exposed to antibiotics.

An important gene that is mutated in more than fifty percent of human cancers[637] is the p53 tumor suppressor, sometimes called "the guardian of the genome." Mice born with p53 loss-of-function mutations are at much higher risk of developing cancer than normal mice are[638]. On average, they develop tumors in just 4.5 months[639], and they are more susceptible to the deleterious effects of carcinogens.

The encoded p53 protein is a transcription factor that is activated by the ATM protein in response to DNA damage[640] and by a few other triggers. It regulates several hundred genes and is capable of halting the cell cycle to allow time for DNA-repair machineries to restore genome stability. It also promotes apoptosis[641] in severely malfunctioning cells. This discovery led to the invention of the first clinically successful anticancer gene therapy, called Gendicine[642-644]. Gendicine involves the transfer of a functional p53 gene copy into a patient's cancer cells with the help of genetically modified adenoviruses. The idea behind this treatment is that the lost antitumor activity of this gene can be reestablished in affected cancerous cells, and they will either stop dividing or undergo programmed cell death.

Loss of p53 function may have its own peculiar benefits, such as increased resistance to acute radiation toxicity[645]. While impaired programmed cell death promotes cancer in the long run, in the short term it translates into the survival of cells with damaged DNA. One could consider p53-deficient mutants to be prime candidates to be supersoldiers under conditions of nuclear escalation, but unfortunately, people born with even one damaged p53 copy suffer from Li–Fraumeni syndrome, which is characterized by early onset of a wide spectrum of tumors[646].

Genes similar to p53 may explain Peto's paradox, the observation that larger animal species don't tend to suffer from increased lifelong cancer rates compared with smaller animals[647]. This is viewed as a paradox because you could fit the cells of more than a hundred thousand mice into a single elephant, so, since it takes only one sufficiently mutated cell for cancer to evolve, an elephant should be a walking tumor, unless its cells are somehow more cancer-resistant than those of mice.

Indeed, evidence of rapid evolution and gene duplications in various tumor suppressors has been found in large animals, including elephants and whales[648]. Extra copies of genes that control programmed cell death ensure that it takes more mutations, more steps, to turn off programmed cell death. Elephants have evolved increased p53 sensitivity to DNA damage, along with other anticancer mechanisms, to compensate for their larger number of cells. This cancer resistance comes at a familiar price: their cells are more susceptible to death after radiation exposure[649]. (Note: a study published in 2025 found that larger animals do in fact suffer from higher rates of neoplasia than smaller animals, suggesting that larger animals compensate for the increased cancer risk only partially[650]).

There is another interesting hypothetical explanation for the lower-than-expected rates of cancer in the largest of animals. The bigger the size of an animal, the larger a tumor needs to grow to become life-threatening. This allows more time for cancer evolution. As a tumor grows, "cheaters of cheaters" may evolve — cells that are selfish even by tumors' standards. For example, first-generation cheaters can produce signal molecules that locally impair an animal's immune system in order for the cheaters to avoid detection, or they may facilitate blood vessel growth inside a tumor to provide a stable flow of nutrients to the tumor. But this requires an investment of resources. Second-generation cheaters might benefit from the collective efforts of their comrades without contributing anything to the "tumor society," thus gaining further evolutionary advantage. As these hypothetical tumors within tumors — or so-called hypertumors—spread, they can potentially destroy the original cancer[651]. While this is full of karmic justice, so far it is supported only by models, not by real biological data. It isn't as easy to study cancer in whales as it is in mice.

Programmed cell death appears to be the most important anticancer mechanism in nature; it is so vital to life that it can be observed even in some unicellular organisms, such as budding yeast. In the wild, these fungi form colonies of closely related individuals. The death of older and damaged yeast cells can save resources for their younger, nearly genetically identical siblings, and promote colony fitness[652].

But there is a sad story about the death of yeast cells that reveals how even fungi need love and appreciation. These organisms have two mating types, which produce signal molecules that affect the opposite "sex" and trigger them to mate. Prolonged exposure to such sexual pheromones with no mating success induces cell suicide of the most tragic kind[653]: it starts from the "heart" (by which I mean the mitochondria). Mitochondria are necessary for yeast's programmed cell death. If the DNA of these organelles is rendered nonfunctional, the resulting "heartless" fungi can survive the lack of mating even after pheromone exposure.

There is controversy as to whether programmed cell death in yeast is related to mammalian apoptosis[654]. But mammalian cells also rely on mitochondria when choosing whether to live or die. Our apoptosis can be triggered by either intrinsic or extrinsic signals. Extrinsic signals are transmitted to cells via "death receptors" located in their cell membranes[655]. These receptors can become activated by the immune system, either through direct cell-to-immune-cell contact or with the help of locally secreted "death hormones", such as tumor necrosis factor alpha[656]. Intrinsic signals act from within. Mammalian apoptosis can be triggered by viral infections, DNA damage, oxidative stress, hypoxia, or the activation of oncogenes. It is the intrinsic apoptotic pathway in which mitochondria play a crucial role.

Normally an animal cell hangs in a delicate balance between numerous factors that promote or inhibit apoptosis, as if it is constantly deciding whether to live or die and needs constant confirmation that it is doing fine and not turning into something monstrous[657]. The mitochondria act as a sort of "inner judge" in this difficult decision. Here is how it works in the case of DNA damage sensing.

Accumulated DNA damage in a cell is sensed by the ATM protein, which activates p53, leading to decreased activity of the anti-apoptotic protein Bcl-2. As the delicate balance shifts, channels are formed in mitochondria's outer membranes, increasing their permeability[658]. This is followed by a release of mitochondria's cytochrome c respiratory-chain proteins into the cell's cytoplasm. Cytochrome c binds to a protein called APAF1, and seven such pairs combine into a wheel-like structure called an apoptosome. If you watched the TV series *Mighty Morphin Power Rangers*, imagine that this is the part where individual robots form the Megazord. If you are more into Japanese cartoons, picture Grendizer. But the apoptosome looks more like a spinner, and its function is to facilitate a destructive biochemical cascade. The next steps involve several important protein-cleaving enzymes called caspases, which modify and activate one another, amplifying the death signal. Isn't nature amazing? Cells evolved such a complex beautiful mechanism just to kill themselves.

Apoptosis is a tightly regulated process. A cell is disassembled neatly, without spilling its components. Its membrane shows irregular buds, which can grow in size. The nucleus becomes fragmented, and the insides of the cell are distributed between multiple vesicles called apoptotic bodies. These are then actively engulfed by the membranes of surrounding specialized cells and ingested through a process called phagocytosis. All of this is required to contain damaged and potentially hazardous molecules present in dying cells, but also to prevent inappropriate immune responses[655]. In its tight regulation apoptosis stands out from cell death via necrosis, which is a less regulated and usually more "accidental" process associated with loss of membrane integrity, random DNA degradation, and inflammation that can affect nearby healthy cells.

Why mitochondria are involved in the programmed cell death of so many species is an exciting evolutionary puzzle, not that this phenomenon is universal. Well-described apoptotic pathways of roundworms do not depend on cytochrome c release from mitochondria, and the role of these organelles in programmed cell death in fruit flies is still not fully understood[659]. However, there is evidence that mitochondria are involved in apoptosis in organisms more evolutionarily distant from us than flies and worms, such as the long-lived hydra[660]. This suggests that mammalian apoptosis resembles ancient versions, and that some animals have lost or reconfigured several components of apoptosis.

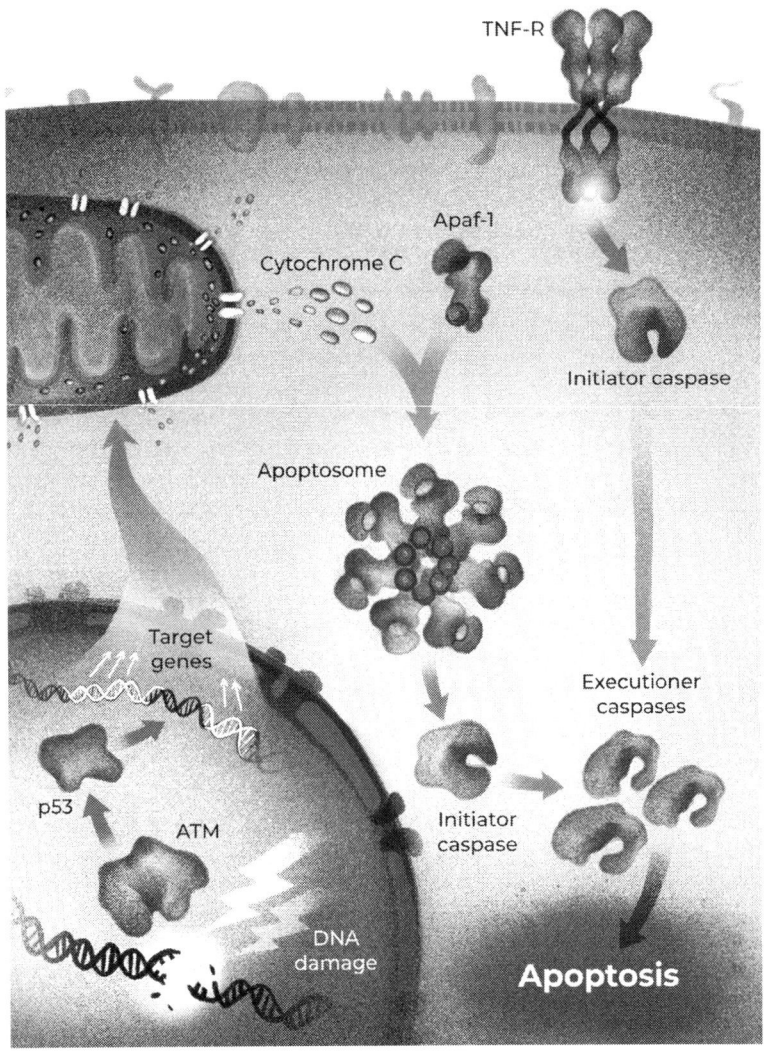

Mitochondria have their own, circular, DNA and are genetically related to bacteria. The nuclear DNA of protists, plants, fungi, and animals is more closely related to that of archaea[661]—a group of unicellular organisms that are somewhat morphologically similar to bacteria, but distinct in their biochemistry and metabolism. It is widely accepted that at some point the bacteria-like ancestors of mitochondria became endosymbionts of our archaea ancestors and that this biological revolution played a critical role in the appearance of complex eukaryotes with aerobic respiration.

Whether the relationships between ancient proto-mitochondria and archaea were mutually beneficial from the start, or whether parasitism or other forms of exploitation initially took place, is still unclear[662]. One hypothesis is that apoptosis could have evolved from an antibacterial defense mechanism that sensed undesirable invaders and killed infected cells before disease could spread to related colony members. Later, as a more endosymbiotic relationship between a particular archaea host and its resident bacteria was established, that mechanism of pathogen-mediated death may have been reconfigured to serve a new, yet similar, role.

We don't know if this was the exact case, but I find it remarkable that the crucial process of animal programmed cell death is, in many ways, controlled by resident bacteria-like entities. It is also surprising that mitochondria themselves can undergo programmed death, via mitoptosis, when damaged[663]. This process involves extensive fragmentation of these organelles, and the remnants of dead mitochondria can be either extruded from a cell in the form of mitoptotic bodies or subjected to autophagy[664]. This mechanism exists in parallel with the utilization of mitochondria through mitophagy, which we discussed earlier[665].

Apoptosis serves many functions in mammals. It is not just a mechanism that protects against cancer, but an important factor in embryonic development. Without it we would be born with tails[666], with membranes between our digits[667], and with malfunctioning nervous systems[667] and various autoimmune disorders[668]. That is, if we would be born at all. The majority of mice genetically engineered to lack caspase-9, a protein critical for apoptosis, die prenatally, with abnormally enlarged and malformed fetal brains[669] (I guess that's the kind of research Acme performed in their genetics lab in the animated series *Pinky and the Brain*). Very few mice engineered to lack both intrinsic and death-receptor-induced apoptotic pathways survive to adulthood[670].

Apoptosis is also vital for the functioning of our adaptive immune system—a system that is capable of improving itself as we encounter novel pathogens, provides us with long-lasting immunity, and acts as an additional layer of defense against cancer. The potential of this evolutionary invention is immense! Humans have around twenty thousand regular protein-coding genes, but millions of different T-cell receptors[670]. These are proteins by which immune T cells can identify various foreign substances and invaders, as well as certain damaged and mutated cells. This vast diversity exists thanks to the process of genetic recombination: the random rearrangement of specific gene segments in T cell DNA. As a result, genetically unique immune cells are formed, each capable of binding to its own specific set of potential targets — such as bacteria and viruses. To undergo recombination, precursors of functional T cells must travel from bone marrow to the thymus (this is why the cells are called T cells), a small gland of our lymphatic system located in our upper chest, behind the breastbone. Only after doing so, they become part of the active-immune-system combat force.

This isn't the only important thymus function. The gland contains cells that present a diverse array of short amino-acid fragments called peptides that are derived from a wide range of proteins that our bodies normally produce[671] — our unique "protein footprint." T cells that have undergone recombination "examine" these peptides, and they self-destruct if an interaction occurs—like a samurai who commits seppuku after failing his master (or after attacking him, more like). Without this mechanism of negative selection there would be a large quantity of autoreactive, self-targeting T cells. Due to its random nature, the recombination process that creates T-cell variability simply can't guarantee in advance whether a particular receptor will help fight a pathogen or make a cell rebel against the body—hence the necessity for such extreme diversity and a mechanism of "natural selection" that sorts our good, bad, and ugly cells.

That is not the only function of apoptosis in our immune system. It is also used to remove T cells that are no longer needed after a pathogen is defeated. Memory T cells can avoid this process and remain intact to protect us against reinfection. While individual memory T cells have a median lifespan of just about half a year[672], our cellular immune memory can last for years or even decades after infection or vaccination[673,674]. This is because memory T cells can divide and produce clonal populations that survive after the deaths of individual cells[675]. If a pathogen, or its components, is introduced on multiple occasions, as in the cases of booster vaccinations and reinfections, more memory cells will be formed.

Similar genetic rearrangements and apoptosis of autoreactive cells are also a prerequisite for the development of our adaptive humoral immunity, represented by B cells. The role of these cells is to produce antibodies—secreted protein molecules capable of binding molecules present on the surfaces of viruses, bacteria, and other pathogens. The areas that antibodies recognize and bind on pathogens or other targets are called antigens.

While T cells fight in close quarters, B cells can fight at a distance—although the reality is more complicated, due to the complex interactions between different immune cell types, which can activate or inhibit one another depending on circumstances. The development of B cells starts in the bone marrow and continues in the spleen, lymph nodes, and other secondary lymphoid structures[676]. Mature B cells actively divide after being exposed to their antigens, and they can turn into effector B cells, whose role is to secrete antibodies in large quantities. They can also turn into memory B cells, which will ensure a quicker and more efficient future response to similar invaders.

Remarkably, when B cells go rogue against our bodies, they don't immediately commit apoptosis, but first try to repair themselves genetically by editing their receptors[677]. There is a preference for this mechanism during the early stages of B cell development in the bone marrow, which acts as a sort of correction facility. Unfortunately, none of these mechanisms are perfect, and some B cells produce autoimmune antibodies that cause health complications. The most famous (thanks to *House, M.D.*) of these is probably lupus, an autoimmune disorder that commonly comes with fever, fatigue, pain, stiffness and swelling in the joints, and a butterfly-shaped rash on the face.

Immune cells use apoptosis both to kill and to be killed. T lymphocytes that recognize and kill cancerous, infected, or otherwise damaged cells are called killer or cytotoxic T cells. One mechanism of killing involves the secretion of specialized granules that contain digestive enzymes along with proteins capable of making pores in a target cell's membrane. The enzymes then trigger apoptosis[678]. An alternate process is set in motion by the interaction between cytotoxic T cells and special death receptors on the surfaces of their targets and utilizes the extrinsic apoptotic pathway. Basically, a cell is "ordered" to commit suicide. A killer T cell determines that a cell is damaged or infected by "noticing" unusual or foreign components on its membrane. Special MHC proteins facilitate this examination by exposing small peptides derived from internal cellular components on the surface of a cell[679]. Thus, even a sneaky virus hiding inside a cell may trigger an immune response. That's how thorough the immune system is, but it isn't perfect, and it has a tendency to deteriorate with age, increasing risks not only of infectious diseases, but also of cancer.

The COVID-19 pandemic demonstrated just how much aging contributes to vulnerability to infections and to immune-system decline[680]. There is a seeming paradox that aging is characterized by both increased risk of autoimmune diseases such as rheumatoid arthritis[681] and inflammation and, at the same time, immune deficiency[682]. Likewise, aging is associated with both an increase and a decrease in programmed cell death, depending on cell type[655]. Both issues can be translated into a simplified formula of dysregulation: cells that should die don't, while cells that should be removed have increased chances of survival.

One might think malfunctions of the adaptive immune system are unlikely to play a central role in aging, because it arose 500 million years ago, in jawed fish[683], and most animals don't have lymphocytes. The number of diverse cells that need to be created and self-destroyed for these mechanisms to function properly is not affordable for smaller creatures, but this doesn't prevent such animals from aging normally. However, experiments on mice reveal just how important adaptive immune cells can be. For example, mice lacking perforin—the protein used by cytotoxic T cells to make holes in the membranes of their targets—suffer increased accumulation of senescent cells, as well as chronic inflammation, fitness reduction, older appearance, and shortened lifespan[684].

Mice with artificially induced dysfunction of mitochondria in their T cells (this occurs naturally during aging) display multiple signs of premature senescence and increased mortality[685]. The malfunctioning T cells were shown to induce cytokine storms—increased production of pro inflammatory molecules such as interleukin-6 and tumor necrosis factor alpha[686,687]—in peripheral tissues. This can lead to multiple organ failure and even death. Cytokine storms played a major role in the poor clinical outcomes of aged patients during the COVID-19 pandemic[688]. This illustrates the point that a properly functioning immune system serves us well, but a dysregulated one can kill us.

Many factors can cause T-cell diversity and functionality to decline as we age. One important contributor is thymic involution. The volume of human thymic epithelium, which is necessary for T-cell maturation, continuously declines from infancy to the end of life[689,690]. As thymus output drops with older age, the majority of T cells begin coming from their preexisting counterparts[691]. While the total number of T cells can be maintained by this process, their variability gradually diminishes.

Even though thymus involution seems to be a negative feature, it is universal among vertebrates—including mammals, birds, reptiles, and amphibians[692]—so it demands an evolutionary explanation. One hypothesis is that it is another example of antagonistic pleiotropy: a mechanism that increases our survival rates early on but contributes to immune senescence later in life. As we age, the thymus can become compromised by infections[693]. If it does, T cells capable of fighting disease might be mistakenly regarded as autoreactive and be removed via apoptosis, leading to a weakened immune response. One strategy to overcome this issue would be to create a reservoir of diverse populations of long-lived T cells during early life, when we have the greatest chance of being infection-free.

An alternative hypothesis is that thymus involution serves no evolutionary purpose. The thymus ages like all other organs, but given the long time lag between loss of thymus function and negative consequences for the immune system, we can afford for it to deteriorate earlier. Even if the thymus is lost later in life, the immune system will still be robust for decades—possibly even up to the end of the shorter lifespan of our ancestors.

Remarkably, the long-lived naked mole-rats have recently been shown not to experience thymic involution up to the age of eleven years[694], and they have additional ectopic thymuses in their neck area (the ectopic thymus is similar to the original thymus in its cellular composition). This is truly remarkable, especially considering that thymic involution starts after just several weeks of life in mice[695]. Given the fact that naked mole-rats are subterranean animals that don't spread many diseases, it is tempting to speculate that the lack of thymic involution could be an anticancer rather than anti-infection adaptation[696].

Indeed, naked mole-rats are so resistant to cancer that the discovery of tumors in individual animals is scientific news worth publishing[697]. Imagine how wonderful it would be to live in a world where this was the case for humans. Whether the enhanced thymus function of naked mole-rats is responsible for cancer resistance is, however, debatable because of the existence of other suggested anticancer mechanisms in these animals. One is hypersensitivity to contact inhibition, the ability of cells to stop dividing after sensing that they have achieved high density (through a p53-mediated and p16-senescence-associated pathway[698]). Another is the more-than tenfold increased half-life of the anticancer p53 protein in the embryonic fibroblasts (and possibly other types of cells) of naked mole-rats in comparison with their human and mouse counterparts[699]. What is definite is that we can learn much about combating cancer by studying these cute "saber-toothed sausages".

In any case, the thymus is very important at the beginning of life. So, what happens if it's not present? The treatment of certain rare congenital heart diseases requires early-life partial or complete surgical removal of the thymus. This procedure saves lives, but unfortunately it reduces T-cell receptor variability[700] and is associated with an increased risk of infection and autoimmune diseases in the following years[701]. There is also accumulating evidence that the thymus continues to function in adults[702], and that its functionality can be a biomarker of health status[703]. A functioning adult thymus becomes especially important for people subjected to diseases or treatments that reduce the T-cell repertoire—such as chemotherapy and certain chronic infections[704], especially HIV—and perhaps for people of extreme old age. Thus, reduction of thymus involution may be added to the pool of strategies to combat aging.

Experiments on adult mice have revealed that thymic functionality and T-cell production can be boosted by several hormones. Some of them belong to a group called fibroblast growth factors (FGFs)[705]. Fibroblasts are common cells of connective tissues; they synthesize the extracellular matrix, and thus contribute to the structural environment within our bodies. There are many FGFs that regulate biological processes such as growth, development, differentiation, and wound repair via binding to receptors expressed on the surfaces of specific cell types. For example, fibroblast growth factor 7 (FGF7) exhibits its protective effect on thymic epithelia by stimulating cell proliferation and DNA repair[706].

Fibroblast growth factor 21 (FGF21) is normally secreted by the liver during fasting and appears to have pro-longevity effects. Genetically modified mice with increased FGF21 levels have median survival times increased by 30% and 40% in males and females, respectively[707]. In the study that first reported this effect, more than 90% of normal mice died by age three, and roughly 60% of transgenic mice were still alive at that time. FGF21 was initially thought to prolong life by reducing insulin-like growth factor signaling, but it was later also found to delay thymic involution[708].

There are other molecules that may act against thymic involution and its presumed negative effects. One, called ghrelin, or the "hunger hormone," increases our drive to eat. Its concentration in our bloodstream is highest when we are just about to take a meal. Adult mice with ghrelin injections have experienced improved thymocyte numbers and T-cell diversity, while ghrelin-receptor-deficient mice have displayed increased thymic involution[709]. Hunger may be beneficial, but malnutrition accelerates thymic atrophy in these animals (although this effect can be partially counteracted by injections of leptin, the "satiety hormone[710]").

There is also a signaling molecule, called interleukin-22 (IL-22), that the body uses to promote proliferation and survival of thymic epithelial cells. Administration of IL-22 enhances thymic recovery after damage induced by total-body irradiation in mice[711]. Finally, there is evidence for a negative role of sex hormones such as testosterone on thymic involution[712]. This could be the reason why the rate of thymic involution is greater in men than in women. Perhaps this could also partially explain sex differences in lifespan[713] and why neutered animals often live longer than their counterparts[714]. And perhaps not just nonhuman animals! A small study published in *Current Biology* compared historical Korean eunuchs with noncastrated men of similar socioeconomic status and found a surprising 14.4-to-19.1 year difference in lifespan[715]. In case you wonder: I'm not suggesting castration as a solution for aging.

But maybe there are better ways to reduce thymic involution. One important gene in the development and function of the thymus is called *FoxN1*[716]; it is related to the *Fox* transcription factors that we discussed in earlier chapters. In mice and humans, *FoxN1* mutations can cause a lack of thymus and a severe primary immunodeficiency[717], leading to early death from infections. If that isn't enough, *FoxN1* mutants also suffer from congenital alopecia (hair loss) and nail dystrophy, or the so-called "nude" phenotype. Some abnormalities of nervous system development have also been reported.

Remarkably, while levels of *FoxN1* in the thymus decline in postnatal mice, they remain at the neonatal level for long-lived adult naked mole-rats[694]. Mice genetically modified for gradual loss of *FoxN1* genes in their cells experience accelerated thymus involution, such that their thymus phenotype at 3 to 6 months resembles that of 18-to-22-month-old mice. Meanwhile, delivery of extra *FoxN1* gene copies by vector injection into the thymus increases the gland's size and the total number of thymocytes[718], suggesting that a gene-therapy approach against this component of aging is possible.

The thymus-involution problem may also be solved by cell transplantation. Perhaps you've heard about the conspiracy theory among QAnon followers that Hollywood elites stay youthful by harvesting a drug they call adrenochrome from the blood of children? Well, it's real! Well . . . it will be real. Maybe. And it's not really a drug. But the authors of one study extracted thymic cells from mouse embryos and day-one postnatal mice and injected them into the thymuses of middle-aged mice. Young thymic epithelial cells were able to engraft and maintain functionality as the size of the thymus increased due to treatment[719]. Obviously, this approach is not translatable to humans in its current form. This is why an alternative and more realistic approach was recently invented.

Fibroblasts can be easily isolated from any animal or person and genetically engineered for increased *FoxN1* activity. This results in the creation of induced thymic epithelial cells that can be transplanted. This approach was tested on eighteen-month-old mice and led to the proliferation of both transplanted cells and native thymic epithelial cells in the recipient, presumably because of the improved microenvironment. As a result, the treated thymus grew in size and weight[720]. Furthermore, the authors observed a reduction in circulating pro-inflammatory cytokines such as interleukin-6, a finding that can be interpreted as an indicator of decreased autoreactive T cell–induced self-damage to tissue. Thus, we have several possible avenues to pursue in the battle against age-related decline in thymus function.

Aside from thymic involution, there are other processes leading to immunosenescence, some of which are not unique to the immune system. These include mitochondrial dysfunction, telomere shortening, genetic and epigenetic alterations, accumulation of damaged or misfolded proteins, and cellular senescence[682]. This means that treatments that reduce general aging phenotypes may also benefit the immune system[721]. For example, the antidiabetic drug metformin, which has some anti-aging potential[722], promotes memory T-cell formation and immunity against subsequent infections and tumors in mice[723]. Unfortunately, as we discussed in the "false Grail" chapter of this book, it is unclear whether people without diabetes can benefit from this drug.

Telomere shortening is another problem of aging for the adaptive immune system. After all, B and T cells and their progenitors have to divide many times, especially in the face of infection, when large volumes of their clones are produced. Indeed, genetically acquired short-telomere syndromes usually come along with the disruption of immune functions[724]. Conversely, high-performing centenarians, characterized by freedom from age-related diseases, have greater telomere length in their immune cells. The telomerase activity and proliferative potential of their T cells is also greater than in low-performing centenarians and old controls[725]. So perhaps the telomerase-activating gene therapies we discussed in the "immortal cell" chapter can also benefit the immune system.

As we age, the performance of our immune system diminishes, but the need to deal with precancerous mutant cells is ever increasing. Malignant evolution within us tends to invent new ways to avoid cell-cycle arrest and apoptosis, whether internally or externally induced. As a result, cancer-related mortality skyrockets as we follow a fate as old as multicellular life: the fate of the cheater-infested bacterial colonies mentioned at the beginning of this chapter. Even in the best-case scenarios of technological success with anti-age treatments, the extremely long-lived humans of the future will encounter cancer one day or another. Evolution can't be stopped. Cheaters always win. Period. Or is this not necessarily true?

Fortunately, life-saving "police" interventions from outside the body can tackle the corrupt biological systems that refuse to fix themselves. And scientists have been working hard on developing sophisticated anticancer "bioweapons."

One such approach involves training immune cells to attack specific types of cancer cells. Chimeric antigen receptor (or CAR) T-cell therapy involves the extraction and purification of T cells from a patient's blood, followed by their genetic modification[726]. Viral vectors provide these T cells with a tailor-made gene that encodes a receptor capable of binding specific antigens on the surface of a patient's cancer cells. The genetically modified immune cells are injected back into the patient, where they begin to remove cancerous cells.

Several generations of CAR T-cell therapy have been developed[727]. Not only has the efficiency of the approach improved over time, but safety measures have been developed. For example, the T cells can be modified to include inducible suicide genes, allowing the withdrawal of the fighting force once the patient's cancer is defeated[728]. Alternatively, T cells can be supplied with "on switches[729]" ensuring that they will

197

work only when special activation molecules are administered during treatment. This allows treatment to be easily stopped in case of severe adverse effects, offsetting some of the current dangers of cell therapy[730]. Furthermore, though initially developed to fight cancer, CAR T-cell therapy was recently reconfigured to remove senescent cells in animal models, which might prove helpful in the general quest against biological aging[731,732].

Another promising anticancer approach comes from a discovery made by immunologists James P. Allison and Tasuku Honjo, which earned them the 2018 Nobel Prize in Physiology or Medicine. Programmed cell death protein 1 (PD-1) is normally situated in the membranes of activated T cells, B lymphocytes, and a few other immune cell types[733]. Along with other so-called checkpoint proteins, it acts as a "brake" for the immune system, preventing it from becoming overzealous, acting as a safeguard that prevents our body from turning into a police state. It can signal T cells to reduce their activity and proliferation or even to undergo apoptosis[734]. Mice genetically engineered to lack PD-1 suffer from various autoimmune disorders due to the lack of this important negative-feedback mechanism[735].

Unfortunately, successful cancers often evolve the means to exploit this self-limiting feature of our immune system. They acquire mutations that allow them to produce checkpoint proteins and thus gain "diplomatic immunity" for any crimes that they may commit against our bodies. Basically, they trick the immune system by pretending to be part of it.

A solution comes in the form of a class of drugs called checkpoint inhibitors. They prevent PD-1 and other checkpoint proteins from binding their receptors. Many such drugs have been developed and approved for cancer therapy over the last years[736], although issues with their toxicity persist[737]. It appears to be difficult to strip away the diplomatic immunity of cancer cells without collateral damage. However, checkpoint inhibitors can be very effective, especially in combination with one another; there are case reports of complete tumor disappearance following a single treatment[738]. But there is one more idea worth sharing.

Remember we talked about the progress of cancer being an evolutionary process? This evolution can be extremely fast paced. One reason for this is the genomic instability of cancer cells. One known contributor to this instability is the propagation of self-copying selfish DNA elements called retrotransposons[739]. These genetic elements played a great role in generating the diversity of mammalian genomes in the past, resulting in the "endless forms most strange" that we observe today[740]. The copying and switching of different parts of genes and regulatory genetic elements provided evolution with multiple genetic combinations to select from—new genetic ideas, most worthless, but some useful. Mutations favored by natural selection were passed down, along with the mechanisms that brought them into existence and other products of those mechanisms. As a result, our genome contains a "graveyard" of ancient DNA relics—or, more precisely, a sealed crypt.

Most DNA sequences of retrotransposons have degenerated and lost the ability to code any proteins. Those that remain intact are usually silenced by epigenetic mechanisms in mammals, "chained" and repressed to protect our cells from harm. When tumor cells break these seals, they unleash the dormant potential of ancient genetic recombination mechanisms upon their genomes. For example, the activation of LINE-1 elements (the only active and autonomous human retrotransposon class) can lead to various kinds of DNA damage in cells, including mutations in the p53 tumor suppressor gene[741].

The good news is that our bodies can produce antibodies against proteins encoded by some of the functioning LINE-1 elements[742], allowing the immune system to specifically target and eliminate cancerous and precancerous cells that lack proper epigenetic silencing. In fact, it is now widely accepted that precancerous cells are generated and eliminated on a regular basis without us even noticing. Cancer becomes a real threat when it evolves mechanisms that allow it to evade the immune system. Some researchers even argue that we can estimate the number of pre-cancers that we have defeated by looking at the number of antibodies for specific proteins encoded by LINE-1 transposable elements.

This knowledge may allow us to improve existing anticancer therapies. Here is one very interesting suggestion. Retrotransposons rely on enzymes called reverse transcriptases to copy their DNA. As usual, DNA is turned into RNA, and then the RNA is turned back into DNA and inserted into the genome. The enzymes responsible for RNA-to-DNA conversion can be targeted with drugs called reverse transcriptase inhibitors, which are commonly used to treat viral infections. Preliminary research suggests that such inhibitors can reduce the adaptive ability of cancer[743,744] and allow animals to survive longer after anticancer treatment. The repression of retrotransposons may be beneficial not just in treating cancer, but in preventing other diseases and even aging[745]. There are ongoing trials testing whether reverse transcriptase inhibitors can be repurposed to treat age-related cognitive decline and even Alzheimer's disease[746,747], given that retrotransposons can be active in neural tissues[748] and that neurons from affected brain regions show increased DNA variation[749].

As you read this, the evolution of cancer within the bodies of so many people is running against the evolution of our science. For every trick that a tumor invents, we must come up with ever more complex, selective, and well-tolerated treatments. Next-generation anticancer therapies heavily rely on analyzing the exact genetic backgrounds of tumors and their individual pathological mutations. More and more interventions are selected on a case-by-case basis from a wide array of tools and their combinations, including the use of our own immune system. For me, even this very broad vision of a possible future inspires hope that cheats will be defeated, our benign cells will prevail, and our inner utopia will be preserved. But will we defeat aging when we provide all necessary conditions for the good cells of our body to prevail? It might not be enough. As you will see in the next chapter, aging is not just about cells.

Crossing the River Styx of malignant transformation, where time flows like a relentless current, carrying cancer through its stages of evolution. Nine figures stand as guardians, discerning the benign from the malignant, forging a new path—one untouched by death.

CHAPTER 10
Fragile, but not that fragile

> I am the bones you couldn't break.
> —*Placebo* ("*Battle for the Sun*")

Summary: in this chapter we discuss the most long-lived components of our bodies, how durability can become a weakness, and how the extracellular matrix influences the aging of surrounding cells.

In Greek legend, Theseus, the mythical king and founder of Athens, sailed to the island of Crete and rescued fourteen Athenian children from the Minotaur, a half-man, half-bull monster. Every year, the Athenians commemorated this legendary event by taking the ship of Theseus on a pilgrimage. This ship required regular maintenance, and this circumstance led to a philosophical debate: if all of the ship's parts are replaced over centuries, is it still the same ship? The answer depends on our definition of what it means to be "the same."

Similarly to the ship of Theseus, our bodies undergo constant maintenance and change. New cells take the place of dead ones and have their components frequently renewed. This has led to a popular, yet false, belief that the human body replaces itself completely every seven years or so. Indeed, most cells of our body undergo turnover; however, this can't be said about all our cells. For example, the majority of the neurons in our brains are as old or almost as old as we are.

Surprising confirmation of this came thanks to twentieth-century nuclear-weapons tests[750]. Nuclear detonations led to an increased atmospheric level of heavy carbon isotope C-14, followed by its decline after the 1963 Partial Nuclear Test Ban Treaty. Altered carbon in the form of carbon dioxide was captured by plants, through photosynthesis, and ended up in contemporary humans, incorporated into their newly synthesized DNA strands. As cells divide, the accumulated C-14 within them becomes diluted, and the amount that remains can be used as a measure of a cell's age. In a way, nuclear tests set up a massive experiment on humanity without anyone's consent. Based on this improvised radiocarbon-dating technique, it was determined that the gray matter of the human cerebellum is almost as old as its owner. The same study also estimated that the average age of nonepithelial cells in an adult human's intestine is 15.9 years, while skeletal muscle cells have an average age of 15.1 years.

Similar studies have revealed that while some heart muscle cells are apparently replaced during our lives, the majority of them are not[751], and cardiomyocyte exchange rates are less than one percent per year in adulthood[752]. Instead, during aging, cardiomyocytes tend to undergo polyploidization, a multiplication of their chromosome copy numbers that increases metabolic output and reduces the cells' susceptibility to gene-disabling mutations[753]. Perhaps these are reasons why primary heart cancer is very rare[754].

The topic of human neurogenesis remained a subject of heated debate for a rather long time. It's almost like a detective story that shows the remarkable ability of science to self-correct and remove internal contradictions. Until recently, it was widely accepted that loss of neurons in adult human brains was irreversible, based on the long-lasting idea that new neurons are never formed. This led to a weird situation in science, because evidence for neurogenesis in adult rat brains had existed since 1962[755], and by 1977 scientists knew that some new neurons are formed in at least two areas: the dentate gyrus of the hippocampus, and the subventricular zone that provides neuron precursors that migrate into the olfactory bulb[756,757]. By 1983, neurogenesis was identified in the vocal-control nuclei of adult birds[758,759], and distinct precursors of neurons were isolated from the brains of mice by 1992[760]. It would be truly remarkable if humans were the only exception to the rule of neurogenesis. The problem was that, although new neurons are easy to detect in lab animals, which can be subjected to genetic engineering, labeling techniques[761], and dissection at an experimenter's convenience, for human subjects are much more difficult.

In 1998, a group of researchers seeking adult human neurogenesis exploited the fact that some cancer patients were receiving a synthetic nucleoside analogue called bromodeoxyuridine for diagnostic purposes. Bromodeoxyuridine is a tag that can be incorporated into and label newly synthesized DNA[762], and thus mark all sorts of newly formed cells, including rapidly dividing malignant ones. Postmortem analysis of brain tissues from patients who received bromodeoxyuridine revealed labeled cells that had morphological characteristics of neurons[763] in their dentate gyri—the same regions in which neurogenesis was earlier established in rats[764]. By 2013, further studies using nuclear-bomb-test-derived C-14 measurements in brain cells estimated that as many as seven hundred new neurons are formed in the adult human hippocampus each day[765].

In 2018, these results were challenged by researchers who used a different approach to identify young human neurons[766]: a set of proxy markers in the form of proteins that are produced more or less exclusively in immature neurons or their progenitors[767] and can be tagged by specific antibodies. The authors of this new study claimed that they observed hardly any labeled cells in postmortem human brains and concluded that neurogenesis in the adult human hippocampus is nearly absent.

Soon these results were challenged by even newer studies, which found that the presence of such proxy markers of neurogenesis is strongly dependent on the conditions of brain-tissue preparation. When all is done correctly and studied brains are fresh, detection methods align and there is conclusive evidence for human hippocampal neurogenesis. Remarkably, it occurs even in people who have reached their ninth decade; but its rate drops sharply in patients with Alzheimer's disease[767,768]. It's true that the majority of our neurons are as old as we are, and we lose these cells faster than they are replenished, but our brains do have the capacity to form new neurons.

In any case, it seems that at the cellular level, the paradox of the Ship of Theseus is not entirely relatable to us, and we are never fully replaced. However, the question of full-body replacement becomes more intriguing at the level of proteins, which usually undergo constant synthesis and degradation within our bodies. The median half-life of diverse proteins is estimated to be around forty-three minutes in yeast[769], and eight to nine hours[770] in human cells[771]. Such measurements may vary depending on experimental methods and on organisms and cell types used, but most results fall within a range from hours to several days[772,773]. Interestingly, average protein turnover rates are inversely correlated with the lifespans of mammals, being low in long-lived species such as humans, naked mole-rats, and bowhead whales, and much higher in mice[774].

Protein degradation is usually an active process. Cells have various tools for the disposal of proteins, among which we have already discussed autophagy and the protein shredders called proteasomes. Timely protein replacement is almost as vital as novel synthesis, and reduced degradation of proteasome targets has been shown to reduce the lifespans of certain animals, such as *Caenorhabditis elegans*[775]. Proteins typically have high rates of turnover, but there are outliers with lifespans measured in months, years, and even decades.

Crystallins have among the longest lifespans of the proteins in our body, exceeding seventy years. They account for roughly ninety percent of the proteins in the lens of the human eye[776], where they are present in a soluble form within lens fibers, specialized cells that are transparent and firmly packed. While new crystallin-containing lens fibers can be added to the outside of a lens throughout life, the center of the lens contains cells that were present when the lens was formed[777]. Lens fibers lack organelles and nuclei, which are necessary for RNA and, consequently, protein synthesis; thus, crystallins are the same age as the cells they reside in. This fact was recently reconfirmed by radiocarbon dating: C-14 levels in the crystallins of an eye lens core matched atmospheric levels at the time of a person's birth[778]. Unfortunately, this doesn't mean that these proteins are invulnerable. Over time, crystallins tend to accumulate damage (for example, due to exposure to ultraviolet light[779,780]) and aggregate, leading to lens opacity and cataracts—the most frequent cause of blindness, and a common example of age-associated pathology.

This doesn't mean, however, that the lens has no potential for regeneration. The most widely used treatment for cataracts is surgery to remove the old biological lens and install an artificial one in its place[781]. This is an effective procedure, although not without possible complications. One problem is postoperative disorganized growth of residual lens epithelial stem cells, which may lead to increased lens opacity and reduced vision[782]. Sometimes these cells are intentionally disrupted during eye surgery to improve treatment outcomes.

Recently, a group of researchers found that if a cataractous lens is removed through a very small wound opening, preserving the abovementioned stem cells and surrounding membrane, natural lens regeneration can occur. This was first tested on rabbits and macaques and later shown in a clinical trial of human infants with congenital cataracts[783]. The human lenses regenerated within three months and were functional, though imperfect due to insufficient regenerative potential[783]. Still, the reported outcome was better than that of traditional surgery, and most likely the method can be improved. While good solutions for cataracts exist and better ones are on the way, it is still reasonable to protect our eyes and crystallins. One possible solution is wearing eyeglasses or contact lenses made from materials that block ultraviolet light[656].

Next among the extremely long-lived proteins, we have elastins and collagens, which are known for their vital role in the extracellular matrix, a three-dimensional network of various molecules that surrounds cells and provides them with biochemical and structural support. For human elastins, the mean measured carbon residence time is around 74 years[784], while the estimated half-life of the stiffer collagens is 15 years, 117 years, and up to 215 years, for skin, cartilage, and intervertebral discs[785] respectively[786]. Collagen half-life is much shorter in rats, but still reaches 45, 74, and 244 days in muscles, skin, and the gut[787]. The interspecies differences for collagen half-life are apparently influenced by variations in collagen production[788], as well as by substrate specificity and tissue distribution of enzymes that degrade matrix proteins[789].

Collagens can also be found in blood vessels, hair, bones, tendons, ligaments, and corneas, and on the surfaces of cells and in other parts of our bodies. They are the most abundant mammalian proteins, comprising roughly thirty percent of total protein mass[790], and a prime component of artery walls[791], serving a load-bearing function. Collagens typically form triple helical strands with elastic properties that allow them to deform reversibly, but their main role is to provide support for numerous tissues. For this purpose, collagen molecules can be cross-linked in a tissue-specific manner, which alters their rigidity and other mechanical properties. They are like a glue that holds different parts of our bodies together.

Not only do collagens contribute to the organization and shape of our tissues, but they also interact with various cellular receptors and regulate cell behavior, proliferation, and differentiation[792]. Fibroblasts—the most common cell type of our connective tissues—secrete collagens and use them to guide their migratory pathways. They are our very own Samuel Porter Bridges, the protagonist of Hideo Kojima's computer game *Death Stranding*, who is tasked with making the United Cities of America whole again after a global cataclysm (he also uses his excretions to succeed in the task with which he is entrusted).

If the main role of collagens is to provide structure and rigidity, elastins are essential in providing stretch, recoil, and elasticity. Wrinkles, aged appearance, and skin[792] fragility are associated with damage to these extracellular-matrix[793] components. Elastin in the form of elastic fibers is the second-most abundant protein in the walls of arteries. There it persists throughout our entire lives, but unfortunately it undergoes modifications that lead to blood-vessel stiffening with age. This can lead to hypertension[794], increased cardiovascular mortality[795], and other circulatory problems. Genetically modified mice that lack elastin die within several days of birth and are characterized by a smaller and thicker aorta[796].

As with crystallins, the extreme lifespans of elastins and collagens are both a blessing and a curse. It's much easier for our cells and enzymes to replace proteins entirely than to fix them. Doing so would be especially metabolically inefficient with large molecules such as collagen and elastin. Imagine that the only way to remove a scratch from your car was to disassemble the car and manufacture a new one. You would probably just let the scratch be. But over time, small scratches and dents would accumulate until you couldn't ignore them anymore.

Similarly, long-lived proteins face problems that short-lived ones don't. They are vulnerable to slowly accumulating damage that comes from an unexpected source: sugars.

Sugars such as glucose can be attached to proteins, either with the help of specialized enzymes or spontaneously, via opportunistic chemical reactions[797]. An enzymatic addition called glycosylation is typically well regulated[798], and the modified proteins serve various beneficial functions. A nonenzymatic addition called glycation is more problematic, occurs of its own accord, and involves several stages. First, an amino acid of the protein (typically lysine, but sometimes arginine) reacts with a sugar, forming a freely reversible Schiff base. This product can then transform into a more stable form called an Amadori product, which interacts with itself and other molecules, eventually forming irreversible cross-links, and as a result: advanced glycation end products, or AGEs[799]. These can increase the stiffness of our muscles, thicken arteries, and interfere with the normal functioning of tissues[800]. They are very difficult for an organism to remove, due to their chemical stability. What's worse is that they can even cause inflammation, by interacting with certain specialized receptors, contributing to illnesses such as atherosclerosis, arthritis, diabetes, and even Alzheimer's disease[801].

Here is a metaphor for AGE formation. Imagine that you are on your first date with someone you've just met. Despite the flirting and initial attraction, your current connection is easily reversible. It wouldn't be hard for one of you to leave and never speak to the other again. You go on subsequent other dates with the same person and enjoy casual, obligation-free sex. It's still possible that you will stop seeing each other, but the bond may strengthen, leading perhaps to a serious relationship. As you spend more time together, the chances of marriage incrementally increase. Even marriage is a reversible condition, but now your separation may require a special "enzyme" in the form of a court-issued divorce decree. Otherwise, in the blink of an eye you may find yourself "cross-linked" by shared possession of a house, a car, and several children. My point is that sugar can be very dangerous!

Sugar-induced cross-links can form between various molecules in our bodies, but their accumulation is especially high in long-lived proteins, because conversion into stable glycation products takes time. Human tissues accumulate many types of AGEs, but the most abundant and important type is glucosepane—a sort of sugary handcuff, a cross-linked product made between the protein amino acids lysine and arginine via glucose[802]. The amount of glucosepane in the body dramatically increases with age, and especially in people with type 2 diabetes, as a consequence of higher blood glucose levels[803].

The accumulation of long-lived proteins cross-linked by sugar is an important component of human aging[804]. Within cells, undesirable proteins can be dealt with thanks to autophagy and proteasome degradation. In small amounts, protein-shredding proteasomes can also be found[805] in the extracellular space and can assist in removing some long-lived molecules, including collagen[805]. But the most studied specialized enzymes that normally perform such functions belong to a group called zinc-dependent matrix metalloproteinases.

These proteins participate in the normal turnover of the extracellular matrix as part of tissue maintenance, regeneration[806], and development. But they also play a role in some pathological processes, such as exaggerated cardiac remodeling following myocardial infarction[807]. Mutant mice with collagens resistant to these enzymes have markedly shorter lifespans; develop features of premature aging, such as hypertension[808]; and suffer from severe wound-healing delays[809]. Over time our collagens become stiffer and more resistant to digestion by the enzymes. This happens to such an extent that a person's age can be determined quite accurately based on how quickly his or her collagens can be dissolved[810].

FRAGILE, BUT NOT THAT FRAGILE

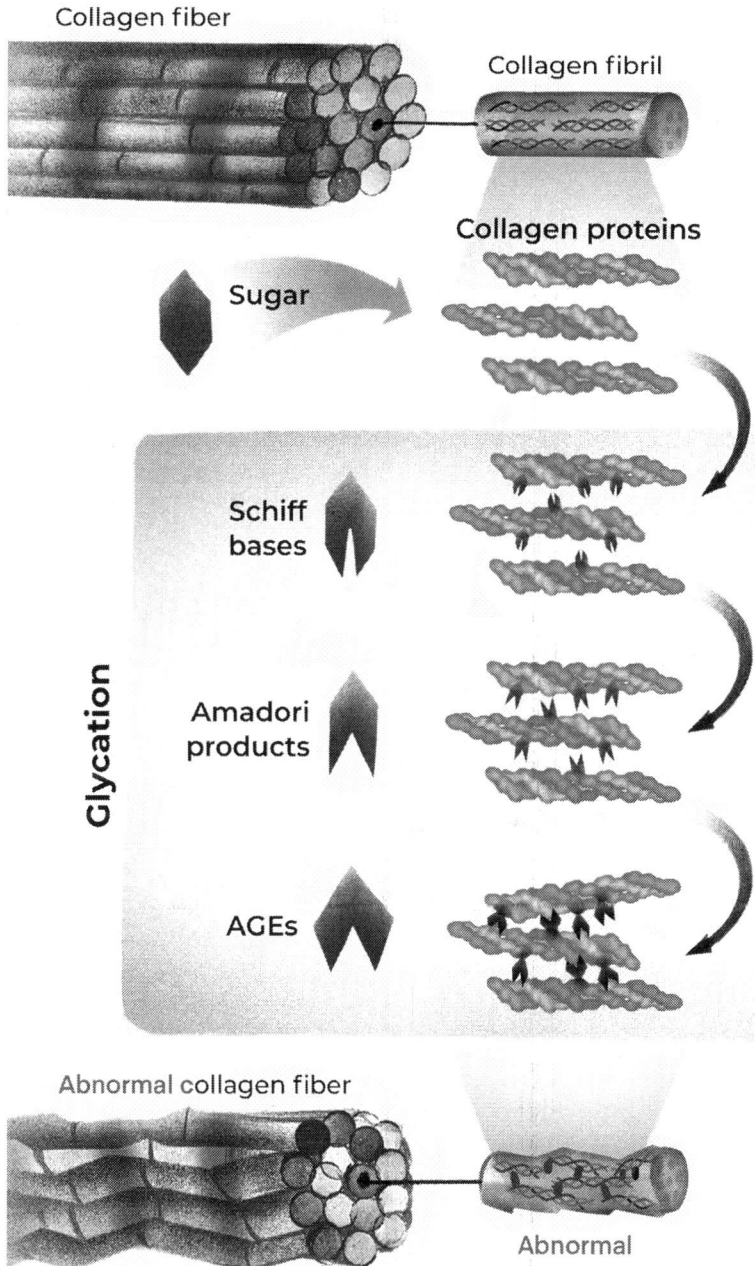

A number of studies[811], some of which have involved the incubation of collagen with sugars, have found evidence that cross-link formation can cause said resistance, but with an interesting caveat: cross-linked collagen becomes vulnerable to digestive enzymes when stretched by mechanical deformation[812]. Perhaps this also applies when artery-wall collagen is subjected to temporary increased blood pressure during physical activity. This could explain some of the health benefits of sports. While this explanation is hypothetical, it is true that exercise reduces arterial stiffness[813,814] in humans and reduces AGE formation in animal hearts[815] and arteries[814]. That alone is a reason to increase your daily motion.

There is another problem with AGE accumulation besides altered mechanical properties. Cells can sense external AGEs, via specialized receptors called RAGEs[816], and respond by promoting inflammation[801] and metalloproteinase activation. Unfortunately, these defense mechanisms have a tendency to backfire. AGE–cross-linked collagens' increased resistance to matrix metalloproteinases can lead to a situation in which these enzymes preferentially remove normal matrix proteins instead. This can further damage the extracellular environment.

Evolution has not fully prepared our bodies to tackle these issues, probably because our ancestors rarely survived to an age when such issues would have become real problems. Only recently have we started living long enough to reach the expiration dates of our long-lived proteins. Surely, in a few million years, nature will provide us with improved extracellular-matrix regulation and remodeling, especially given the low rates of extrinsic mortality that we currently enjoy thanks to science and technology.

Unfortunately, we require more immediate solutions. One paradoxical short-term idea is to use metalloproteinase inhibitors to prevent excessive matrix damage[817]. Aside from potentially conferring anti-aging effects, such inhibitors might be useful in various fields of medicine, including oncology. Certain cancers use metalloproteinases to destroy their extracellular environment in order to grow faster, invade tissues, and improve blood flow toward tumors. Metalloproteinase inhibitors can prevent that from happening.

But in the longer term, we need interventions capable of selectively removing AGEs. In 1996, a group of researchers discovered an AGE cross-link breaker with the chemical name N-phenacylthiazolium bromide[818]. It was later shown to reduce AGE formation in the blood vessels of diabetic rats and to reduce vascular hypertrophy when given as an early treatment[819]. A more stable modification of this chemical, called alagebrium, along with other "AGE breakers," went through a number of clinical trials aimed at improving cardiovascular outcomes[820]. These studies showed that alagebrium has the potential to reduce arterial stiffening[821], sparking further interest in this and similar compounds.

However, other studies have complicated the issue. It turns out that while N-phenacylthiazolium bromide and alagebrium can break some types of AGEs, they are not effective in breaking the most relevant and abundant type, at least not in the skin and tendon collagens of the diabetic rats that were used as a model[822,823]. Today, the predominant view is that existing AGE breakers are better suited for preventing AGE formation than for removing stable AGEs[824]. Preventing AGE formation would be a good thing, but the mechanism of action of AGE breakers has also been called into question. Unfortunately, due to financial issues, clinical trials of alagebrium were terminated before it could be licensed as a drug, and its therapeutic potential in humans is still unclear[825]. Hopefully research in this direction will resume.

Let's make a small detour. While extracellular-matrix aging is perhaps not the hottest topic in contemporary academic research, the picture becomes completely different when we look at the public sector. Social networks are infested with millions of #collagen posts. Collagen supplements are advertised to improve skin healing, elasticity, hydration, and "glow"; to reduce wrinkles, "support synthesis of collagen in the skin," "restore and maintain youthful levels of collagen[826]," and even delay aging. Most popular posts and videos come from biased, financially backed accounts that recommend specific brands of collagen without citing any evidence for their claims[827].

Several meta-analyses have found evidence in favor of using collagen-containing supplements for wrinkle reduction and improved wound healing[828], skin elasticity, and hydration[829]. However, collagens' praise should be taken with a grain of salt; many studies of collagen-containing supplements are industry funded, with risks of bias, and some have severe methodological flaws, such as improper blinding or lack of blinding (study participants know whether they are taking the supplements or not, which influences behavior). Not all research findings are true[830], and considering the methodology of studies is very important.

Recently, this point was illustrated in a journal called *Clinics in Dermatology*, which published a parody article called "Chicken soup for the skin[831]!" by Dr. Leonard Hoenig. Mocking flawed collagen-supplement research, the author claimed to compare experimental outcomes of eating chicken soup and eating "placebo soup." One conclusion is clearly a reference to p-hacking (misuse of data analysis to find patterns that can be presented as statistically significant[832]): "Skin elasticity of the lower part of the left eye did not show any significant differences, but that of the middle forearm extensor in the chicken soup group increased after 4 weeks (P < .01)."

The author then explains that "although the participants did not feel the improvements in skin elasticity and pigmentation, the authors thought that chicken soup was therapeutically effective in increasing skin elasticity and decreasing facial pigmentation," highlighting the problem of experimenter bias, which is especially relevant for study protocols without proper blinding.

The study proceeds to quote a twelfth-century physician on the benefits of chicken soup against leprosy based on ancient Greek and Roman medicine, citing teachings of Hippocrates and Galen, who believed that illnesses come from imbalances among the four humors (blood, phlegm, yellow bile, and black bile)—a joke about the misguided popularity of ancient and "traditional" medicines (which, coincidentally, are also used to promote collagen and other supplements).

This reminded me of the *South Park* episode "Red Man's Greed," in which a spirit journey initiated with the help of paint thinner helps a character discover the traditional American cure for SARS: Campbell's chicken noodle soup, cold medicine, and Sprite. Finally, the article notes that "chicken soup consumption is usually well tolerated" and provides a recipe for one, along with an image of the prepared dish.

Mockery of collagen supplements isn't entirely fair. It is true that our gastrointestinal tract does not absorb large proteins. It digests collagens in the same way it digests all proteins (including those from chicken), by reducing them to their simplest building blocks—amino acids. Some small peptides made from two or three linked amino acids may survive this process and be transported inside our bodies, but that's probably it[833]. And in any case, only single amino acids can be used to produce the proteins that function in our bodies. So in a sense, chicken is no worse than collagen supplementation, and the source of collagen does not appear to matter[834]. But individual amino acids (such as glycine, which is abundant in collagen) can produce biological effects, as discussed in the "false Grail" chapter, so maybe some of the purported benefits of collagen are actually benefits of glycine.

The human extracellular matrix contains many molecules besides collagen and elastin. Most of them fall beyond the scope of this popular-science book, but a group of long-lived proteins called laminins surely deserves to be mentioned[835]. Here is a fun fact: these proteins are typically depicted as a cross (a form that allows them to bind to other matrix and cell-membrane proteins), leading some Christians to believe that it's a mark of God. As Colossians 1:17 reads: "He is before all things, and in him all things hold together." But seriously, laminins do demonstrate the extracellular matrix's vital importance.

Do not confuse laminins with nuclear lamin proteins, which help maintain chromosomal organization in the nucleus and are linked to premature aging[836] and Hutchinson–Gilford progeria syndrome. Laminins help maintain tissue structure and function. They are essential components of the basal lamina, a layer of extracellular matrix sitting on top of the epithelial-cell layer that creates it. The membranes of epithelial layer-cells contain another kind of protein called integrin, which sticks out and binds to laminins and other extracellular matrix components, such as collagens, linking the whole thing together. Different tissues use different laminins, and mutations in their genes can lead to outcomes ranging from muscular dystrophy to developmental problems in the nervous system, kidneys, or lungs, and even death[837].

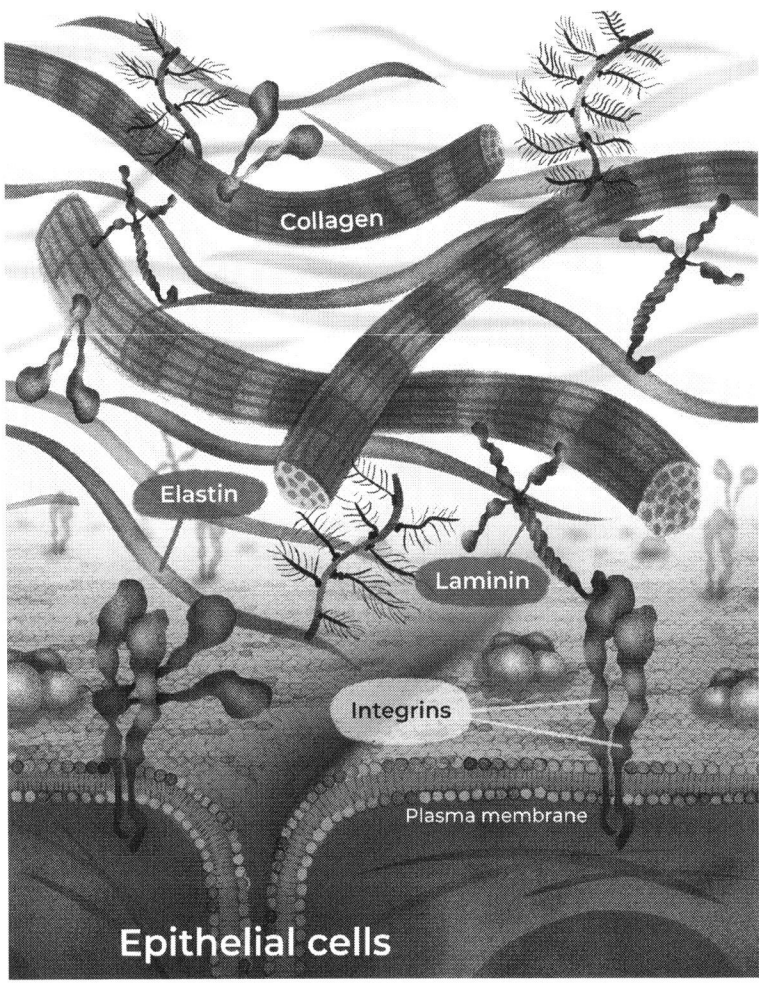

Junctional epidermolysis bullosa is one genetic condition that is caused by a laminin disorder. It leads to extremely fragile skin that easily blisters, erodes, and becomes infected, and is susceptible to cancer. This condition seemed absolutely incurable for centuries, until genetic engineering was invented. In 2015, the first treated junctional epidermolysis bullosa patient, a seven-year-old boy who had lost more than eighty percent of his outer skin layer to the disease, had some of his skin cells removed and genetically modified with a functional copy of the damaged laminin gene. These new cells were cultured and used to replace his old skin. The boy recovered[838].

One of the reasons why laminins are so important is because they are involved in the maintenance of stem cells and controlling the differentiation of progenitor cells, which are capable of turning into cells of other types. For example, ultraviolet light damages skin laminins, leading to loss of skin stem cells and causing skin photoaging. This is the kind of damage normally dealt with by matrix metalloproteinases[839].

There are other examples of this connection between cells and their surrounding matrices. For one, scientists collected bone marrow mesenchymal stem cells from young and old mice and placed these cells respectively on old and young cell-free extracellular matrices. Interactions with the young matrix markedly improved the ability of old stem cells to create bone structures and increased their telomerase activity and their ability to proliferate[840]. Likewise, stem cells responsible for muscle regeneration have lower replicative potential when their surrounding extracellular matrices stiffen due to AGE accumulation[841].

This complicates radical life extension. As it turns out, not only do we need to repair cells; we need to repair what's around them. In a personal communication, Roman Litvinov, a researcher who specializes in extracellular-matrix aging, shared his vision of a future solution to AGE accumulation that he calls "in vivo reassembly of the body." It starts with the use of "AGE-formation blockers" to protect laminins and other components of the basal membrane that are long-lived, but not as long-lived as collagen or elastin, and have more potential for renewal. The rejuvenated basal membranes can be populated with mesenchymal stem cells that will now maintain a healthy phenotype as well as the ability to form and structure the matrix around them. At this stage, the old parts of the matrix can be removed by targeting AGEs with antibodies or with ubiquitin that facilitates proteasomal degradation. A new matrix will then be synthesized by fibroblasts. Unfortunately, this plan might prove more challenging in practice than in theory.

Another possible solution is inspired by evolution. When a species experiences natural selection that favors increased lifespan, the long-lived molecules of its members should evolve to support the increasing upper age limit. This explains why human collagens tend to be more resilient than those of mice. Recent decreases in human extrinsic mortality rates provide the necessary background for ongoing lifespan evolution. Perhaps more beneficial forms of collagen and elastin are already spreading in our populations. Normally, only the next generations will truly benefit from this slow ongoing evolutionary process. Instead of waiting for this evolution to occur, we can search for longevity-associated mutant versions of collagen and other long-lived proteins[842], or even design them artificially. We can then use gene-therapy techniques to replace the regular versions of corresponding genes in our mesenchymal stem cells with their improved versions. For example, we could find or create collagens that are less resistant to dissolving proteases when glycated. This could reduce the volume of AGEs and alleviate some age-related disorders.

Or perhaps there are other mechanisms that protect against glycation. A 2025 study published in eLife analyzed glucose levels and glycation rates across 88 bird species. Surprisingly, the researchers found that long-lived birds had higher glucose levels, yet their glycation rates did not increase as much[843]. This suggests that these birds evolved specialized mechanisms to resist glycation. If we can identify these mechanisms, they could become valuable tools in our quest to combat aging.

While such approaches may take years to develop, one thing is clear: the aging matrix is a problem that we will have to deal with one way or another, and the field welcomes original suggestions. Until we have answers, we might consider other ways to reduce AGE formation. For example, a drug called acarbose blocks an intestinal enzyme that digests larger carbohydrates into glucose. Acarbose treatment has been shown to reduce the serum levels of AGEs in patients with diabetes[844]. It's generally considered safe[845], and according to epidemiological studies, it reduces the incidence of type 2 diabetes, although it's still unclear whether it reduces all-cause mortality[846]. The drug provides 16% to 17% increase in median lifespan for male mice and 4% to 5% for female mice[222].

Let's conclude that some of the most distinct characteristics of aged phenotypes, including skin abnormalities, cataracts, and increased risk of cardiovascular diseases, can be at least partially attributed to extremely long-lived molecules that are difficult to replace. It might seem counterintuitive that the most durable and long-lasting parts of our bodies could also be their Achilles' heel (coincidentally, collagen from the heel adjacent to a rabbit's Achilles tendon is a model for studying the stiffening effects of glycation[847]).

Here, I'd like to introduce the concept of *antifragility*, a term coined by mathematical statistician Nassim Taleb (whom I'm not personally a fan of, though I find this particular idea compelling). Antifragility describes a system that not only withstands mistakes, damage, failure, and volatility but actually benefits from them. Consider the difference between a large rock and an animal. At first glance, the rock seems more resilient—it is unchanging, capable of enduring forces that would harm a living creature, and will likely outlast the animal. If dropped, it would crush the creature beneath it. Yet, unlike the rock, life has the ability to adapt and improve over time. Through Darwinian evolution, organisms become more diverse, complex, and better suited to their environments. Meanwhile, the rock undergoes only slow, inevitable erosion, with no capacity for growth or improvement.

Living cells within our bodies constantly react to internal and external changes. Despite their vulnerability, they can activate or suppress autophagy, transform into other cell types, increase DNA repair, and produce protective proteins as a reaction to stress. Our adaptive immune system is capable of self-improvement in the face of novel pathogens. It learns through a sort of internal natural selection that favors cells producing the most efficient antibodies and cellular receptors.

The cells of our body come from a line that has existed for billions of years and has undergone countless gradual improvements. Immortal cell lines can be maintained in laboratories, and many normal cells are capable of outliving the organism that originally contained them. Mouse erythropoietic stem cells serially transplanted from older to younger animals can produce functional erythrocytes for a hundred months—double the maximum lifespan of these mammals[848]. Likewise, T cells can divide and function even after ten years of mouse-to-mouse transplantation[849]. Long-lived extracellular proteins may be durable and capable of withstanding time, but like the rock, they erode and must rely on cells for maintenance and renewal, just as cities require humans as much as humans require cities. Making these vital components of our bodies more antifragile in addition to durable could improve our lifespan and healthspan.

But the most important system that has both fragile and antifragile properties is the human brain. On one hand, our aging neurons accumulate irreversible damage and cannot be easily replaced. But on the other, the architecture of the brain allows for impressive plasticity. It physically learns from mistakes and adapts to its environment by forming new neural connections and removing those that are unused. We start with little knowledge about the world, but when faced with obstacles we learn to overcome them. Preserving the brain's function and our knowledge, experience, and ability to think is perhaps the most important goal of radical life extension, and the topic of the next chapter.

Our body is not the ship of Theseus—some damage cannot be undone. Yet, ten individuals are replacing what was once deemed irreplaceable, ensuring the ship can sail on without end, transforming it into the ship of Theseus.

CHAPTER 11
The ageless brain

> Your head will collapse, but there's nothing in it,
> And you'll ask yourself,
> Where is my mind?
>
> —Pixies ("Where Is My Mind")

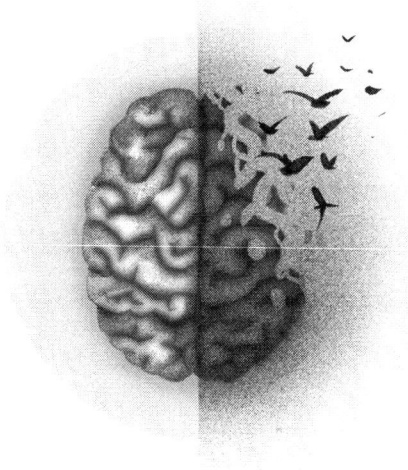

Summary: in this chapter we discuss head transplants, cybernetic bodies and uploading one's consciousness, how the aging body affects the aging brain, the rejuvenating potential of "young blood," and the mechanisms of Alzheimer's disease's progression.

In 1926, Soviet scientist Sergei Brukhonenko designed a blood-circulation and -oxidation apparatus called the "autojector." Using this invention, he was able to keep a dog alive for two hours after its heart had stopped[850]. Two years later, the researcher demonstrated something completely outrageous: the ingenious apparatus sustained a living dog's head that was severed from the rest of the body. The head was able to react to its environment, open its mouth, and even swallow a piece of cheese—which immediately came out on the other side.

This grotesque experiment was surpassed by Soviet transplantologist Vladimir Demikhov, who, in 1954, managed to attach one dog's head onto another dog's neck, creating a two-headed animal. Several such surgeries were performed, and while most dogs died within a few days, one reportedly survived for almost a month[851].

These were not the first experiments of their kind. In 1908, a similar head transplant was performed by French surgeon Alexis Carrel and American physiologist Charles Guthrie, although the resulting animal had to be euthanized within a few hours. Regardless, the possibility of creating a living Cerberus or CatDog was never so close to being in our grasp (not that I advocate creating any such creatures).

Aside from the difficult surgical operations, a major challenge in the first head transplants was the rejection of the newly attached parts by the body's immune system. In 1970, as effective immunosuppressants were becoming available, American neurosurgeon Robert White and his team achieved the first monkey head transplantation[852,853]. The transplanted head could chew and swallow food, and track objects with its eyes[854]. It even bit an experimenter. However, it couldn't control the body, and the monkey died within nine days.

Some members of the public regarded animal head transplantation experiments as barbaric. However, it is difficult to deny that the abovementioned researchers advanced transplantology and that their ideas saved many human and animal lives. Vladimir Demikhov performed the first experimental coronary bypass surgery[855] and contributed greatly to clinical heart and lung transplantation[856], while Alexis Carrel was awarded the 1912 Nobel Prize in Physiology or Medicine for his work on vascular sutures and the transplantation of blood vessels and organs.

These studies inspired and were inspired by the idea that someday human head transplantation will be possible, as both optimistic and pessimistic discussions of this looming technology spread throughout our societies and cultures. In 1925, Russian writer Alexander Belyaev published a horror story called *Professor Dowell's Head*, which was about a lab assistant who sustained the detached head of his mentor on life support for the purpose of extracting scientific knowledge. The assistant then used this knowledge to transplant the head of one young woman onto the body of another.

Not all fictional uses of this technology have been so grim. In the lighthearted animated series *Futurama*, head-in-a-jar technology keeps celebrities alive. Even the heads of dead people can be revived and sustained. Undesirable consequences still come, such as Richard Nixon's head obtaining a robot's body and exploiting legislation stating that "n*obody* can be elected more than twice."

Actual human head transplantation would be very difficult to achieve, although some people believe it's possible. In 2015, Chinese surgeon Ren Xiaoping and his colleagues performed a head-to-body transplantation on mice. They described this research as "laying the groundwork for human head transplantation[856]," although only twelve of forty pairs of animals survived more than twenty-four hours. In the same year, Italian neurosurgeon Sergio Canavero attracted media attention by claiming at a TEDx conference and in multiple interviews that he would be the first person to transplant a human head. Previously he had published a peer-reviewed article highlighting the various challenges that such surgery would face, and his proposals to overcome them[857].

Canavero suggested the use of hypothermia and ultrasharp blades to reduce tissue damage; polymers that supposedly improve nerve healing; immunosuppression, and so on. Indeed, the problem of immune rejection can be addressed with modern drugs that allow patients to live with transplanted body parts, even limbs, for decades[858]. But connecting a brain to another body's peripheral nervous system and restoring motor and sensory functions seems especially problematic at the current technological level. Thus, even if a head transplantation is successful and the patient survives, he or she will most likely be physically disabled.

Some scholars have called Canavero's proposal "exciting, provocative, problematic, and evidently contentious[859]." Others have emphasized that the "surgical, ethical, psychosocial, and immunologic hurdles associated with body-to-head transplantation are enormous[860]." Meanwhile, Canavero and Xiaoping coauthored a paper describing the fusion of two deceased human cadavers, a "rehearsal" surgery performed in 2017[861]. To date, this is as far as any such transplantation has gone. And we are surely far from making such a procedure viable or acceptable.

Perhaps the issue of paralysis following head (or should we say *body?*) transplantation can be overcome by nonbiological means. One group of researchers managed to connect the motor cortex of a rhesus monkey to its disconnected spine via an electronic device[862]. A microelectrode array was installed in the monkey's brain to monitor the activity of its neurons. That information was encoded and wirelessly transferred to a receiving device inserted into the spinal cord below the monkey's injury. As a result, the monkey regained control over its previously paralyzed legs.

In 2008, another group of researchers taught monkeys to use a computer–brain interface to control a robotic arm[863]. Since then, this technology has been successfully translated for use in quadriplegic humans. It has even been improved by the addition of sensory inputs from the robot arm via stimulation of the subject's somatosensory, or "body feeling," cortex[864]. So perhaps we should consider using robotic bodies for detached human heads or isolated brains? This idea might be especially preferable given that a healthy body of a brain-dead donor is an expensive rarity that can be used to save not just one life but many lives through multiple organ transplants. In contrast, robotic bodies could in theory be manufactured on any scale, like cars. You could even own several.

CHAPTER 11

When it comes to brain–computer interfaces, most people probably think of Elon Musk's company Neuralink, given the wide media attention it has received. The technology involves implanting numerous electrodes into the brain. While this might help people with quadriplegia, cerebral palsy, and maybe blindness (Musk hopes), I believe there is an even more impressive and less invasive approach, which would allow one to manipulate neurons.

A method called optogenetics involves the genetic modification of cells with genes encoding light-activated proteins, such as those naturally found in some species of algae. If the modified cell is a neuron, you can activate or deactivate it with a flash of light, with great temporal and spatial precision[865,866].

Optogenetics has been used in animal models to control physical movement[867] remotely, to stimulate auditory pathways[868] and restore auditory-driven behavior[869], to modify and recall memory[870,871], to alleviate ventricular arrhythmias in the heart[872], to control the urinary bladder[873], and even to induce penile erection[874]. Just give your girlfriend a blue flashlight, and enjoy the fun. It's not surprising that the journal *Nature* proclaimed optogenetics the "Method of the Year[875]-2010." In 2021 optogenetics was used to partially restore vision to a patient with neurodegenerative eye disease[875].

Just as it's possible to produce neurons that are activated by light, it is possible to make neurons emit light[876] with the help of bioluminescent proteins, such as those found in fireflies or in the glowing jack-o'-lantern mushroom[877]. Imagine a futuristic scenario in which the active neurons emit light signals when they fire. These signals are received and processed by a device installed at the very surface of the brain, analyzing your brain activity with cellular precision. The device can then signal specific groups of neurons with short, focused beams of light of a different color. Much larger regions of the brain's surface might be involved in this communication than with the Neuralink device. Think about connecting minds, or experiencing The Matrix, or the "brain dances" from the video game *Cyberpunk2077*, which allow one to relive someone else's experience (unlikely, but not completely impossible). We just need to live long enough to witness this technology.

Futurama's head-in-a-jar technology has also had some traction in reality. Research on mollusks such as the sea angel has shown that an isolated nervous system can survive for days and can even participate in "imaginary behavior." For example, while sending signals that would normally result in swimming upward, the sea angel will pause for motionless relaxation[878]. Scientists have also found a way to keep a completely isolated guinea pig brain alive for several hours[879,880]. The isolated mammalian brain continues to function, allowing its neural activity to be registered and recorded. This means that, in theory, it could one day control some kind of device.

Another speculation, this one more popular, is that we will achieve immortality by uploading our consciousness to some electronic medium. Notable proponents of this futuristic idea include philosopher Nick Bostrom, and Ray Kurzweil, who predicted that such technology will appear by 2045. Theoretical physicist Michio Kaku has said that, "at least in principle, maybe it's possible to transfer our consciousness, and at some point even become immortal[881]." The idea has become influential in contemporary science fiction, inspiring several episodes of *Black Mirror*, and Amazon Prime's series *Upload*.

My favorite movie that addresses this subject is *Chappie*, which is about a robot with experimental artificial intelligence that mimics the human abilities to think, feel, and learn. Warning: spoilers ahead. The robot gets captured by good-hearted gangsters (played by the hip-hop band Die Antwoord), who teach him many things, including how to engage in criminal activity. After learning about the issue of death, Chappie uses his superior robotic intelligence and an Internet connection to effortlessly solve the problem of consciousness, and he invents a method to transfer his mind into another robot or a computer. This technology is later used to save the minds of his human friends and place them in robotic bodies.

The idea that artificial intelligence will someday become so smart that most of our problems will become trivial for it is called technological singularity, and it was predicted based on observation of exponential computational-power growth, formulated as Moore's law.

Unfortunately, we aren't anywhere close to achieving consciousness transfer or copying an entire brain. However, there have been some notable achievements. In 2014, members of the OpenWorm project created a toy robot controlled by a neural network copied from a *Caenorhabditis elegans* worm. This was possible because the nervous system of this animal consists of only 302 neurons, and all of their connections have been mapped. Remarkably, the robot displayed some roundworm behavior, such as obstacle avoidance; however, it is difficult to say to what extent this behavior matched that of a real worm[882,883]. Just recently, another group of researchers mapped the entire network (or *connectome*) of a fruit fly larva's brain[884], which consists of 3,016 neurons with 548 thousand synapses. As for mammals, one of the most impressive achievements is a model of a mouse's primary visual cortex, with approximately 230 thousand neurons. However, this model is not an exact partial copy of an individual mouse's brain[885]; rather, it relies on general neurobiological knowledge. The greatest-yet achievement in human-brain digital reconstruction involved the three-dimensional structure of a cubic-millimeter region comprising 50 thousand cells with 130 million connections[886], obtained via electron microscopy.

While technological advancement will likely continue, there may be some fundamental issues with the concept of consciousness transfer. In the middle of the twentieth century, Polish sci-fi writer Stanisław Lem proposed a teleportation paradox in his book *Dialogues*. The idea is that, in the future, a machine could scan your body atom by atom, destroy it, and assemble an exact copy at a distant place, effectively transporting you at the speed of light. Given that our personalities are defined by our brains, which are made up of atoms, there would be no way for an observer to discern the new you from the old you. But would it truly be you?

If you think that there is no problem with this, consider what happens if the machine malfunctions and creates multiple copies of you in various locations. Which of them would you become? In what location would you expect to arrive? What if the original was not destroyed but was merely scanned and re-created? Likewise, if a perfect model of your brain could be uploaded to a computer, what would prevent it from being copied and placed in several different virtual environments simultaneously? Which environment would you observe? Would you have achieved true immortality or would you have merely perished and been replaced by an identical copy that will live on with your memories? Perhaps a solution to these problems is in slow stepwise integration, by which the parts of your brain are continuously and functionally replaced, like the parts of the ship of Theseus. But what would be the advantage of such an approach if, objectively speaking, the result is the same?

Maybe there are certain physical limitations, such as the uncertainty principle, that prevent the creation of an identical copy or a perfect model of a brain. Even so, one could argue that a perfect copy is not necessary. We are certainly quite different today than we were during the first days of our lives. This is not only because of the constant exchange of atoms and molecules in our bodies, but because of the formation of and loss of connections between our neurons. Nonetheless, we feel as if our identities are preserved. But is this feeling an illusion? Or perhaps the whole idea of the self is an illusion? I won't pretend that I know the answers to these questions, but let's discuss a bit further.

Some time ago I authored a sci-fi novel called *The Harvard Necromancer*, in which the protagonist proposed a metaphor to explain the teleportation paradox. I don't necessarily consider this metaphor to be accurate or convincing, but I will mention it nevertheless, for the purpose of entertainment. The novel is about a scientist who performs a series of rigorous experiments that, to his surprise, reveal the existence of a kind of "magic" involving genetically engineered animals with human genes. The scientist tries to understand the laws behind this newly discovered phenomenon, which appears to contradict modern physics and biology. At one point he mentions a made-up concept called "the soul as blockchain."

Blockchain is a technology used in cryptocurrency exchange that is based on the idea of blocks that hold information about transactions. Each block also contains a reference to a previous block. So it is possible to go back in time, step-by-step, and recover the whole history of a series of transactions and, by doing so, establish the ownership of a certain asset[887]. This leads to an interesting fundamental property of the blockchain. Imagine that two people each give you an equivalent of ten dollars in crypto. The amounts are the same, but they are coded as different objects, with different transaction histories. If our universe were similar to a blockchain, then a person would be different from their identical copy because of the different "transaction histories" that have occurred over time. Just as the slices of two different three-dimensional objects can be identical in two dimensions, two people can be the same in the present but different in their past and future.

I believe that there are many interesting questions to ponder and argue about regarding the definition of identity, but our brains are not getting any younger. To give ourselves more time to experience the joy of thinking and debating, it would be beneficial to understand how the biological human brain ages and what can be done about it. There appears to be no easy solution to the problem of death, and no matter what new technologies await us, it's best that we arrive in the future at our maximum intellectual capacity.

Fortunately, it appears that the brain is not the weakest link in our gradual deterioration, and that at least part of the aging brain's problems stem from the rest of the body. After all, some of the greatest risk factors for cognitive decline are cardiovascular diseases, and metabolic conditions such as diabetes and obesity[888]. A large meta-analysis revealed a reduction of more than thirty percent in the risk of cognitive decline among people who engage in regular physical exercise compared with those who report a sedentary lifestyle[889].

Aside from improving vascular function[890], physical exercise has been shown to increase the release of brain-derived neurotrophic factor (BDNF) in both humans and mice[891]. This protein plays a crucial role in the survival, differentiation, and interconnection[892] of neurons during development[893]; and mutations that disrupt its secretion, transport, or processing have been linked to poorer episodic memory[893], as well as reduced volume[894] of and activity[895] in the hippocampus, which is responsible for long-term-memory formation. Genetic variations in BDNF have also been linked to changes in cognitive performance[896] and rates of cognitive decline[897].

Neurons are some of the most long-lived cells in mammals, and their survival is often limited by their environment. In 2013, a group of scientists transferred embryonic cerebellar precursor cells from a short-lived mouse strain, with a twenty-six-month maximum lifespan, into the developing brains of a rat species capable of surviving up to thirty-six months. The donor mice were genetically modified to produce a green fluorescent protein so that their cells could be easily visually identified in recipients' bodies. Eventually, many colored mouse neurons were found, even in the oldest rat cerebellums, to have survived well past the maximum lifespan of their original species.

Some of these cells were Purkinje neurons, which play a primary role in coordination, control, and the learning of movements, and are progressively lost in aging mammals. Mice of the studied strain normally lose Purkinje neurons at alarming rates, but the lifespans of these mouse cells were drastically increased inside the brains of rats[898].

Another way to test the effects of body aging on the brain is through experiments involving heterochronic parabiosis—a surgical connection between the circulatory systems of old and young animals. This allows blood components to transfer freely between a pair. In 1999, the writers of *Lexx*, one of my favorite sci-fi TV series, were apparently inspired by the idea of parabiosis when they created an antagonist named Brizon. He was a dying former "Bio-Vizier" who forcefully connected himself to the liver of the young female character Xev to prolong his life in a gruesome and memorable way. Yes, this is another unfortunate example of our culture portraying those who seek life extension as evil and misguided.

In any case, early studies of heterochronic parabiosis reported some life extension in old rats compared with untreated single controls, but the effect was significant only in females[899]: 835 versus 740 days estimated expected length of life. The beneficial effects of shared circulation can be masked by the detrimental health consequences of the surgery and by alterations in the lifestyles of the treated animals, such as restricted physical activity, so comparison is usually made between heterochronic (different age) and isochronic (same age) parabionts. According to the study, old males and females connected to same-aged rats had lives that were, respectively, roughly 120 and roughly 150 days shorter than those of old rats from heterochronic pairs.

This comparison is somewhat problematic because age-related diseases that develop in one old animal of an isochronic pair may reduce the lifespan of its connected counterpart. As an extreme example, consider the spread of metastatic cancer or inflammation signaling. One could say that old animals connected to young ones have comparatively longer lives simply because the alternative of being connected to another old animal is an even bigger hazard.

To reduce the negative effects of long-term parabiosis, some more recent studies have used short-term parabiosis with separation. One of these studies found that this intervention reduces the lifespan of young mice attached to older counterparts, but there was only a slight trend toward increased longevity of the latter, and it did not reach statistical significance[900]. However, another study did find lifespan extension of old female mice that were temporarily connected to their young counterparts, and it even demonstrated a reduction of biological age measured via epigenetic biomarkers of aging and gene activity. The greater improvements achieved in the second study can perhaps be attributed to the different mouse strains and sexes used, and to variations in the quality of surgery.

It should be made clear that nobody is considering connecting old people to children as part of some dystopian anti-age therapy. Nor should we become vampires. But parabiosis experiments in animals may help elucidate whether a young body produces beneficial blood components, or perhaps deleterious ones that it's better at removing. And in any case, the majority of current parabiosis research is primarily exploring potential avenues for brain rejuvenation.

In 2014, a group of researchers reported that old mouse parabionts that spent five weeks connected to young ones formed more new neuron progenitors than old isochronic parabiont controls[901], and that they had better cognitive performance. The serum of young and heterochronic old parabionts contained elevated levels of a growth factor called GDF11, and injections of this factor promoted neurogenesis in aged animals. Further research showed that GDF11 does not effectively pass the blood-brain barrier but instead promotes neurogenesis indirectly, by improving cerebral vasculature[902] and enhancing blood flow to the brain[903].

Other potentially beneficial blood molecules have also been found[904]. For example, animals that exercise have increased plasma levels of BDNF and a liver enzyme called GPLD1. Artificial increase of GPLD1 levels in adult mice leads to increased BDNF levels, to cognitive improvements, and to increased neurogenesis. These effects were suggested as an explanation of the finding that injections of plasma derived from mice that exercise led to improved learning[905] and increased hippocampal neurogenesis in recipient mice, while the plasma of sedentary mice was no better as a treatment than saline. Unfortunately, despite the mentioned beneficial effects, plasma injections have so far failed to improve the lifespan of model animals[906].

While some researchers focus on the cognitive benefits of young plasma[907], others focus on the properties of young blood cells. Cells that have gotten special attention are macrophages—white blood cells that can clear dead cells. Macrophages are known for their role in rejuvenation. For example, injections of these cells can lead to improved wound healing, especially when the macrophages come from young donors[908].

It is widely known that the aging brain faces impairments in signal transmission, caused by the gradual loss of the insulation of the brain's "wiring." That insulation comes in the form of a lipid-rich material called myelin, which sheathes the axons and dendrites through which neurons connect to one another. Myelin sheaths are produced by Schwann cells in the central nervous system and by oligodendrocytes in the periphery. Brain injury, aging, and neurodegenerative diseases can lead to demyelination and reduced cognitive and motor function[909]. Multiple sclerosis is probably the most common demyelinating disorder, and it leads to vision problems, extreme fatigue, trouble moving, and odd feelings of tingling and burning.

One study of heterochronic parabiosis showed that the ability of old precursor cells to differentiate into oligodendrocytes and enhance insulation is improved when old mice are connected to young ones[910]. This effect was mediated by young macrophages, which were more capable of removing old myelin debris, allowing new myelin to form. A more exotic approach also helped boost oligodendrocyte function in mice—the infusion of young cerebrospinal fluid into aging brains[911]. The beneficial effects could be mimicked by infusions of a single component of the fluid, a growth factor called Fgf17, whose concentrations drop in mouse neurons, and in aged human cerebrospinal fluid and plasma.

There is also accumulating evidence for the existence of detrimental components of old blood, which are worth diluting. For example, young mice exposed to old plasma or heterochronic parabiosis experience reduced synaptic plasticity and impairments to their cognitive functions, such as spatial learning and memory[912]. One suggested explanation is the spread of inflammation signaling. Aged plasma contains increased levels of chemokines with pro-inflammatory properties[913], such as CCL11 and prostaglandin E2 (PGE2). The administration of CCL11 to young mice reduces neurogenesis[912], while the prevention of PGE2's ability to act upon its receptors improves spatial memory, hippocampal function, and neural plasticity in aging mice[914].

Despite their limitations, studies of parabiosis have provided a number of exciting hypotheses regarding the effects of various blood components on the aging brain, some positive (such as BDNF, GDF11, and young macrophages) and some negative (such as pro-inflammatory cytokines). However, researchers from UC Berkeley reported that neuroinflammation and animal cognition[915] can potentially be improved by a much simpler procedure called neutral blood exchange. Neutral blood exchange involves replacing about half of an animal's plasma with saline with five percent albumin—a soluble protein that is common in serum and is used to transport various substances[916]. We have yet to find out whether the beneficial effects of this procedure can be translated to humans, and if it can prolong lifespan. Hopefully some "young blood" research will be confirmed in future clinical trials[917].

Now that we have established that brain aging is connected with the aging of the body, let's discuss the more specific aspects of age-related cognitive decline which is probably the most unnerving part of getting older. Just watch Anthony Hopkins's Academy Award–winning performance in *The Father*, a 2020 movie that meticulously paints the grim details of senile memory loss. According to systematic reviews, approximately five to nine percent of people aged more than fifty years are diagnosed with some form of cognitive impairment that hinders their ability to remember, think, make decisions, or perform everyday activities. Two to five percent suffer from Alzheimer's disease, and another one to two percent from vascular dementia, characterized by reduced blood flow to the brain. We are all potentially at risk. Even dolphins develop Alzheimer's-like pathologies[918].

CHAPTER 11

One day a person is a great scientist or a brilliant actor. Soon he or she is struggling to speak or remember the simplest of words. Peter Falk suffered from Alzheimer's and Robin Williams from Lewy body dementia, and Bruce Willis developed frontotemporal dementia. By the age of 100, the risk of all-type dementia reaches the brutal mark of 65%[919]. The risk of developing Alzheimer's disease rises from 15% at age 70 to around 65% at 105[920]. With the human population growing and the average lifespan increasing, the number of people with dementia is projected to rise from 57 million in 2019 to 153 million in 2050 unless effective treatment is found[921].

Federal funding of dementia research in 2022 exceeded 3.5 billion dollars in the United States alone. Decades of research have yielded more than 200 thousand published peer-reviewed scientific articles on this topic. Hundreds of clinical trials were performed. Unfortunately, most of them failed[922,923]. At the time of this writing, the only exceptions are in the form of several drugs that provide modest symptomatic benefits for cognition and behavior without altering the progression of Alzheimer's disease. A drug called aducanumab was approved by the US Food and Drug Administration in 2021 despite the controversial and contradictory results of its clinical trials[924,925]. Another drug, called donanemab, showed promising results in recent randomized clinical trials. Unfortunately, even donanemab does not prevent the progression of Alzheimer's disease, but is reported to merely slow it down[926].

We will focus on Alzheimer's disease because it is the most common form of dementia, but also because the science behind this disorder is fascinating. It's almost like a true-crime detective story with multiple suspects, some of whom might turn out to be innocent bystanders who took the fall for the actual bad guys.

Alzheimer's disease starts slowly, with initial symptoms including mild learning problems and short-term memory deficit. Over time neurodegeneration progresses, leading to significant impairments in speech and long-term memory and to an inability to perform even the simplest daily tasks. This is eventually followed by death. Autopsies of Alzheimer's disease patients typically reveal brain atrophy that is most pronounced in the hippocampal region[927], with two special features. The first hallmark is extracellular deposition of aggregated protein fragments called amyloid beta peptides (amyloid plaques) localized in the affected brain areas. The second hallmark is an aggregation of hyperphosphorylated proteins called tau in the bodies of neurons[928] (neurofibrillary tangles, or NFTs—not to be mistaken for non-fungible tokens, which are cryptographic assets).

These observations, along with evidence for the toxicity of some amyloid beta forms[929], have led to the predominant amyloid-cascade hypothesis of Alzheimer's disease. According to this hypothesis, abnormal accumulation and aggregation of beta amyloids leads to neuronal dysfunction, causing the formation of neurofibrillary tangles and, eventually, neuron death.

Let's get a closer look at one of our primary suspects. Amyloid beta peptides are produced by cleavage of amyloid precursor proteins (APPs), which reside within the membranes of various cells, including fibroblasts, endothelial cells, oligodendrocytes, and, most importantly, neurons[930]. APPs can be cut, with variable outcomes, by three membrane-bound enzymes, called alpha, beta, and gamma secretases. When APPs are cleaved by alpha secretase benign non-aggregating products are formed and that's the end of the story[931]. But if this doesn't happen and APPs are cleaved by both beta and gamma secretases, aggregation-prone amyloid beta peptides are formed and released into extracellular space. Because alpha secretases compete with beta and gamma secretases for APPs, the stimulation of alpha secretase activity has been proposed as one of many potential Alzheimer's disease treatment approaches[932]. Other suggestions include the use of antibodies such as the abovementioned aducanumab and donanemab against amyloid beta aggregates, and reduction of amyloid beta production with the help of beta and gamma secretase inhibitors[933].

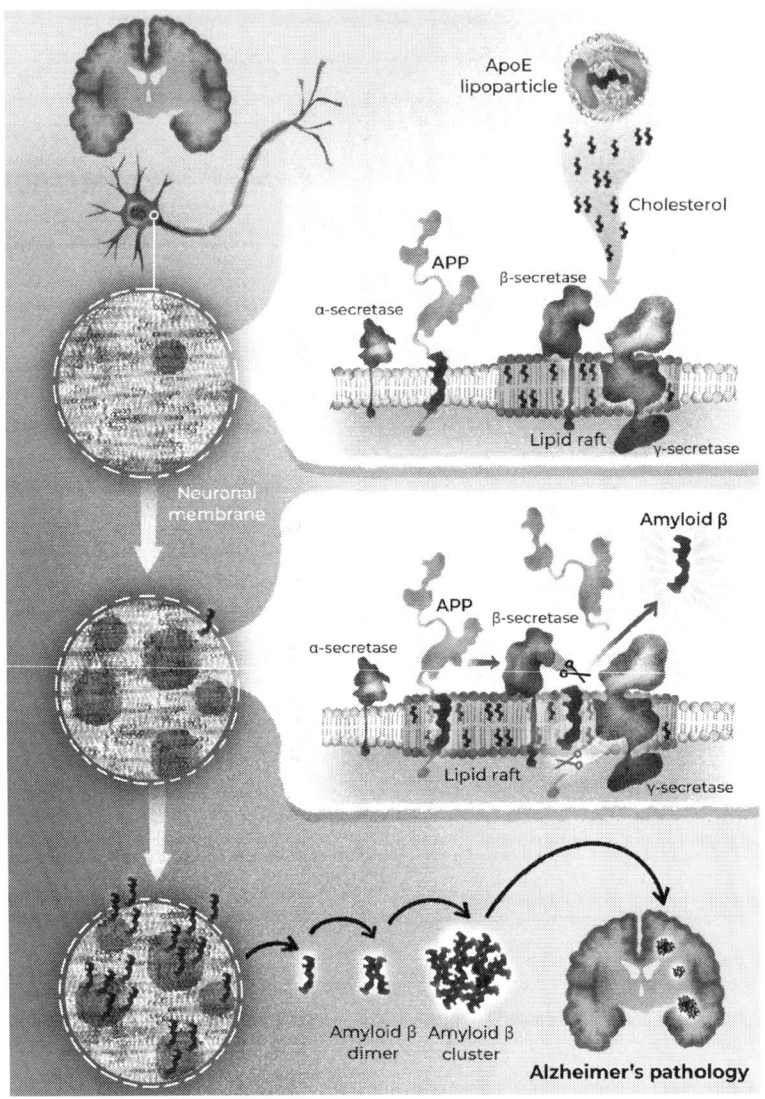

Although the amyloid-cascade hypothesis of Alzheimer's disease has so far failed to produce a truly effective treatment, it's supported by a number of genetic studies. Approximately five percent of Alzheimer's cases are early onset and begin before the age of sixty-five. Five to ten percent of these cases are explained by mutations in the APP gene and the PSEN1 and PSEN2 genes that encode protein components of the gamma secretase complex[934]. So, there is a direct connection between some mutations in key genes involved in beta amyloid formation and cognitive decline. However, the majority of genetic variability in early-onset Alzheimer's disease remains unexplained[935,936].

Meanwhile, the greatest known risk factor for the more common sporadic Alzheimer's disease is a genetic variant in the apolipoprotein E (APOE) gene—the ε4 allele[937]. Carriers of two copies of this allele are at a striking fifteen-fold increased risk of the disorder[938]. Genetic testing for this heritable feature is readily available to the public, and what a relief I felt when I found out I have two "normal" (ε3) APOE variants!

Apolipoproteins are proteins that bind lipids, such as cholesterol, that are insoluble in water, and transport them through our blood, lymph, and cerebrospinal fluid. The APOE ε4 allele is associated with high levels of cholesterol, which leads to the accumulation of fat in the arteries and with atherosclerosis, resulting in increased risk of coronary heart disease, stroke[939], and mortality. The connection between APOE gene variants and cognitive decline through impairments of the circulatory system is straightforward, but the connection with full-blown Alzheimer's disease and beta amyloid formation is a mystery that researchers are still trying to unfold.

The remarkable story of what *is* known starts with cellular membranes. A cellular membrane is a structure that separates the contents of a cell from the surrounding environment, and it consists mainly of molecules called phospholipids, which are organized in two layers. The "heads" of phospholipids are hydrophilic. This means that they "like" to be near water, and thus, they form a membrane's surface. The "tails" of phospholipids are hydrophobic. Due to their "fear" of water, they prefer to mingle among themselves in the internal part of the double layer, protected from H2O by the hydrophilic heads.

The second-most abundant lipids in membranes are sterols such as cholesterol in animals. Cholesterol molecules also have hydrophobic and hydrophilic parts and tend to cluster together. As a result, some parts of the cell membrane are especially enriched with cholesterol and form tightly packed domains called lipid rafts[940]. Some membrane proteins have a preference for these rafts and float with them, while others tend to localize in less organized areas with lower levels of cholesterol.

Amyloid beta–producing beta and gamma secretases are among those proteins that have a preference for lipid rafts, while the "benign" alpha secretases avoid such regions. APP can be found all over the membrane, but when concentrations of cholesterol increase, APP tends to move onto lipid rafts, where it becomes spatially co-localized with beta and gamma secretases. This increases the frequency of interaction between these proteins and, hence, the production of amyloid beta. This is where APOE comes into play. Neurons have impaired cholesterol synthesis, and they rely on APOE to transfer these lipids from supporting astrocyte cells. In short: APOE brings cholesterol, cholesterol builds lipid rafts, and lipid rafts promote APP cleavage and amyloid beta formation.

All of this was uncovered thanks to modern visualization techniques, which allowed the detection of lipid rafts and their spatial co-occurrence with labeled membrane proteins[941]. Although the described regulatory mechanism is elegant, the authors of the study did not observe any difference between the dangerous APOE ε4 variant and the normal ε3 allele in terms of their ability to load cholesterol onto neurons—so the mystery of APOE ε4's connection with beta amyloid accumulation wasn't solved after all. Perhaps there are other mechanisms involved. For example, researchers have shown that the "bad" APOE ε4 variant can cause altered cholesterol localization in oligodendrocytes, resulting in reduced insulation of neurons, impairing their function and survival[942].

Another puzzle is why treatments based on the amyloid-cascade hypothesis have failed to yield therapeutic results. Perhaps some of the research is wrong. The plot thickened recently, as duplicated images and other signs of misconduct were identified in an influential 2006 paper[943] (it has more than three thousand citations) that strongly supported the causal role of amyloid beta in Alzheimer's disease. The paper's authors claimed to have purified a specific isoform of amyloid beta from the brains of cognitively impaired mice with amyloid plaques. When transferred into young mice, these isoforms caused memory defects. But there were accusations that this evidence was fabricated, along with some other research by Sylvain Lesné, the lead author of the paper[944]. This, however, does not invalidate the amyloid-cascade hypothesis, or other key findings in the field. In fact, independent researchers have shown that injections of amyloid beta into mouse brains can cause cognitive decline that is not observed after "empty" injections[945-947]. Some studies even found that cognitive decline in mice can follow amyloid beta injections into the gastrointestinal tract, suggesting that the potentially toxic protein can somehow redistribute itself[948]. Finally, there is evidence that defects in the molecular machinery responsible for protein clearance via proteasome degradation can be found in all Alzheimer's patients' brains postmortem and that this leads to extracellular amyloid beta deposits[949].

Alzheimer's disease's being caused by a perpetuating maladaptive protein is easy to conceive of because protein aggregates are involved in other brain pathologies, such as Huntington's[950] and Parkinson's disease[951,952]. We know just how damaging abnormal proteins can be from the case of prions—infectious proteins, discovered by neurologist Stanley Prusiner, that are the causative agents behind brain disorders such as scrapie in sheep; mad cow disease, or bovine spongiform encephalopathy, in cattle; and Creutzfeldt-Jakob disease in humans[953]. Kuru, which most likely started as a variant of Creutzfeldt-Jakob disease, was transferred among the Fore people of Papua New Guinea through cannibalistic practices that involved the consumption, mainly by women and children, of dead ancestors' brains[953]. Prusiner received a Nobel Prize for showing that such diseases are caused by abnormal proteins that can convert normal versions of themselves, which are abundant in an organism, into pathological forms, creating a chain reaction of undesirable molecular transformations[954] that can cause inflammation and apoptosis[955]. The discovery of prions was revolutionary because their mechanism of propagation is completely different from that of viruses, bacteria, and other known pathogens that rely on replication of nucleic acid–based genomes.

Despite years of research, it is still not exactly clear how amyloid beta causes the death of brain cells—although some of its toxic forms were found to induce apoptosis in cell cultures of neurons[956]. At the same time, surprising evidence emerged that APP has protective properties. The protein can be found in all studied vertebrates, and similar proteins are present in numerous invertebrates, including cnidarians such as the hydra, implying an evolutionary history dating back hundreds of millions of years[957]. It's safe to assume that the presence of this protein provided an important evolutionary advantage for the ancestors of modern animals. For instance, APP seems to play a role in the migrations of neurons[958]. But mice lacking APP or beta secretase genes develop normally and are mostly asymptomatic under normal conditions, aside from slight memory problems[958].

The true importance of APP and amyloid beta becomes evident under abnormal conditions, such as ischemia or hypoxia. Under such conditions APP and beta amyloid production is usually elevated in the brain, and mice lacking either APP or beta secretase suffer increased brain damage and mortality[959]. Amyloid beta has been shown to help the brain recover from injury by repairing blood-brain barrier leaks and protecting against infections[960,961]. This could explain why so many clinical trials involving potential Alzheimer's treatments that target beta amyloids not only failed, but sometimes produced adverse effects as brain swelling and bleeding[962,963], reduced cognition, and increased hippocampal atrophy.

Let's remember that blood-circulation disorders are one of most important risk factors for Alzheimer's disease and a few other kinds of dementia. Furthermore, Alzheimer's disease is usually accompanied by damage to small brain vessels[964]. In view of this, it seems reasonable to suggest that beta amyloid formation could, at least in some cases, be a response to brain damage, not its underlying cause. This new theory predicts a correlation between beta amyloid content and the severity of brain damage, but with a reversed arrow of causation.

Other observations now also start to make sense: the link between APOE ε4 and Alzheimer's disease through atherosclerosis and other vascular conditions; the link between physical exercise and reduced dementia risk through improved cardiovascular outcomes and increased levels of BDNF, which improve the brain's vascularization (Alzheimer's disease is associated with reduced BDNF levels[965]); the association between moderate alcohol use and reduced risk of Alzheimer's disease[966] and early-onset dementia in large cohort studies[967]; the link between traumatic brain injury and accelerated development of Alzheimer's-like dementias with increased amyloid beta production[968].

In 2023 the journal *Nature* published an article confirming that, at the molecular level, human organs do not necessarily age at the same rate. The aging of individual organs was analyzed by looking at the blood plasma levels of proteins that originate from specific body parts. The study found that nearly 20% of the studied population experience accelerated aging of at least one organ, and that this condition is associated with a 20%-to-50% greater mortality risk, but most importantly, the study showed that accelerated vascular aging was predictive of Alzheimer's disease's progression[969].

This raises the question of whether the predominant theory of Alzheimer's disease is correct. Perhaps, instead of targeting beta amyloids, we should be focusing on improving the brain's blood vessels. If we do, BDNF could well be a promising candidate for Alzheimer's treatment[970].

Another theory is that the progression of Alzheimer's disease is influenced by infections[960]. Once again, the idea is to reverse the arrow of causality. Let's consider amyloid beta not as the primary cause of brain damage but as a defensive response to it. There are multiple lines of evidence supporting this theory. First of all, amyloid beta has antimicrobial properties[971], and certain infections of the nervous system, such as herpes simplex, can lead to increased extracellular levels of amyloid beta[972]. Second, treatment of infected cells with antivirals can reduce the formation of amyloid beta in neuronal and glial cell cultures[973]. Third, meta-analyses consistently show that people with the herpes virus (and people with any of several other infections) are at an increased risk of Alzheimer's disease[974,975]. In 2023, a large study published in the *Journal of Alzheimer's Disease* reported that people vaccinated against herpes zoster infection, pertussis, or pneumococcus were at a reduced risk of developing Alzheimer's Disease[976]. Several periodontal pathogens have also been recently linked with increased risk of dementia[977]—some extra motivation to visit your dentist.

This theory of infection does not contradict the vascular theory of dementia; a brain's aging and cognitive decline can have interacting causes. Moreover, we know that certain infections can damage the brain's blood vessels. For instance, the widespread neurotropic parasite *Toxoplasma gondii* is known to damage endothelial cells in the brain[978], and it is simultaneously a risk factor for Alzheimer's disease[979]. A practical implication is that we might be able to reduce the burden of Alzheimer's disease and other dementias through sanitation, antimicrobial agents, antivirals, and vaccinations[980].

Another important observation is that Alzheimer's disease is associated with sleep disturbances[981]. Both insufficient and excessive sleep are associated with poor cognitive function[982]. While the biological function of sleep remains somewhat enigmatic[983], mammalian sleep is linked with the clearance of various (sometimes toxic) metabolites, including amyloid betas[984], from the brain. Slow-wave sleep causes a synchronous rhythmic dilation of blood vessels, and the resulting change in blood volume increases the circulation of cerebrospinal fluid[985] and promotes waste removal[986]. Experiments show that sleeping human subjects are better at removing injected contrast agents from the brain than their waking counterparts[987]. One potential explanation for sleep disturbances in Alzheimer's patients is reduced nighttime production of the sleep hormone melatonin in comparison with age-matched controls[988]. Although clinical studies show that melatonin supplementation alone cannot reverse Alzheimer's disease, receiving melatonin-based therapies and maintaining seven to eight hours of sleep nightly might improve quality of life and prevent early stages of cognitive decline[989,990].

Amid all these theories, we almost forgot about the other potential "culprit"—the aggregating protein tau, which appears excessively in the brains of Alzheimer's patients. To understand tau's normal role, we must first discuss how molecules and organelles are transported from the bodies of neurons to the most distant parts of neuronal appendages, in journeys that are sometimes greater than one meter. For these feats of transportation to be accomplished, molecular highways are maintained within the axons of neurons. These highways take the form of microtubules—rigid, hollow rods that are essential parts of the cytoskeleton. These roads are polarized, with their plus ends growing rapidly and usually reaching into a cell's periphery.

Vesicles, organelles such as mitochondria, protein complexes, and even some RNAs[991] are transferred along microtubule surfaces by motor proteins called kinesins and dyneins, which can essentially walk in a hand-over-hand motion[992], similarly to how humans use their two legs. I recommend watching a short film called *The Inner Life of the Cell*, which shows this magnificent walking process in action. The reality of this movement is even more interesting and nuanced than what is portrayed in the animation. As small molecular machines, proteins such as kinesins are subject to stochastic molecular interactions and can sometimes make random steps in opposite directions, especially if they are carrying heavy loads[993], but forward steps occur more frequently, ensuring that eventually the cargo reaches its destination. Most kinesins are biased to move generally toward the peripheral plus ends of microtubules, while dyneins are responsible for backward transportation.

Both kinesis and dyneins require chemical energy that comes in the form of ATP[994], and disruptions of these molecular transports can lead to severe health problems, including various neuropathologies[995]. Remarkably, some viruses, such as rabies, have learned to hijack the microtubule transportation system for their own purposes and essentially "ride" motor proteins to reach the central nervous system from their place of entry[996].

Tau proteins are involved in the maintenance of the molecular highways detailed above. Under normal conditions, these proteins are believed to bind and stabilize microtubules that dynamically assemble and disassemble within our cells[997]. This suggests that tau may be important for axonal growth—but apparently it's not entirely necessary, because tau-deficient mice are mostly normal, except that they experience increased muscle weakness, impaired balance, and a few behavioral abnormalities that come with aging[998,999].

Under pathological conditions such as Alzheimer's disease, levels of tau become elevated, causing "traffic jams" on the molecular highways by inhibiting the kinesin-dependent transportation of vesicles and organelles toward the ends of axons[1000]. This can lead to various problems, such as lack of mitochondrial renewal at a neuron's far end, causing additional damage to a cell.

Hyperactivity of tau is normally reduced by phosphorylation, the addition of phosphoryl (PO_3^{2-}) groups to certain amino acids of this protein[1001]. This leads to detachment of tau from microtubules and reduction of "traffic jams"; however, phosphorylated tau tends to aggregate and form the neurofibrillary tangles observed in Alzheimer's disease. While, for some time, the predominant view was that the aggregated versions of tau were the primary cause of cellular toxicity, there is accumulating evidence that soluble tau may be at fault[1002], causing the abovementioned transportation deficits of mitochondria and other important cellular components. In this scenario the neurofibrillary tangles could be by-products of pathology, not its cause, explaining why therapies that attempt to reduce tau aggregation have failed as much as therapies that attempt to prevent amyloid plaque formation.

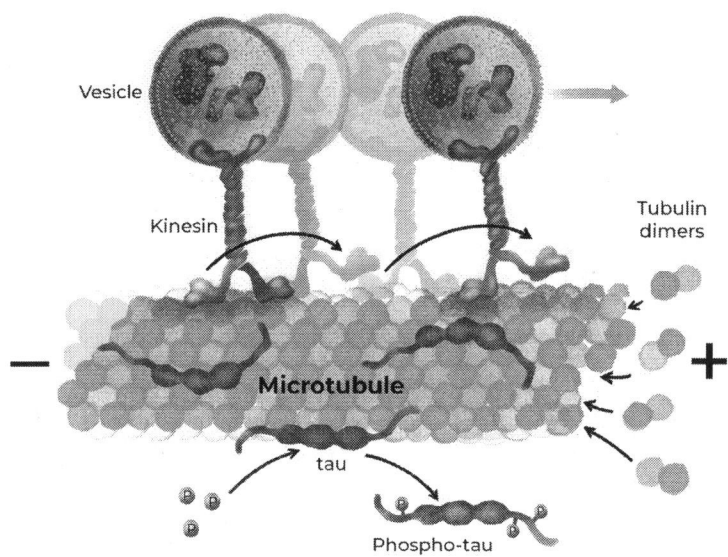

While it's too early to rule out the mainstream hypotheses of the cause of Alzheimer's disease, now might be the right time to dedicate more research to alternative explanations. Most likely, there are many factors that trigger brain inflammation and neuron death. If so, several contributing causes of Alzheimer's disease will need to be addressed simultaneously. Just as there probably won't be a single treatment for general aging, there probably won't be a single treatment for the aging brain. The good news is that, for all we know, a healthy body really helps maintain a healthy mind.

But what do we do if our bodies fail us and our organs malfunction? Perhaps a solution lies in creating new organs and tissues from individual human cells. This is the topic of the next chapter.

A healthy brain depends on a constant flow of oxygenated blood and an intricate transportation network that reaches every distant nerve fiber—much like a modern city relies on a steady supply of water and electricity to function.

CHAPTER 12
The twenty-first–century cure

> Your designer heart still beats with common blood.
> And what if you could have genetic perfection?
> Would you change who you are if you could?
>
> —*Repo! The Genetic Opera*

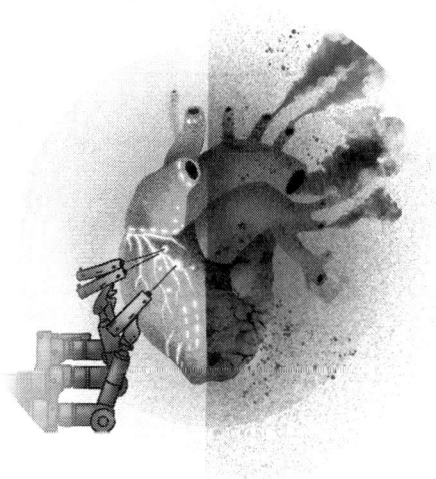

S ummary: in this chapter we discuss 3D bioprinting, chimeras, blastocyst complementation and other methods of creating artificial tissue and organs.

The ethics of human cloning is one of the most controversial subjects in modern biology. Human cloning is banned in more than seventy countries, and it's difficult to imagine any scientific advancement for a technology that causes so much fear for members of our species. Only the fear of genetically modified organisms (GMOs) is comparable. Recently I stumbled upon an article called "Human cloning: unnatural selection" on a religious website, and it provided the perfect textual illustration of some of the myths surrounding this subject.

The article started with the astonishing false claim that cloning requires a pregnant woman whose fetus will be removed and substituted with a clone. This has nothing to do with reality; preexisting pregnancy is neither necessary nor warranted. Then the article mentioned that a human is "not created a second time, but only repeated," probably referring to the religious idea that the soul comes into existence at the moment of conception, which the cloning process bypasses. This idea is theologically problematic given the existence of identical twins, who have to share one soul if the logical conclusion drawn from the idea is correct.

The author painted a "horrible" picture of a future in which clones are bought by "married homosexual pairs"; and "rich immoral women" who "do not wish to put their life in danger during childbirth"; and "madmen" who will "reproduce like monsters from horror movies" after hiring "thousands of illiterate women" to bear their offspring in a "secluded, well-guarded place," "producing geniuses—the highest race of people," according to Adolf Hitler himself. The fate of the "produced people" (as the article named them) will not be enviable, as they will be put to death and their body parts will serve the interests of their originals.

I have briefly mentioned Kazuo Ishiguro's novel *Never Let Me Go*, but I haven't explained how profoundly this novel affected my judgment of humanity's readiness to use cloning technology. As a biologist, I see absolutely no reason for a newborn clone to be treated differently than any other human child, because there would simply be no difference between cloned children and non-cloned children. Hence, the idea that clones will be harvested for organs is as alien to me as the idea that someday the dissection of toddlers will become acceptable. It is as clear as daylight that clones should receive a normal birth certificate and that all human rights should apply to them as much as to anybody else.

However, Ishiguro argues in his novel that humanity is irrational, no matter how much we wish to see ourselves as otherwise. People who believe in a mystical, undetectable soul that is created at conception (like the religious author of the article mentioned above) can paint the moral landscape in such a way that clones appear to be "produced," different, "soul lacking," and thus do not deserve the same sentiments as the rest of us. And there is no objective way to prove them wrong, because a metaphysical soul isn't something we can measure or detect, in clones or in people generally. Thus, it is easy to see how a portion of humanity could turn against clones, or any other group declared "soulless" on religious or spiritual grounds. Perhaps many people fear not the procedure of cloning, but what the existence of clones might reveal about human nature.

It feels weird to have to say that reproductive cloning is not the solution to the current shortage of transplantable human organs[1003]. Of course it isn't. But we do have a very real problem to solve: more than 100 thousand people are on just the United States' national transplant waiting list. The greatest demand is for kidneys, followed by livers and hearts[1004]. As the human population and lifespan increase, the number of people requiring organ transplantation will likely also rise. Fortunately, science is at the brink of achieving much better solutions than those proposed in dystopian movies and prose.

Let's start with a little bit of history. The first successful human organ transplant was performed in 1954[1005], when a kidney was transferred from one identical twin to his brother. In other words, one clone willingly saved another's life. The genetic identities of the donor and recipient played a crucial role in the long-term success of this surgery, because there was no risk of immune rejection and no need for immunosuppressants.

The main factor that influences organ rejection is a series of adjacent genes located on chromosome six and called the major histocompatibility complex, or MHC[1006]. As discussed in an earlier chapter, the proteins encoded by these genes make the constituents of a cell visible to the immune system. This is done through surface presentation of small peptides derived from the cell's proteins. For a cell to be left unharmed by the immune system, it must demonstrate matching MHC molecules and lack foreign peptides, such as those derived from infectious agents.

If all people had the same MHC molecules it would be easy for pathogens to adapt to this defensive measure and spread effectively throughout the whole human population. Thus, evolution favors MHC diversity, and the corresponding genes are highly polymorphic[1007]. Furthermore, we receive two complementing sets of MHC genes, one from each parent. Since all MHC genes are inherited together, two nonidentical siblings will have a 50% chance to have one matching MHC set and a 25% chance of having two or none. This is why sometimes siblings are suitable donors for one another and sometimes they are not.

MHC diversity is such an important evolutionary feature that numerous researchers have investigated whether it plays a role in mate selection. In theory, the ability to choose a partner with MHC sets different from one's own could be beneficial for one's offspring. Such a preference was initially shown in mice[1008] and was linked to olfactory cues, eventually leading to studies involving men and women with varying MHC similarity sniffing one another's sweat and urine samples in the name of science[1009]. Most recent rigorous studies and meta-analyses have not found MHC-biased mate choice in humans[1010,1011], which is a bit anticlimactic.

Anyhow, MHC matching in organ transplants is crucial beyond doubt. Organs for transplantation should be as genetically similar to their recipients' as possible. But most people don't have identical twins or access to cloning facilities, and even if they did, it would be greatly preferable, both ethically and practically, not to rely on other human beings for transplantable organs. What if we could produce new organs outside of human bodies? What if we could make a copy of your own kidney, liver, or heart that would be genetically perfect for you?

Four main paths are unfolding in pursuit of this goal. These are: (1) growing human organs in animals[1012]; (2) creating genetically modified animals (optimally with a few human genes) that will be suitable donors for humans; (3) inducing organ development outside of a living organism, from pluripotent or other kinds of stem cells, controlled by growth factors or other chemical signals; (4) 3D bioprinting of tissues and organs from more or less specialized cells[1013,1014].

Animals with human organs or human cells would be called chimeras. This scientific term is a reference to the mythical fire-breathing monster that was part lion, part serpent, and part goat—like *South Park*'s ManBearPig, known to be half man, half bear, and half pig. The human part is not necessary; a chimera is any creature that consists of cells with different genetic backgrounds, even if they come from the same species.

Chimerism can occur naturally. Perhaps the weirdest example is the sexual symbiosis of deep-sea anglerfish. The males of some species are very small in comparison with the females and can anatomically bind to the latter, establishing a shared circulatory system. Multiple males can join a female at once in this remarkable form of naturally occurring parabiosis. Up to eight males have been observed integrated with one female. Over time the integrated males lose their fins, eyes, and internal organs—except for the testes, which can grow and provide the female with an endless supply of sperm. This incredible evolutionary feat was accompanied by deleterious mutations in crucial genes that are involved in the normal development of the adaptive immune system. Two such genes are *RAG1* and *RAG2*, which are essential for the DNA-recombination process responsible for the diversity of B and T lymphocytes and their receptors[1015]. Anglerfish also lost some, but not all, of their MHC genes. Perhaps these adaptations, together, help prevent females from rejecting male tissues.

In humans, chimerism can occur via the merger of two or more embryos. Such a merger can lead to people being born with different-colored eyes, or with non-matching skin tones on different parts of their bodies. It is even possible for some of one person's cells to carry male XY chromosome sets while others carry the XX female karyotype[1016]. Perhaps the most famous case of human chimerism is that of Lydia Fairchild, who was accused in court of not being the biological mother of her children because of negative genetic parental test results. Lydia was vindicated when further investigation revealed that she was a chimera with reproductive parts that were from an unborn twin and hence genetically different from other parts of her body[1017].

Artificial creation of chimeras has now become reality. The story of growing parts of one animal in another started in 1993, when a group of scientists from Harvard University Medical School came up with an idea called blastocyst complementation[1018]. A blastocyst is a small ball of cells that forms during early pregnancy, a few days after fertilization. The researchers created genetically engineered mice that lacked the *RAG2* gene (one of the genes that anglerfish with sexual symbiosis are deficient in), which led to an absence of functional mature B and T cells, and consequently, an impaired adaptive immune system. They then complemented the blastocysts of these mice with a small number of genetically normal embryonic cells, which resulted in the development of chimeric organisms that had normal immune systems. All mature B and T lymphocytes of those immune systems were derived from the added cells.

The underlying principle of blastocyst complementation is that a genetic mutation that leads to the unavailability of a certain organ or cell type opens a "developmental niche" in the embryo, and that niche can be filled only by foreign cells. This idea was used to create chimeric mice with organs such as kidneys[1019] or lungs[1020], consisting of cells from other mice. In the case of the kidney, a regulatory gene had to be turned off in recipient embryos, and two genes had to be deactivated in the case of lungs.

In 2010, there was a major development in the blastocyst-complementation approach, as a group of researchers at the University of Tokyo used a similar technique to create a rat pancreas inside a mouse[1021]. This time the chimera was made from cells of two different species. The mouse was genetically designed to lack Pdx1, a regulatory protein that plays a critical role in pancreatic development. Mice lacking both copies of the corresponding gene do not have a pancreas, and they die soon after birth. However, if blastocysts of these mice are complemented by rat pluripotent stem cells, functional pancreases are formed, derived predominantly from cells from the second species.

The best part is that, thanks to Shinya Yamanaka's work on cellular reprogramming, any living adult animal, including a human, can become an endless source of pluripotent stem cells. An amazing consequent idea is that cells derived from an individual who needs transplantation can be added to an animal blastocyst, and that animal will grow not just a human organ, but one that is genetically tailored for the recipient. Furthermore, no human embryos will be created or wasted in the process.

Of course, you can't grow human organs inside a mouse, due to the extreme difference in the animals' sizes. But pig organs are sized similarly to ours. In fact, pig heart valves have been used for human transplantation since the 1960s[1022]. The main problem with such xenotransplantations is immune rejection. In the case of heart valves the issue isn't critical; various treatments and techniques, such as pig cell removal, can improve the success rates of transplants, which can remain functional for many years[1023]. But the problem persists for larger organs that rely on functioning cells. One solution is through the creation of human-pig chimeras.

The first such chimera was reported in the journal *Cell* in 2017[1024]. The authors used Yamanaka factors to reprogram human foreskin fibroblasts into an induced pluripotent state. They then added these cells to a pig blastocyst and showed that some human cells integrated and differentiated inside the resulting pig embryo. Unfortunately, the level of chimerism was much lower than what was achieved between rats and mice, perhaps because of the greater evolutionary distance between humans and pigs. The same group later published an article in which they reported creating chimeric monkey embryos with human cells[1024], but they were still far from creating a transplantable human organ inside another animal.

There are several problems that need to be solved before blastocyst complementation can save any human lives. Let's say we somehow manage to create a pig with the Pdx1 mutation and it grows a pancreas consisting of human cells. The vasculature of this organ will still be porcine, which may lead to organ rejection after transplantation. A recent breakthrough revealed a possible solution[1025]. A gene called *ETV2* is a master regulator of blood-vessel development. A mutation that turns this gene off is lethal in mice and pigs because embryos harboring it do not develop vasculature. However, blastocyst complementation can restore vessel formation. This means that it is theoretically possible to create a pig that has both a human organ and human blood vessels. The authors of this study also found that the low level of human-pig chimerization is partially due to human cells' undergoing programmed cell death in a pig environment. By artificially enhancing the activity of the anti-apoptotic gene *BCL2* in human cells, it was possible to increase their number in a pig embryo fivefold.

The outcomes of the competition between human and host cells in chimeric embryos can perhaps be improved by other approaches[1026]. For example, human cells with a deficiency in apoptosis-promoting p53 produced higher rates of human-pig chimerism in a study that attempted to grow pigs with human skeletal muscles[1027]. Another idea is to grow organs in pigs with compromised immune systems, to reduce the targeting of human antigens[1028].

The most effective way to create human organs in animals is still unknown. Perhaps it will be as simple as turning a couple of genes on or off, and we need only to figure out which ones. The case of Yamanaka factors has revealed the tremendous rejuvenative power of just a few regulatory genetic components. It is also possible that the solution will require multiple complex interventions. Whoever finds the answer will save millions of human lives.

While having a renewable source of actual human organs would be an ideal solution, some animal organs are almost as good. In 2022, the story of the first successful pig-to-human heart transplant made headlines[1029-1031]. The organ was of animal origin, but the pig donor was genetically modified to lack several genes that are known to provoke acute immune response. It also contained several human genes chosen to help the heart escape human immune-system recognition and to reduce coagulation and inflammation. The scientists who performed the transplantation also disabled the donor pig's growth hormone receptor to prevent excessive heart growth[1032]. Pig hearts grow continuously, even in adult animals, and this can create a problem for humans, whose chests don't have enough space for ever-growing pig hearts.

Although sophisticated genetic engineering was able to solve the problem of acute organ rejection, the patient died two months after transplantation. The second terminally ill patient who received an experimental pig heart experienced great improvement during the first month of recovery, and even started working toward regaining the ability to walk, but died six weeks after transplantation. The causes of death are not entirely clear; the heart failures did not have typical characteristics of transplant rejection. Maybe the hearts failed because the patients were not in good health before the experimental procedures and were receiving multiple medications. In the case of the first patient, the pig heart was found to be infected with a porcine cytomegalovirus, which could have damaged the patient's organs and contributed to his death.

Humanity has never been so close to achieving an infinite source of transplantable organs, although it may take time for us to make a pig heart completely suitable and safe for transplantation. Scientists are experimenting with genes that can be turned on and off[1033] in donor animals, but testing these solutions poses difficulties. The only experiments so far performed on living humans involved severely ill patients who were not eligible for regular human heart transplants. Several studies involved the transfer of genetically modified pig kidneys into brain-dead human recipients on life support. No acute rejection was observed within several days of the transplantations[1034,1035].

Here is where the news gets good. Studies on baboons have shown that, with the help of pharmaceuticals, a genetically modified pig's heart[1036] and kidney[1037] can remain in a primate for up to 945 and 260 days respectively. In rhesus monkeys the survival time of pig kidneys has surpassed one year[1038]. Perhaps the issue lay with the poor condition of human recipients rather than the organs themselves?

One concern regarding the use of pig organs for xenotransplantation has to do with certain transmissible pig endogenous retroviruses (ERVs). Through husbandry practices and vaccination, it is relatively easy to create pig lines that are free of regular viruses, bacteria, and other pathogens. The ERVs in question are an exception because they are already integrated into the DNA of all domesticated pigs. Pig cells can release these porcine endogenous retroviruses and infect different species, including humans—although this is not typically observed during preclinical and clinical xenotransplantation trials[1039].

To ensure complete safety, genetic engineering once again comes to the rescue. In 2017, a group of scientists including George Church used a genetic editing tool called CRISPR-Cas9, a sort of DNA-cutting molecular scissors, to disable the endogenous viruses in a line of pig cells[1040]. They then used a cloning technique involving somatic-nucleus transfer to create a virus-deactivated line of pigs. The best thing about these and any other genetically modified animals is that once they have been generated, you can easily produce them on any scale through simple biological reproduction to fit any demand. The monetary price of organs can theoretically be brought down to that of pork chops. You can also upgrade the animals with new genetic modifications until an ultimate donor is created.

The CRISPR-Cas9 gene-editing tool deserves a special explanation. Molecular biologists are known to "borrow" nature's inventions with no regard for copyright. We have stolen fluorescent proteins and their associated genes from jellyfish to visualize the cells and tissues of model organisms[1041]. The photoreceptors of unicellular green algae called channelrhodopsins have been used for a technology that allows researchers to optically control the neural activity[866] of genetically modified animals and even alter their memories[871]. A wide range of viruses are now serving humankind as DNA-delivery tools that allow us to treat genetic diseases[1042] and to produce vaccines against pathogens such as SARS-CoV-2[1043], the virus that causes COVID-19.

CRISPR-Cas9 is an ingenious natural invention. Bacteria suffer infection from viruses called bacteriophages. Like other viruses, bacteriophages hijack their hosts to produce copies of themselves, often leading to the destruction of the infected cells. In order to adapt, bacteria evolved a mechanism that is reminiscent of computer antivirus software with a preinstalled system for receiving regular updates. Bacteria proactively incorporate small parts of invading viral genomes into a special region of their own circular DNA called the CRISPR cassette[1044]. The CRISPR cassette is like a database that stores small, harmless fragments of viral genetic material. This database allows the bacteria to access "memories" of previous infections of their ancestors and "record" information on new infections that can be passed on to their descendants.

Bacteria can then produce RNA copies of these viral DNA fragments that can guide the Cas9 enzyme toward recognition and cleavage of full-size matching viral DNA[1045]. This is similar to how police use fingerprints to identify criminals. The CRISPR database update procedure reminds me of the wonderful but unrealistic computer game *Assassin's Creed*, the protagonist of which, Desmond Miles, has access to the memories of his ancestors encoded in his DNA. Memories of human experiences are stored in the neural connections of our brains, not in our DNA, so this wouldn't work in the real world. But if Desmond were a bacterium fighting against the Order of Bacteriophages, access to ancestral "memories" would make slightly more sense.

CHAPTER 12

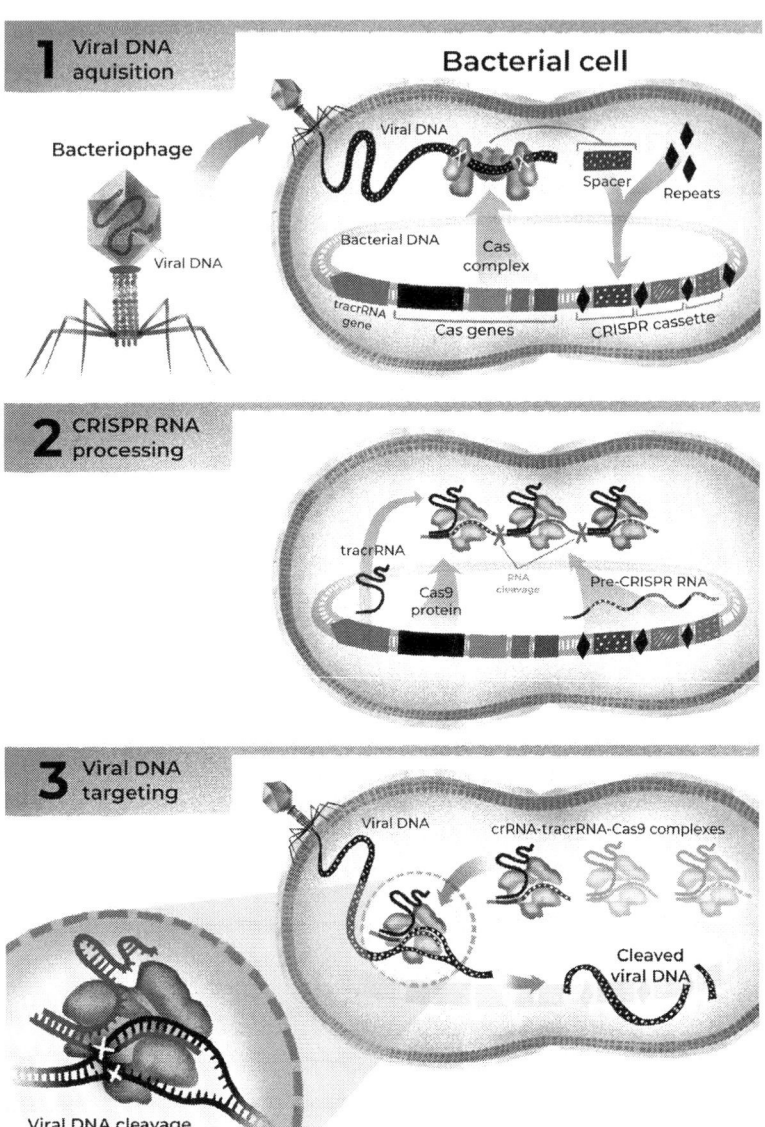

One great thing about Cas9 is that this pair of molecular scissors can be programmed to cut almost any DNA target. By providing specially designed alternative guide RNAs, you can tell Cas9 to make a surgical incision in a specified DNA sequence in the genome of a plant, a human, or another organism instead of a bacteriophage. Cas9 can be used to excise integrated HIV DNA from the genomes of infected human cells[1046] or to protect a plant from a plant-specific viral disease[1047]. The applications are almost limitless.

It is not surprising that Cas9 was used to create the first genetically modified human babies—a controversial event that resulted in the imprisonment of the primary researcher, He Jiankui, in China[1048]. The biological part of the controversy has to do with the fact that Cas9 has the capacity to induce mutations at sites other than intended ones[1049], a problem that researchers are currently trying to address by genetically modifying this instrument of genetic modification[1050,1051]. There was concern that something negative could happen to the modified babies if the genetic editing tool Jiankui used was not perfect.

Treating humans is not the same as performing animal experiments, which you can redo as many times as needed to get desired results. Another issue was that, through genetic modification, Jiankui tried to induce resistance to HIV, which is a good but unnecessary trait to have. The rationale for the experimental intervention was that the experimental babies had an HIV-positive father. But their mother was HIV negative, and simple extracorporeal fertilization is sufficient for virus-free children to be born, so there was no real medical purpose to outweigh the risks.

Endogenous retroviruses are present not just in pigs but in many mammals, including humans. In fact, roughly eight percent of the human genome consists of DNA sequences that are remnants of retroviral infections that occurred over the past millions of years. While most ERVs and their individual components are inactive, some have evolved to have beneficial functions, while others have been implicated in diseases, and even in aging.

One important ERV-derived gene is the one that encodes syncytin, a protein that once formed a retroviral envelope. Syncytin was previously used by a virus to fuse with host cells in our ancestors, but now it is used to facilitate fusion between cells during the formation of the mammalian placenta[1052]. Placental mammals would not exist if not for this ancient viral infection. Another example is Arc, a protein that is very similar to its retroviral counterparts that form the inner shells of infectious particles that encapsulate viral RNAs. Today, Arc transfers RNAs from one neuron to another[1053]. This mysterious function is important in the formation of long-term memories[1054,1055], presumably due to Arc's effects on synaptic plasticity—the ability of neurons to modify the strength of their connections. In our fear of GMOs, humanity often forgets that we are all transgenic organisms, with functional genes of viral origin incorporated into our genomes, the lucky survivors of nature's vast genetic experiments.

Not all of nature's genetic experiments have been kind to us. An example of a still-dangerous ERV is the comparatively recently acquired HERV-K, whose increased activity is linked to the progression of multiple cancers[1056,1057]. In adult cells HERV-K is normally silenced by epigenetic mechanisms, although it becomes active during some stages of embryogenesis[1058]. In 2023, the journal *Cell* published an article revealing that, as our cells age, the epigenetic shackles that normally repress HERV-K start to falter. The awakening of HERV-K can lead to cellular senescence and triggers an immune-system response. With the help of a modified version of CRISPR-Cas9, the authors could activate or deactivate HERV-K. Increased HERV-K activity induced senescence in cultivated human cells, while HERV-K deactivation restored their ability to proliferate and improved a number of aging biomarkers. This means that gene therapy that targets HERV-K might be an effective anti-aging approach. But there is an even better idea for the future. Just as we can create pig organs without pig ERVs, we could create human organs without some human ERVs. This could result in organs that would be less susceptible to aging. Other genetic improvements of human organs are also possible.

Gene-editing tools such as CRISPR-Cas9 are already widely used on human cells in an area of biotechnology that hints at a possible future in which the creation of human organs won't require an entire organism. Vertebrate cells have an immense capacity for self-organization, which has allowed scientists to manufacture organoids—miniature self-assembling versions of organs. These three-dimensional structures consist of multiple spatially ordered cell types and can perform some or all of the original organs' functions[1059].

The "mini-brain" was among the earliest organoids. The first specimen was merely patterned brain cortical tissue that contained functional neurons derived from mouse embryonic stem cells[1060]. In 2014, a proper, complex human "mini-brain" was created[1059]. The researchers who created it started with human induced pluripotent stem cells that grew into three-dimensional aggregates called embryoid bodies in a medium[1061]. These structures were subjected to another medium, which caused undifferentiated cells to turn into neural progenitor cells—cells that form the neuroectoderm—recapitulating the first steps of nervous-system development. These tissues were placed in droplets of Matrigel[1062], a substrate for culturing cells that is similar in composition to the extracellular matrix and contains structural proteins such as collagen and laminin.

The growing organoid absorbed nutrients in a spinning bioreactor and developed into a complex brain-like structure that survived for months. Parts of the structure resembled the cerebral cortex; others were reminiscent of the liquid-filled inner ventricles. Even an immature retina and the protective membranes that form between the brain and the bone of the skull were observed. So was neural activity.

The researchers didn't stop there. There is a medical condition called microcephaly, which occurs when children are born with smaller-than-normal heads and brains, often resulting in intellectual disability. The causes can be genetic or related to the fetus's being exposed to toxic substances, such as alcohol, or to certain infections, such as the Zika virus[1063]. The researchers identified an individual with a severe genetically attributed version of this disorder. From that person's skin fibroblast cells, they created several lines of induced pluripotent stem cells harboring the same pathological mutations and used them for "mini-brain" development. This research found that patient-derived organoids displayed premature neural differentiation when compared with normal brain organoids.

Thus, organoids turned out to be an extremely useful tool in investigating the root causes of human diseases. Infection of human "mini-brains" by the Zika virus later helped uncover some of the molecular mechanisms through which the pathogen compromises brain development[1064,1065]. Human airway organoids have become models to assess the infectivity of respiratory viruses[1066]. "Mini-organs" can also be used to cultivate and study specialized pathogenic bacteria that are difficult to grow in culture[1067]. Meanwhile, CRISPR-Cas9 and other gene-editing tools have been used to induce targeted mutations in lines of pluripotent stem cells in order to study the properties of derived organoids, greatly expanding our knowledge of human gene functions[1068].

It is beyond doubt that human organoids will play an important role in the future of individualized medicine. Drug tests performed on organoids grown from genetically distinct patients can reveal treatments that benefit some people more than others. For example, the drug responses of rectal organoids have been found to predict corresponding therapeutic endpoints for patients with cystic fibrosis[1069]. Likewise, patient-derived tumor organoids can serve as models to identify the best available anticancer treatment options for an individual[1070] and spare him the adverse effects of drugs that are inefficient in his case.

To date, protocols have been established for the creation of organoid types including brain[1071], lung[1072], liver[1073], kidney[1074], pancreas[1075], intestine[1076], bladder[1077] ... and the list goes on[1078]. While the predominant use of organoids is to study organismal development, genetic disorders, and medical treatments (slowly but steadily reducing our reliance on animal studies), some organoids have turned out to be transplantable.

For example, human blood-vessel organoids derived from pluripotent stem cells can form capillaries, arteries, arterioles, and venules when inserted into immunodeficient mice[1079]. These organoids grow and survive for months, have an endothelium that is ninety percent human cells, and establish functional vascular trees in recipient animals. Such chimeric mice were originally created to study the vascular effects of antidiabetic drugs, but one can envision multiple future uses of this technology in tissue design and regeneration.

In one study, intestinal organoids were made up of mouse cells and injected into the damaged colons of other mice via rectal infusion. Some details of this procedure were unpleasant, such as, when "the anal verge [was] sealed immediately with Histoacryl glue to keep the organoid fragments inside the lumen." However, the glue was removed three hours later, and within several weeks, parts of the recipient's damaged colon were rejuvenated by the added cellular mini-structures[1080].

Recently, a group of researchers managed to transfer self-assembled retina-like structures derived from human stem cells into the eyes of immunodeficient rats with a genetic disorder that causes retinal degeneration and blindness. The organoids survived for more than seven months and even developed photoreceptors. This greatly improved the recipients' visual acuity compared with control animals that received sham surgery or were not operated on[1081]. Human studies of similar approaches are ongoing[1082].

Transplantable kidney organoids are also on the way. These mini organs have been shown to survive after transplantation, and to acquire blood vessels with the help of host animal cells. However, the organoids remain insufficiently mature to be of therapeutic use[1083]—a matter of time and additional research.

One crucial obstacle to organoid transplantation, aside from structural difference from normal organs, is organoid size. Recently, a most impressive self-assembling human heart was created from pluripotent stem cells. The organoid could beat and had many structural features of a developing human heart. However, it was only about 1.5 millimeters in diameter. Despite such a limitation, organoids have many uses, and over time we might find just the right conditions to cultivate them into mature, properly sized, usable organs[1084].

Tissue engineering is the fourth and final fantasy in the field of biological-organ manufacturing. Perhaps the most famous example is the creation of artificial human bladders[1085]. Urothelial cells (cells that line the urinary tract) and smooth-muscle cells were obtained from patients and cultivated in a growth medium. Then bladder-shaped scaffolds of collagen and a biodegradable polymer were created specifically for the patients. Via a sterile pipette, lab-grown smooth-muscle cells were seeded on the exterior surfaces of the scaffolds, and they were incubated with the addition of a growth medium. Two days later, urothelial cells were added in a similar way, coating the inside layers of the scaffolds. For some patients the bladders were wrapped in omentum, adipose tissue that enhances regeneration[1086]. Finally, glue made from proteins involved in blood coagulation was used to coat the external layers of the organs. This biological glue was created by combining a human enzyme called thrombin and its substrate—soluble fibrinogen, which is normally converted into insoluble polymerized strands that form blood clots[1087].

The only thing missing was bladder innervation[1088]. The bodies of neurons that innervate and regulate the bladder are located in the spinal cord and in ganglia outside of the organ, and they must grow new connections after transplantation. As the conditions for nerve regrowth are still unclear, the lack of proper innervation remains one of the major challenges for this and many other types of tissue engineering[1089,1090].

Fortunately, not all organs require innervation. Recently, tissue engineering was used to construct small, functional human thymuses. The authors of the study took murine thymuses and removed their cellular components to create thymic scaffolds. They were populated with thymic epithelium progenitors differentiated from induced human pluripotent stem cells. These were added in combination with human hematopoietic progenitor cells (HPCs) from human umbilical-cord blood. HPCs can develop into immune T cells, but they require the thymus's epithelial cells to mature. The artificial thymuses were inserted into immunodeficient mice and allowed to produce diverse populations of human T cells. Thus, the artificial thymuses were functional[1091]. This study complements previous work on mice, in which researchers reprogrammed embryonic cells into thymic epithelium cells by enhancing the activity of the *FoxN1* gene. When these cells were transplanted into a host, a functional thymus was established[1092].

The automation of tissue engineering is made possible by technology known as layer-by-layer 3D bioprinting, which utilizes hardware and software similar to those used in regular plastic 3D printing but using special bio-inks that consist of human cells combined with hydrogels derived from cell-free extracellular matrix. In 2019, a small human heart was made from two types of such bio-ink[1093]. One contained heart-muscle cells, while the other had endothelial cells for blood-vessel formation. The cells for both bio-inks were created by reprogramming the patient's own fat-tissue cells. Once again, no embryos or donors were required.

Researchers used computerized tomography to make a digital model of the patient's heart and a 3D printer to create a small patch of vascularized heart tissue matching a part of the heart's left ventricle. The patch was immunologically, biochemically, and anatomically perfect for the patient. The creation of a complete heart proved to be much more difficult, however. For one thing, the weight of a 3D organ is too great for existing hydrogels to sustain. The solution was to print the organ inside, basically, a transparent supportive Jell-O made from seaweed polysaccharide salts called alginates supplemented with a growth medium. Still, the printed human heart was very small (twenty millimeters by fourteen millimeters).

The size limitations of printed organs are defined by the number of cells that can be grown in culture, but also by the diffusion limits of nutrients that can reach the deeper parts of the synthesized organ. To overcome these limitations, 3D bioprinting solutions are being used in attempts to incorporate microchannels into biological constructions[1094]. Unfortunately, vascularization at the single-cell level remains a challenge[1095].

There has been an unexpected development in this field. One potential solution for the organ-vascularization problem was hidden in plain sight and is based on the unique properties of the parathyroid glands. Our parathyroid glands produce the parathyroid hormone, which is essential for regulating calcium levels in our blood. Without it we would suffer from hypocalcemia, which can lead to muscle spasms, seizures, numbness, and even cardiac arrest. These small glands are situated on top of the thyroid gland, which sometimes requires surgery due to overactive production of hormones, such as in the case of thyroid cancer. The thyroid gland can be substituted with lifelong thyroid hormone medication, but losing all four parathyroid glands can lead to severe complications. To preserve parathyroid function during thyroid surgery, surgeons have been practicing a technique called parathyroid autotransplantation. This involves transferring a patient's parathyroid gland cells into his or her muscles[1096], such as those of the neck or the forearm. Transferring onto the latter is especially attractive because doing so is a simple operation and allows comparing hormone levels between the two arms to ensure that the transplant is working correctly[1097] (although it becomes especially important not to lose an arm afterward).

It turns out that parathyroid cells are very good at surviving transplantation, even into unusual locations. They do so by producing factors that promote the growth of blood vessels (such as vascular endothelial growth factor, VEGF) and other molecules that protect cells from death. This is why several research groups have investigated whether the survival of other transplants can be improved by the addition of parathyroid cells. And it has worked. For example, in animals the co-transplantation of parathyroid cells has greatly improved the survival of newly introduced insulin-producing cells[1098]. Currently a Phase I clinical trial is being conducted at the University of California at San Francisco to test whether such co-transplantation can be used for the treatment of diabetes[1099].

One might ask where we will get the necessary human parathyroid cells from, and one answer is via the blastocyst-complementation technique. Researchers have already been successful at creating rats with a parathyroid gland composed of mouse cells, indicating that soon it will be possible to recreate the human parathyroid gland in other animals[1100]. Another avenue of research is to identify the gene activity signature that enables the high survivability of parathyroid cells and to mimic it in the cells of other transplantable tissues and organs. Either way, it appears that parathyroid cells hold an unexpected regenerative secret, and perhaps our bodies can teach us other lessons.

The manufacture or growth of accessible personalized human organs is a biotechnological holy grail that could greatly shift the limits of the human lifespan and save countless lives. Many different paths have been taken to reach this goal, with various degrees of success. Some approaches are excitingly close to clinical application. Perhaps the artificial creation of new organs is where all our human technology will converge. On one hand we have our biological tools, including genetic engineering, cellular reprogramming, and blastocyst complementation; and on the other we have software and hardware that allow us to create high-fidelity 3D models of tissues and organs, and devices for blueprint-based cellular assembly. For a recent example, an open-source manual for creating a 3D bioprinter from a regular 3D printer with a total components cost of less than one thousand dollars was published[1101], making the technology much more widely available. Claims that humanity would grow organs in human clones or stoop to organ repossession couldn't have been further from the truth. But while creating new, perfectly suitable organs is an important life-saving endeavor, it would be best to keep our existing organs young and functional for as long as possible. And to do this, we might need to combine all that we've learned about aging so far.

New organs don't grow on trees—yet perhaps they should. With the power of modern genetics, who knows what we might be able to create?

CHAPTER 13
Whatever it takes

> I do whatever it takes,
> 'Cause I love how it feels when I break the chains
> —*Imagine Dragons ("Whatever It Takes")*

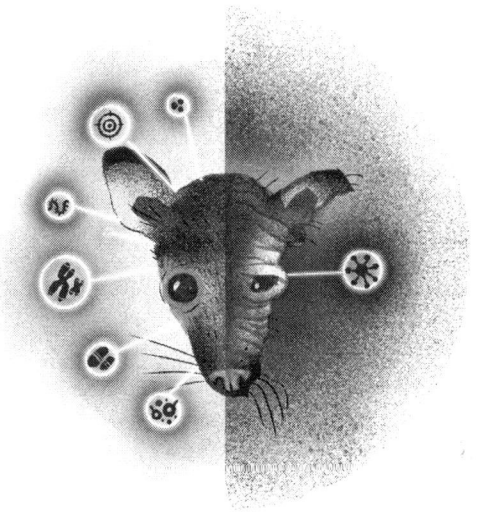

Summary: in this chapter we summarize the hallmarks of aging, the most impressive methods of lifespan extension, and why multiple anti-aging therapies can synergize when used in combination.

Imagine a scientific project code-named Saving Private Mouse, Saving Private Kitten (if you are a cat person), or perhaps Saving Private Platypus (if it pleases you). The best researchers and veterinarians from all around the globe have united for an ambitious project. They must do everything in their ability to keep a single animal alive and healthy for as long as physically possible. Unlimited resources—all the funding and knowledge in the world—are at their disposal. Any dietary or environmental change deemed beneficial will be made. Any disease will be treated with the best available science-based medicine. The most advanced anti-aging interventions, including drugs and gene therapies, will be administered. New organs will be grown and transplanted on demand in any amount. How long will the animal live compared with its untreated counterparts?

No one has ever attempted an experiment even remotely similar to this. This bothers me, because it's such an obvious idea, and because even a combination of just two interventions can increase the median and maximum lifespans of mice by up to an astonishing 69% and 61% respectively[1102], while no individual intervention has caused greater than 50% mean or maximum life extension in mammals in a properly conducted and reproducible study.

The abovementioned successful combination was of calorie restriction and a mutation in a gene called *PROP1*, which leads to abnormal development of the pituitary gland. The mutation is characteristic of Ames dwarf mice, which have deficiencies in growth hormone and several other signal molecules. They have a small body size and achieve approximately 1 to 1.25 extra years of lifespan[1103]. Initially researchers thought that the Ames mouse mutation mimicked the effects of calorie restriction. However, it turned out that calorie restriction can increase the lifespan of Ames mice even further.

The same conclusion was made when calorie restriction was applied to long-lived dwarf mice deficient in growth-hormone-releasing hormone, also known as somatoliberin (yes, there are hormones that affect the release of other hormones). Somatoliberin is produced by the hypothalamus—the part of the brain that links the nervous and endocrine systems together by regulating the pituitary gland. The already dramatically (roughly fifty percent) increased lifespan of these mice also improved further[1104]. For some reason, calorie restriction did not work so well for a third kind of long-lived dwarf mouse that was deficient in the growth-hormone receptor. In this case, only female mice benefited from food limitation, and the effect was small[1105].

Of course, I am not suggesting that we remove or damage our pituitaries (like the optic glands of the female octopuses in an earlier chapter), or that we modify human embryos for genetic abnormalities, or that these approaches would have the same effects on people as on animal test subjects. But we now know that anti-aging interventions can sometimes act synergistically, improving one another when conditions are right. And while there are remarkably few studies on combined anti-aging interventions in mammals, impressive demonstrations of synergy can be found in experiments on simpler organisms[1106].

The lifespan genetics of *Caenorhabditis elegans* is probably the most studied in the world. The SynergyAge database contains a catalog of longevity effects of more than 1,500 mutation combinations, of which more than one hundred have shown positive synergistic effects[1107]. For example, in 2013 a group of researchers combined two longevity-promoting mutations[1108]. The first one was in the famous *daf-2* gene, which is involved in nutrient sensing and encodes the receptor for an insulin-like growth hormone. As mentioned in earlier chapters, worms with this mutation have more-than doubled lifespan. In this study the mean lifespan was increased by 169%. The second mutation was in a gene called *rsks-1*, which encodes a protein that works downstream of TOR, the inhibitor of autophagy. The deactivation of *rsks-1* was previously shown to downregulate protein synthesis and to promote lifespan in worms, presumably by shifting cells into a state that favors maintenance and repair[1109]. The effect of this mutation was modest, resulting in only a 20% increase in longevity. However, the combination of the two mutations in a single animal resulted in a whopping 454% increase in median lifespan compared with normal worms! Some mutants survived for more than a hundred days, while the typical lifespan of *C. elegans* is only around two weeks. The combined effect was greater than what we would have expected by simply adding the numbers for individual genes.

The authors of this study also studied a combination of the *daf-2* mutation with a pharmacological TOR-inhibiting intervention, the drug rapamycin. Rapamycin added 26% to the lifespan of normal worms and 45% to that of already-long-lived *daf-2* mutants, in what appears to be another example of synergy. A version of dietary restriction could also be synergistically combined with the *daf-2* mutation, leading to some worms' surviving for more than 120 days[1110].

Another well-studied model organism is *Drosophila*. Here we can find several examples of prominent pharmacological interventions with synergistic lifespan–promoting effects. In one study, rapamycin, a drug called trametinib, and lithium each prolonged the lifespan of fruit flies by about 11%, but a combination of all three extended the median lifespan by 48%[1111], with some insects surviving for almost 120 days. A combination of rapamycin, two other drugs, dietary restriction (a low-protein, high-carbohydrate diet), low temperature, and photodeprivation provided an even more impressive 185-day maximum lifespan in regular flies, and 213 days in a long-lived mutant strain[1112], which is equivalent to more than 200 human years. This happened even though some interventions increased mortality when used on their own. For example, the triple-drug combination was detrimental for the long-lived mutants, and dietary restriction had a negative impact on the lifespan of regular flies. However, the best result was achieved when all interventions were combined.

A synergistic effect was also observed in a recent study on *Drosophila* during which stem-cell rejuvenation was coupled with the removal of senescent cells[1113]. Once again, two combined methods of life extension worked better than one.

It's worth highlighting that the use of Yamanaka factors to turn specialized cells into induced pluripotent stem cells, or simply to dial back the epigenetic clock of aging, is itself a combined therapy with synergistic effects between its components. The rejuvenation is achieved by the cumulative effect of three or four genes that need to be simultaneously delivered into target cells or activated by other means.

A change in a single gene's activity doesn't provide the combination's effect.

The idea of a universal, single therapy—or "magic bullet"—that can prevent or dramatically slow aging has stumbled upon the harsh reality that aging is a complex process involving multiple components. I would say that if there *is* a universal theory of aging, then it's that *there is no universal theory of aging*. "Aging" is simply a convenient way to describe various forms of damage that accumulate over a lifetime. To understand aging is to understand the various types of damage it involves—and to identify ways to reduce or repair them.

In 2013, researcher Carlos López-Otín from the University of Oviedo in Spain, along with his colleagues, published a review titled *The Hallmarks of Aging*, which has been highly influential in the field of gerontology—it has been cited more than eleven thousand times[513].

The article summarized available knowledge on our physical decline and brought forth a classification of nine hallmarks of aging, the nine "Ringwraiths" that spell our doom. If you defeat one, the others are still deadly. Each hallmark manifests itself during normal aging, and its experimental aggravation or amelioration, respectively, accelerates or slows the aging process in one way or another. By this point in this book, you should already be familiar with most of the hallmarks, but each deserves a few additional comments.

(1) Genomic instability. The DNA of our cells is subject to mutation. One of the most important causes of mutations is internal: the errors associated with DNA replication as incorrect nucleotides are incorporated into growing DNA strands due to stochastic processes and imperfections of DNA-copying enzymes[1114]. Malfunctions of DNA-repair mechanisms can exacerbate these processes and lead to premature aging, such as in the case of Werner syndrome[1115]. Not surprisingly, the rates of such mutations have an inverse relationship with mammalian lifespan[1116].

Another internal cause for mutations is the awakening of endogenous retroviruses (ERVs) that lay dormant in our

DNA. As mentioned earlier, these ancient genetic relics of viral origin comprise around eight percent of our genome[1117]. Most of them have long lost the ability to copy themselves or integrate their copies into DNA, either due to mutations or because they have been silenced by epigenetic mechanisms. However, due to cellular aging, epigenetic suppression of ERVs may erode, allowing some of these elements to escape and wreak havoc on our genes. Recently, it was shown that senescent human cells that fail to suppress ERVs can even produce virus-like particles that can transmit the senescent phenotype to normal cells[558]. Fortunately, antibodies against these particles can prevent this process.

There are other kinds of selfish genetic elements in animal genomes that may contribute to aging[1118]. In the "rise and fall of living utopias" chapter we discussed retrotransposons and their role in cancer. Molecules that inhibit these elements, such as reverse transcriptase inhibitors have shown potential to improve healthspan and lifespan in some short-lived mouse models[1119,1120] by improving genome stability.

There are also exogenous factors that damage our DNA, including some regular viruses, ionizing radiation (such as X-rays and gamma rays), ultraviolet radiation, and mutagens such as those found in tobacco smoke and automobile exhaust[1121]. While some studies have reported that minuscule exposure to DNA damage from ionizing radiation can increase the lifespan of mice[1122] and even decrease the mortality of nuclear shipyard workers[1123], perhaps by stimulating some DNA-repair mechanisms (this concept is called hormesis), subsequent research has found only insignificant effects of the smallest radiation doses, and negative effects of larger ones[1124–1126]. Contrary to many works of popular fiction, such as *Teenage Mutant Ninja Turtles* or *The Incredible Hulk*, reptiles, people, and other animals exposed to radioactive material acquire not superpowers, ninja abilities, or a hunger for pizza, but cancer.

Mutations in nuclear and mitochondrial DNA, as well as the disruption of the nuclear lamina that is responsible for the proper three-dimensional arrangement of our chromosomes, contribute to genome instability and aging[1127].

(2) Telomere attrition. We already discussed this hallmark of aging in detail in the chapter entitled "The immortal cell." As we age, our cells undergo numerous divisions, and the ends of our chromosomes shorten. This can lead to cellular senescence, reduced cellular renewal, and genomic instability[1128]. The enzyme telomerase can counteract this process. Aside from the mentioned studies in which the lifespan of mice was increased by up to 41.4 percent via telomerase reverse transcriptase activation, there is additional evidence linking aging with telomere attrition.

In a 2019 study published in *Aging Cell*, researchers compared two groups of centenarians. One aged much better than the other, as evidenced by better cognitive and physical performance and a smaller burden of disease. The healthier group had longer average telomere length, a lower prevalence of critically short telomeres, and much greater telomerase activity in immune T cells[725]. Other studies have found that shorter telomere length in immune cells is associated with increased risk of critical illness due to certain infections, such as COVID-19[1129]. Also, a high frequency of cells with extremely short telomeres was found to be a marker of other health problems, such as infertility[1130].

(3) Epigenetic alterations. We have already stressed the importance of proper gene activity regulation via epigenetic mechanisms such as DNA methylation and histone modification. As an organism ages, this control becomes loosened; genes that should be silenced can become activated, and vice versa. Measurements of accumulated epigenetic changes can be used to estimate an organism's age rather accurately[1130], and Yamanaka factors can rejuvenate cells through epigenetic mechanisms.

Several kinds of histone modification can greatly affect gene activity. Two especially well studied ones are the attachment and the removal of –COCH3 groups; these processes are called acetylation and deacetylation. Acetylation of histones leads to the unwinding of coupled DNA, making it more accessible to enzymes that transcribe genetically stored messages into RNA, promoting the synthesis of locally encoded proteins. Deacetylation of histones has the opposite effect. It's like opening and closing specific pages of a cookbook, making the recipes for certain dishes more visible and available.

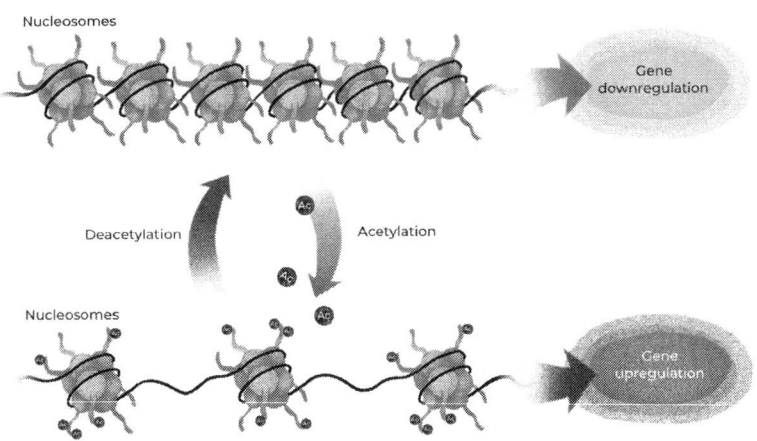

An enzyme called SIRT6 is known to regulate the activity of various genes through the epigenetic mechanism of histone deacetylation. As a matter of fact, it participates in DNA repair[1131,1132] and appears to play an important role in turning off some of the transposable genetic elements that cause genomic instability[1133]. SIRT6-deficient mice develop a number of abnormalities and die in three to four weeks[1134], while male mice with artificially enhanced SIRT6 activity reportedly have longer-than-normal lifespans[1135]. In mammals, SIRT6 presumably works downstream of the longevity-promoting gene *FOXO3a*[1136], which may explain some of the beneficial effects of the latter gene's activation. Enhanced SIRT6 activity has also been shown to improve telomerase function[1137], and more efficient SIRT6 proteins are typically found in longer-lived species[1131]. Meanwhile nicotine suppresses SIRT6 and accelerates senescence of some cell types. This includes spermatogonial stem cells that produce sperm[1138], which might explain some of the negative reproductive consequences of smoking[1139].

A comparison between extremely long-lived people and normally aging people with no long-lived relatives identified a rare SIRT6 variant associated with an increased chance of becoming a centenarian, and greater ability to repress transposable elements and stimulate DNA repair[1140]. A British biotech start-up called Genflow Biosciences is currently developing a gene therapy against aging that will consist of viral vectors to deliver this newly discovered longevity-associated SIRT6 variant to mammalian cells.

The example of SIRT6 shows how hallmarks of aging can be tightly interconnected: nutrient sensing can affect *FOXO3a* activity, which can cause epigenetic changes via SIRT6, which in turn is relevant for both telomere attrition and genomic instability.

(4) Loss of proteostasis. Our cells contain thousands of proteins that need to be synthesized in the right places and in the right amounts, correctly folded into complex three-dimensional structures, transported to target destinations, maintained in their functional conformations, and degraded when they become damaged or unneeded[1141,1142]. Proteostasis is the dynamic regulation of all these processes in order to preserve the functionality of cells, tissues, and the entire organism. By no means is this simple. A protein's folding starts as soon as the first stretches of its amino acids emerge from ribosomes during synthesis and start forming complex three-dimensional structures. Protein-producing molecular factories sometimes need to slow or temporarily halt production in order to more the time for certain steps of this process to finish correctly[1143].

Special proteins called chaperones assist protein folding and prevent harmful aggregation. Chaperones' molecular sizes, structures, and exact functions can vary. Some large chaperones are shaped like barrels and can incorporate and seal off any unfolded or misfolded proteins in their cavities.

Sometimes these barrels even have molecular lids! This allows a protein to "brew" in an environment that is safe from outside molecules that could interfere with the formation of its final functional structure. This becomes especially important under conditions of increased temperature; hence, some chaperones are known as "heat shock proteins," or HSPs. The production of HSPs in our cells can increase not only in response to temperature changes, but also due to infections, inflammation, exposure to harmful substances, and other factors that can negatively affect proteostasis[1144]. Chaperone production also occurs in nonstressful conditions; some participate in protein degradation by marking proteins for autophagy or sorting them to enter proteasomes—the cells' main "protein shredders[1145]."

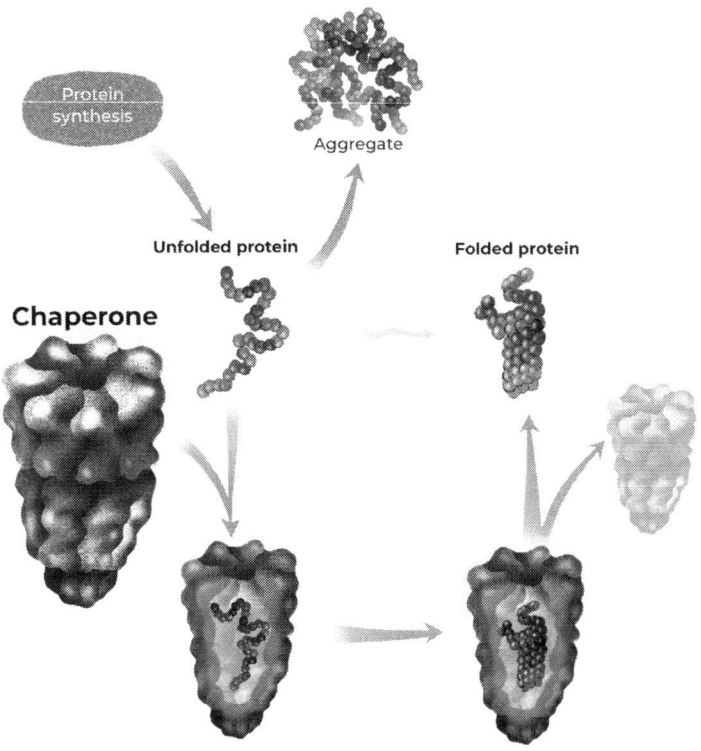

Synthesis of chaperones is controlled by other proteins, among which I've decided to highlight heat shock factor one (HSF-1), a master regulator of heat shock response that promotes heat shock protein synthesis. Increased activity of HSF-1 significantly increases the lifespans of studied model organisms such as *C. elegans* (by up to 40%), while its deactivation reduces lifespan[1146]. Mice lacking HSF-1 are around twenty percent quicker to die when infected with prions—the aggregating misfolded disease-causing proteins[1147]. Meanwhile, there is accumulating evidence that HSF-1 plays a critical role in the progression of other diseases related to protein aggregation, such as Huntington's disease[1147]. It can also protect cellular components from oxidative damage[1148].

It has also turned out that one of the most studied human longevity-associated mutations in the *FOXO3a* gene enhances the ability of HSF-1 to promote *FOXO3a* activity in response to stress factors[1149]. While HSF-1 has other functions beyond promoting the synthesis of heat shock proteins[1150], there is no doubt that aging is associated with the accumulation of misfolded or damaged proteins and that chaperones provide an important mechanism of counteraction[1151].

(5) Deregulated nutrient sensing. We have already discussed several important molecular pathways that are regulated by nutrient intake and affect aging, such as the insulin-like growth hormone (IGF) pathway and the TOR pathway[1152,1153]. Calorie restriction, methionine restriction, TOR inhibitors, mutations leading to reduced IGF-1 signaling—all of these interventions can counteract some aspects of this hallmark of aging in model organisms. Another example is the liver-derived fibroblast growth factor FGF-21, which is produced in response to fasting. Genetically modified mice that make more of this hormone live longer[707,708].

There is a simpler example of why nutrient sensing is so important. It is well-known that excess energy intake with too little energy expenditure leads to visceral obesity, which in turn causes low-grade chronic inflammation and increased risk of developing insulin resistance, diabetes, and vascular diseases[1154,1155]. This is because our fat tissues not only store fatty acids but also control our lipid and glucose metabolism and produce hormones and signal molecules, including inflammatory cytokines such as interleukin-6 that affect other cells of our body. Too much or too little of these signals can lead to undesirable health consequences. One would think that keeping a normal body mass index (BMI) would solve the problem, but it turns out that aging is a comparable risk factor for type 2 diabetes. Older organisms suffer from impaired nutrient sensing and intercellular communication, and as a result, a seventy-year-old person with a normal BMI has approximately the same risk of developing type 2 diabetes as a thirty-year-old with a BMI at the high end of the spectrum[1156].

But there is another side to this story. Aside from greater insulin resistance, aging is associated with lower levels of growth hormone and IGF-1. From this perspective, some pro-longevity changes in nutrient sensing actually occur with age[137,1157].

(6) Mitochondrial dysfunction. As we already know, mitochondria are the main energy generators of eukaryotic cells. They are responsible for respiration—the use of oxygen to convert macronutrients into energy, which is then stored in the chemical form of ATP. ATP is then used by our cells' enzymes and molecular machines as a sort of currency. But mitochondria also act as regulators of various cellular processes, including programmed cell death, and play an important role in several aging-related disorders, such as Parkinson's disease[1158,1159].

For a long time, one of the most prominent theories in gerontology was the free-radical theory of aging, which blamed our deterioration on the toxic reactive oxygen species (ROS) produced by mitochondria. The idea was that this form of oxygen damages all sorts of molecules, including proteins, lipids, and DNA, especially in mitochondria themselves[1160], leading to the dysfunction of these organelles and other manifestations of aging.

Consistent with this theory, mice engineered to have more mutations in their mitochondria have a shorter lifespan and suffer a number of health problems, including hair loss, osteoporosis, and reduced fertility[1160]. The lifespans of mice, flies, and a few other organisms can be increased by these organisms' having their mitochondria produce more ROS-neutralizing enzymes[1161], while mice lacking some of these enzymes have shorter lifespans[1162]. Mild inhibition of mitochondrial respiration can also improve the lifespans of various model organisms, including roundworms, flies, and mice[384,1163–1165].

The main objection to the free-radical theory of aging is that in some cases oxidants have beneficial effects on lifespan. This is because they can activate various protective genes, including *FOX* family[1166–1168]. When I was a university student there was much anticipation for a drug invented by the dean of our faculty, Vladimir Skulachev, an expert on aging and mitochondria. He designed special antioxidants that were linked to positively charged ions. This allowed them to be efficiently delivered straight into the negatively charged matrices of mitochondria[1169]. While the idea is clever, and some reports show that these ions can protect animals against certain types of stress factors[1170], independent researchers have so far been unable to show that these antioxidants cause any increase in animal lifespan under normal conditions[440].

This, of course, does not invalidate the role of mitochondria in aging, but merely indicates that mitochondrial dysfunction needs to be addressed differently. For example, through the stimulation of mitophagy—the process of removing damaged or nonfunctional mitochondria[1171]. Recently, researchers in the laboratory of Vadim Gladyshev at Harvard Medical School found that most lifespan–extending interventions in mice are associated with increased activity of genes involved in mitochondrial protein synthesis, and that this increased gene activity appears to be one of the most consistent biomarkers of mammalian longevity both across and within species[137].

(7) Cellular senescence. In the "Programmed cell aging" chapter we discussed the different types of cellular senescence, and their positive and negative sides. Although senescent cells are heterogeneous and ambiguous by definition, some of them

can provoke inflammation by producing pro-inflammatory molecules such as interleukin-6. Their endogenous retroviruses can become derepressed, and their autophagy can be compromised, while their mitochondria can overproduce reactive oxygen species.

The main argument for the causal role of cellular senescence in aging is that the introduction of senescent cells is linked to inflammation and reduced lifespan, so targeting and eliminating them via genetic and pharmacological interventions can improve the lifespan and healthspan of certain animals. One well-studied senolytic combination is quercetin plus dasatinib. While quercetin apparently does not significantly increase mouse lifespan on its own[1172], it improves the efficiency of dasatinib when the two drugs are used in combination[1173]. This is yet another example of the benefit of combining therapies.

(8) Stem cell exhaustion. Different tissues and organs have different strategies of renewal and repair. Some organs rely on preexisting specialized stem cells, while others rely on cells' giving up their specialization to increase their proliferative ability in response to injury[1174]. In any case, the body requires an influx of new multiplying cells, and aging reduces their availability[1175]. Partial or transient reprogramming of cells by Yamanaka factors has led to increased tissue repair in a number of animal models and has even increased the lifespans of rapidly aging mice. Just recently, this finding was reported in normal mice as well[531].

One interesting piece of evidence for the significance of stem-cell exhaustion comes from the analysis of the mortality rates of patients who received bone marrow transplants[1176]. Bone marrow contains two main cell types[1177]. Hematopoietic stem cells give rise to all our blood cells, while mesenchymal stem cells are precursors of skeletal, cartilage, and fat-tissue cells[1178]. It turns out that the epigenetic age of transplanted hematopoietic stem cells is inherited from the donor[1179], so a person receiving stem cells from someone younger than him can benefit from the donor's youthful potential and live longer[1176].

The observed mortality difference between recipients of young and old hematopoietic stem cells was independent of disease, recipient age, sex, and other variables. It's unclear, however, whether the young hematopoietic stem cells were beneficial or the old ones were detrimental for survival. The first interpretation is consistent with an experiment involving the transplantation of young bone marrow cells into old mice, which led to approximately 28% maximum-lifespan extension[1180]. However, upon closer examination of this study I found that the control mice were few and lived less long than expected for the strain, so perhaps the reported results are too optimistic (unfortunately, this is a common problem in animal studies). A more recent study showed a more modest 12% median lifespan extension for this procedure in mice[1181], which is still an impressive result.

The restoration of the stem cell pool can have other surprising beneficial effects. Apparently some stem cells produce extracellular vesicles through which they can communicate with surrounding senescent cells and suppress their production of pro-inflammatory molecules[1182].

(9) Altered intercellular communication. Aging is associated with changes in the production of various hormones and signal molecules. For example, as we age we suffer from increased levels of interleukin-6 and other pro-inflammatory molecules associated with poorer physical and cognitive performance[1183]. At the same time, BDNF signaling is reduced, potentially leading to reduced brain plasticity[1184].

Aged mammals also suffer from reduced vascular endothelial growth factor (VEGF) signaling. Many cells produce this growth factor to stimulate the formation of blood vessels, and aging is associated with capillary loss and reduced blood flow to tissues and organs. One way to increase the circulating amounts of VEGF is via gene therapy. This intervention has been found to increase the median lifespan of both male and female mice by a staggering 49%[1185]!

But this solution is perhaps imperfect. The problem with VEGF comes not from its decreased production with age, but from increased production of an inactive soluble form of its receptor. Soluble receptors bind VEGF and prevent it from targeting its normal membrane-bound receptors in responding cells. This appears to be a regulatory mechanism to prevent excessive VEGF signaling; VEGF sometimes does its job too well. For example, the progression of Alzheimer's disease was recently linked to increased production of soluble VEGF receptors and reduced cerebral blood flow[1186]. This could explain why anti-VEGF injections that are used to treat other conditions, such as neovascular macular degeneration, are associated with mild cognitive impairments in humans[1187].

There is another dark side to VEGF signaling. Tumor cells sometimes "learn" to produce VEGF to increase their blood supply and their ability to grow[1188]. Hence, some anticancer drugs aim to reduce VEGF signaling[1189]. Unfortunately, tumors can become resistant to these interventions[1190], pushing researchers to test ever more complex combinations of anticancer drugs.

The original nine hallmarks of aging were published in 2013, and approximately ten years later they were expanded with an additional three in a new article coauthored by Carlos Lopez-Otin[1191]. The added hallmarks also deserve our attention.

(10) Chronic inflammation. This hallmark, also known as "inflammaging," appears to be one of the strongest predictors of death. The term was coined by professor Claudio Franceschi in 2000[1192]. Studies involving supercentenarians (people aged 110 years or more) and other extremely aged groups reveal a strong association between mortality[1193] and levels of inflammation markers[1194,1195], including interleukin-6. This is not surprising given the role of inflammation in cardiovascular diseases[1196], the leading cause of human death.

The master regulator of inflammation is called nuclear factor kB, or NF-kB. Like *FOXO3a*, it's a transcription factor, a protein that can enter a cell's nucleus, bind DNA, and selectively activate genes. Unlike *FOXO3a*, NF-kB induces mostly pro-inflammatory genes, such as those encoding interleukin-6 or tumor necrosis factor alpha[1197]—a potent inducer of programmed cell death. It plays an important role in the activation of multiple immune-cell types. Unfortunately, as we age, activation of NF-kB becomes more and more pronounced contributing to the aging phenotype.

By increasing or decreasing NF-kB levels in the hypothalamus of mice via gene therapy, it is possible to shorten or lengthen, respectively, their lifespans by up to 25%[1198]. But perhaps there is a simpler solution for this hallmark of aging. In 2024 the journal *Nature* published a study that showed an up-to-25% increase in murine lifespan following the deactivation of just one pro-inflammatory molecule, interleukin-11[1199], which is related to interleukin-6[1200]. This is a good addition to our list of potential anti-aging tools.

(11) Dysbiosis. In 2004, American biotechnologist Craig Venter and his team published the results of a bold project, the sequencing (reading) of the Sargasso Sea "genome"[1201]. The Sargasso Sea is not a living organism, like the fictional ocean of Stanisław Lem's *Solaris*, but that does not prevent us from applying the methods of DNA analysis to environmental samples from the sea. You can collect seawater and extract the DNA of all the organisms found in it. Then you can indiscriminately cut that DNA into pieces with enzymes, and with the help of sequencing techniques you can analyze the nucleotide sequences of the fragments. Finally, you can attribute the DNA sequences you found to known organisms or classify them as unknown.

Craig Venter's team discovered DNA belonging to many previously unknown sea creatures, and brought this method, called metagenomics, into mainstream science. Metagenomics involves the qualitative and quantitative analysis of all microorganisms living in a sample—such as the surface of your eye or skin, or the inside of your intestine—by the amounts of their DNA[1202].

Today this method is widely used to analyze the hundreds of bacterial species residing, cooperating, competing, and sometimes exterminating one another[1203] within our guts[1204]. Thanks to metagenomics it has become much easier to analyze how microbial composition changes with age, disease status, place of residence, and eating habits. It turns out that some features of the microbiome, such as its compositional uniqueness[1205], are associated with increased longevity, while the spread of certain bacterial groups is linked to inflammation or poor survival. It is, however, difficult to determine whether human health is driven by microbial content or vice versa, or whether a third factor, such as diet, influences both human health and microbial content.

Gerontologist Ilya Mechnikov was a vocal proponent of the idea that a healthy microbiome is crucial for human longevity. He advocated consuming more dietary fiber (long before the idea became mainstream[1206]) and drinking fermented bifidobacteria-containing milk in order to achieve a healthy lifestyle. While some epidemiological studies have found lifespan benefits of fermented (but not regular) milk, in line with Mechnikov's ideas, the effect on adult human mortality appears to be relatively small[1207].

Bifidobacterium usually dominate in the guts of newborns[1208] but aren't good at long-term colonization. As we age, their relative abundance drops significantly[1209] and other microbial groups take over[1210]. Bifidobacteria have known antimicrobial properties, and there is some, although limited[1211], evidence that probiotics containing these and several other groups of commensal bacteria can reduce mortality of preterm, low-birth-weight infants[1212] and protect children from pathological microbial species[1213].

Aside from using probiotics, there is another method of altering microbiome composition. Fecal microbiota transplantation from a healthy donor to a patient is a promising treatment for several gastrointestinal disorders, including long-term colon and rectum inflammations[1214]. Thus, microbial interventions may have beneficial effects on lifespan by reducing systemic inflammation signaling[1214]. For example, transplantation of microbiota from young mice into older counterparts has resulted in increased bifidobacteria content, reduced markers of inflammation, and improved fertility[1215]. In another study, microbiota transferred to mice from long-lived humans reduced a number of aging biomarkers in comparison with a control group of animals that received microbes from typically aging people. Unfortunately, no data on lifespan changes was provided[1216].

Some researchers believe that gut microbiota may influence lifespan by affecting conditions such as obesity. The authors of one study took a pair of twins who were discordant in weight (one was lean and one obese) and transferred their microbiota to mice. The receivers of the obese twin's microbiota eventually gained significantly more adipose mass than control mice[1217].

Our microbial cohabitants can affect our bodies by producing metabolites—including signal molecules, vitamins, and essential amino acids—that our own bodies can't produce[1218], making the microbiome an interesting potential target for anti-aging interventions and other health-promoting interventions. On the other hand, germ-free mice living in sterile conditions have been found to have lifespans that are the same as or even longer than normal mice with nonpathogenic microbiomes, somewhat undermining the idea of lifespan–promoting microbes[1219,1220]. Perhaps "good" microbes are good simply because they keep the "bad ones" away through competition for a place to live. Unfortunately, even though more than a hundred years have passed since Mechnikov proposed his ideas, we are still far from understanding or even quantifying the role of the microbiome in human aging.

(12) Disabled macroautophagy. We have already discussed the importance of autophagy, and its negative regulator TOR, in aging. But there are also positive regulators of this process. A protein called ATG5 is essential for the formation of autophagosomes. It's so important that mice lacking a functional copy of the corresponding gene die within a day after birth[1221]. Genetically modified mice with an ATG5 deficiency that is limited to their neural cells suffer progressive neurodegeneration,

accompanied with accumulation of intracellular protein aggregates in their neurons[1222].

At the same time, a genetic enhancement that moderately and ubiquitously increases ATG5 production in mice provides an approximately seventeen percent increase in lifespan[1223]. To further test the relationship between autophagy and mortality, scientists created genetically modified mice in which ATG5 activity could be turned off or on via the dietary presence of a regulatory molecule called doxycycline. Such mice could be born normally, allowing researchers to observe what happened when autophagy was turned off later in life. Doxycycline-induced temporary inhibition of autophagy in adult mice led to severely increased mortality and tumor incidence, and restoration of autophagy partially reversed the damage[236].

Further evidence for the importance of this hallmark comes from a naturally occurring molecule called spermidine, which promotes autophagy and mitophagy, improves heart function, and increases mouse lifespan. Genetically modified mice lacking ATG5 in their heart cells do not benefit from the cardiac effects of spermidine[237], suggesting that improved autophagy is indeed the longevity-promoting mechanism of action for this molecule.

Aside from these classic and widely accepted hallmarks, some scientists have argued that a few more should be added to the list.

(13) Stochastic nonenzymatic modification of long-lived macromolecules. The addition of this hallmark was recently proposed by researchers Alexander Fedintsev and Alexei Moskalev[804]. They emphasized that age-related stiffening of the extracellular matrix can lead to cellular senescence, stem-cell exhaustion, blood-vessel calcification, and inflammation. These things happen due partially to activation of cellular receptors (RAGEs) that detect advanced glycation end products (AGEs). The "Fragile, but not that fragile" chapter of this book had a detailed explanation.

(14) Splicing dysregulation. This hallmark was proposed by a group of researchers following a 2022 aging meeting

in Copenhagen[1224]. Each of at least half of our genes can produce more than one type of messenger RNA and more than one type of protein[1225]. In fact, a single gene can encode a large number of different proteins. The current record holder is *Drosophila melanogaster*'s *DSCAM* gene, which can generate more than 38 thousand different protein isoforms. Note that the entire fruit fly genome contains only around 15 thousand genes. "DSCAM" is short for "Down syndrome cell adhesion molecule." In humans, this molecule is located on chromosome 21, and an extra copy of that chromosome (which comes with an extra copy of the *DSCAM* gene) causes severe impairment of neural development.

In *Drosophila*, the variability of DSCAM proteins allows each neuron to have an "individual identification tag" that helps growing nerve fibers discern self from nonself and avoid "short-circuiting[1226]." This is why the gene needs to make so many different products. This variability is achieved through a process called alternative splicing. The fruit fly's *DSCAM* gene's full-length RNA is segmented into regions called exons, which are interspaced with introns. Splicing is when the introns are cut out by a complex structure called a spliceosome, and the remaining exons are combined into a final, mature RNA, which is used for protein synthesis. Alternative splicing means that there are possible variations in what parts are cut out and kept in. The same part of an RNA can act as an intron in one scenario and as an exon in another, producing remarkable protein diversity.

The complexity of *DSCAM* is an evolutionary innovation of insects. The *DSCAM* of humans and other vertebrates has only one isoform. But splicing is universal among animals and other eukaryotes, and aging is associated with changes in its regulation[1227] (as well as dysfunctions of the RNA-synthesis machinery in general[1228]). Because of splicing dysregulation a rare isoform can become predominant, and vice versa. Previously we discussed how declining VEGF signaling leads to reduced vascularization and tissue blood flow with age. Alternative splicing is partially responsible for this. It happens that one gene encodes both the functional membrane-bound VEGF receptor and its inactive soluble form. Splicing regulation and dysregulation cause increased production of the soluble receptor that binds VEGF and prevents it from interacting with its normal membrane-bound version[1229], disrupting the formation of blood vessels.

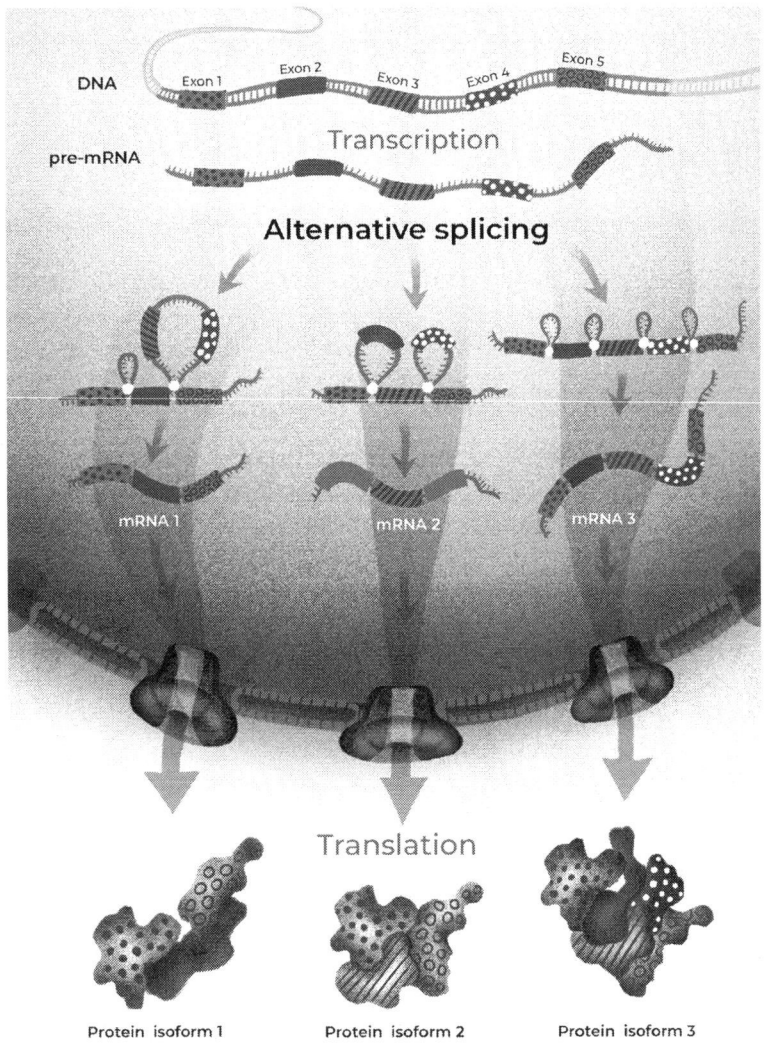

Splicing can really complicate some genetic interventions, including those that target aging. For example, if you want to increase the activity of a gene, you have to consider whether the correct splice isoforms will be produced. For example, VEGF itself has many isoforms, with different binding affinities[1230], obtained through alternative splicing, so one might ask which isoform is a better drug candidate.

Now that we have established and described the complexity of aging, we can return to this chapter's main concern: the possibility of synergistically combining multiple anti-aging interventions—preferably interventions that act upon different hallmarks of aging. So far, there has been a surprising dearth of research on the effects of combined therapies on normally aging mammals.

My colleagues and I recently published a review titled "targeting multiple hallmarks of mammalian aging with combinations of interventions[1231]." We found just over a dozen relevant studies in which lifespan changes were measured. Another dozen studies assessed aging-related pathologies or biomarkers. Several combinations were tested on mouse models of progeria-accelerated aging. Most of these combinations were tested in just a few treatments.

We noticed that most researchers stick to targeting just a few mechanisms of aging or biological processes (processes or mechanisms about which they probably feel they are most knowledgeable, or that they consider to be most important). Such studies provide valuable results but do little to address our primary concern.

Genetically modified mice with increased activity of two tumor suppressors (p53 and ARF) live longer than mice with enhanced activity of only one[1232]. Likewise, mice with increased activity of two enzymes that combat reactive oxygen species (peroxisomal catalase and superoxide dismutase) have a longer median lifespan than those with increased activity of just one such enzyme[1161]. The use of the cholesterol-lowering simvastatin and the antihypertensive drug ramipril has provided a positive effect on mouse lifespan in combination, but not as individual therapies[1233]. The TOR-inhibiting drug rapamycin has been found to work better in combination with other potential TOR inhibitors, such as phenylbutyrate, the antidiabetic drug metformin[143], and the glucose-metabolism regulator acarbose[453].

Only a few studies have broken this pattern. Telomerase activation has had a great effect in combination with tumor suppressors[503], with a net result of forty percent increased lifespan, accompanied by cancer resistance, improved physical condition, increased telomere length, and reduced frailty. In 2022, a gene-therapy combination of three longevity-associated genes (FGF21, αKlotho, and a soluble receptor for TGFβ1) protected mice against four aging-related features: obesity, type 2 diabetes, heart failure, and renal failure[1234]; unfortunately, this combination's effect on lifespan was not reported. FGF21 is involved in nutrient sensing and glucose metabolism. αKlotho has several mechanisms of action, including reduced inflammation signaling[1235]. Genetic deficiency of αKlotho causes a severe premature-aging syndrome in mice[1236], while increasing its activity provides up to thirty percent increase in mouse lifespan[1237]. The final gene in the mix encodes a soluble receptor that inhibits a protein called TGFβ1 by binding it, similarly to how soluble receptors bind VEGF and prevent it from acting. Increased TGFβ1 activity reduces blood flow to a mouse's brain and causes Alzheimer's-like cerebrovascular abnormalities[1238]. In this study, a combination of two treatments (FGF21 plus the soluble receptor) was most effective.

Fortunately, the idea of testing combined anti-aging interventions is gaining increased attention, and additional studies are underway[1106,1239]. As I was finishing this book, Aubrey de Grey—a visionary in the field of radical life extension—reported the results of a combination therapy in mice. This approach involved rapamycin, the senolytic agent Nav-Gal, hematopoietic stem cell transplantation, and *telomerase reverse transcriptase* (TERT) gene therapy, targeting at least six hallmarks of aging.

Although a peer-reviewed research paper had not been published at the time of writing, de Grey shared the results on the social platform X. In female mice, the highest maximum lifespan was observed with rapamycin and the combination therapy. Surprisingly, in males, the combination therapy did not outperform the untreated control group. Even more unexpectedly, the longevity-promoting effects of TERT gene therapy seen in previous studies were not reproduced; in fact, the therapy was associated with a reduction in male lifespan.

These inconsistent results highlight the complexity of aging biology and underscore the need for rigorous, sex-specific, and independently replicated studies to evaluate both the efficacy and potential risks of multi-targeted interventions. Aubrey de Grey is currently seeking funding for Robust Mouse Rejuvenation Study 2.

It appears that we have reached a peculiar point in history, when the amount of potential anti-aging treatments that deserve to be tested far exceeds the current practical (and financial) capabilities of the scientific community. With additional resources we could ensure that every study on every anti-aging drug, supplement, or intervention is properly conducted and independently replicated in animals, followed by human clinical trials. Meanwhile our knowledge about longevity-linked genetic variations could be translated into potential gene therapies, which could be tested alone and in combination in search of an optimal solution to the problem of aging.

In the meantime, we need to prioritize allocating resources to strategies that are most likely to work. In doing so we should consider the safety, cost, and effectiveness of interventions, the comparative difficulty of their application in adult humans, their potential adverse effects, and possible interactions between therapies. One strategy could be to combine as many treatments as possible, starting with the ones that have provided the greatest increase in mammalian lifespan, while being practically feasible. Judging from existing studies in mice, the top of the priority list should include VEGF gene therapy (49% median lifespan increase), methionine restriction

(42%), TERT gene therapy (41%), rapamycin plus acarbose (28% to 34%[453]), follistatin gene therapy (32%), interleukin-11 deactivation (20%-to-25%), and perhaps bone marrow transplantation (12% to 31%).

All of these interventions were mentioned earlier, except for follistatin gene therapy. Follistatin is a protein that promotes muscle growth and is capable of countering age-related decline in muscle mass. You might have seen photographs of incredibly lean, ultramuscular Belgian Blue cattle on the Internet. They are often presented as "GMO cows," although they are really a product of selection. Belgian Blue cows and bulls have a disruptive mutation in a gene encoding a protein called myostatin, which is a muscle-growth inhibitor. Enhanced follistatin activity has similar effects to the disruption of myostatin, but it is easier to manipulate and was shown to prolong murine lifespan. Although the FDA has not approved it, follistatin gene therapy is currently available for approximately $25,000. Until recently, individuals could access this treatment by traveling to Prospera, a special economic zone in Honduras, a Central American country known for its relaxed regulations. This is what biohacker Bryan Johnson famously did at the end of 2023. But I would advise waiting for proper clinical trials of anti-aging gene therapies to be concluded before one attempts any such treatment.

Another strategy for combined therapies could be to affect as many hallmarks of aging as possible, regardless of the resulting effects on lifespan. For example, the abovementioned combination doesn't cover cellular senescence, which can be corrected by a combination of dasatinib plus quercetin. Perhaps a molecule like alagebrium could be added to reduce the formation of advanced glycation end products, despite a lack of studies confirming its effect on longevity. Antidiabetic drugs such as acarbose could be used to reduce blood glucose levels for the same purpose. Gene therapies that increase SIRT6 or *FOXO3a* levels could additionally be tested as factors that improve genome stability. Likewise, heat shock factor 1 (HSF-1) could be activated to maintain proteostasis.

A third strategy could be to preferentially or additionally include molecules with good safety profiles that are already consumed by humans, such as caffeine or spermidine. Then we have the amino acids taurine and glycine. Taurine supplementation has increased the median lifespan of male and female mice by 10% and 12% respectively, has increased the lifespan of roundworms by 10% to 23%[229], and has been found to improve glycemic indexes in humans[1240]. Glycine has provided a small yet consistent (4% to 6%) and statistically significant increase in lifespan for both female and male mice, according to the Interventions Testing Program[224], and appears to mimic methionine restriction.

It's difficult to imagine what will happen to an animal's lifespan if all mentioned interventions are combined, including those previously unavailable, such as gene therapies. The only way to find out is through trial and error, with the help of science, as in the comic strip about two Stone Age scientists discovering the best way to solve disputes:

"Big rock is most fundamental particle in universe."
"No! Big rock is made up of small rocks."
"To Collider!"

The bad news is that even murine studies take years, and they are no substitute for human clinical trials. After all, people are not mice, and we are already benefiting from many adaptations for exceptional lifespan. Even the structure of mortality is different for mice and humans, with cancer being the primary cause of death for the former and cardiovascular diseases for the latter. Mammals display much variation in their physiology and metabolism; what's good for one can be bad for another. It is normal for a giant panda to absorb more than fifty milligrams of cyanide per day from its bamboo diet; such a dose is nearly lethal for humans[1241]. And humans safely use isoniazid to treat tuberculosis[1242], while even in small doses this antibiotic is potentially lethal for dogs, causing seizures, and gastrointestinal and cardiovascular abnormalities[1243].

Even when human clinical trials of combined anti-aging therapies have been approved and funded, it might take decades to accumulate sufficient mortality data to draw meaningful conclusions. Proposed solutions include studying animals and patients that are already extremely old at baseline, and using huge sample sizes (with dramatically increasing study costs). A third solution is to use proxy measures of aging, including biomarkers such as DNA damage accumulation, epigenetic clocks, gene-activity clocks, protein glycation indexes, measurements of glucose sensitivity, telomere shortening, cognitive and physical decline, disease burden, and other indicators of health. All of these strategies can be combined to obtain data as quickly as possible.

To ensure our survival, we need a transformative expansion in both the quantity and quality of scientific research on longevity-promoting interventions. We don't need just one Interventions Testing Program but multiple programs running parallel efforts, alongside numerous human clinical trials for every potential anti-aging therapy that clears preclinical stages. Additionally, an even larger scale of studies is essential to explore promising combinations of these interventions, including advanced gene therapies. This revolutionary shift in research priorities must begin now if we hope to benefit from its breakthroughs. To achieve this goal, we must transform how our societies view death, challenging and moving beyond the prevailing 'culture of death' that limits our progress and existence.

Humanity is waging the wrong battles. By uniting our efforts, we could strive to overcome death itself—just as we must combine therapies to address the complex hallmarks of aging. It's time to break boundaries and cross the streams!

CHAPTER 14
Cancel death culture

> Get your head out of your ass.
> Listen to the goddamn qualified scientists.
> Just look up.
> Turn off that shitbox news,
> 'Cause you're about to die soon, everybody!
>
> —*Ariana Grande & Kid Cudi ("Just Look Up")*

Summary: in this chapter we discuss anti-aging movements and organizations, what needs to be done to defeat aging, and how everyone can contribute.

In 2005, philosopher Nick Bostrom's The Fable of the Dragon-Tyrant appeared in the Journal of Medical Ethics[1244] and later as an animated film with millions of views[1245]. The fable tells of a kingdom terrorized by a dragon that devours thousands daily. No weapon can harm it, and people eventually accept their fate, surrendering loved ones year after year.

As centuries pass, scientists finally discover a material that can pierce the dragon's scales. While the king debates building a weapon, advisers argue that the dragon is a natural part of life, controlling overpopulation and creating jobs, even providing an entry to a blissful afterlife. They claim that fearing the dragon is selfish and that life's brevity brings meaning. But a young boy speaks of his lost grandmother, moving the crowd to recognize the simple truth: the dragon is a monster, and they don't have to accept its rule.

The king orders the project, but before the weapon is launched, a young man begs him to spare his father from one last sacrifice. The king can't risk exposing the plan but regrets not acting just a few days sooner. Bostrom's tale suggests that to defeat aging, our own "dragon," science alone won't suffice. We need voices like the one of this boy to challenge society's acceptance of death and to rally around a life-extension effort grander than any in history. The first dragon-piercing materials have been invented, and now it's time to forge them into a weapon—but to do so may require a life-science project that would eclipse the grandness of the Large Hadron Collider project, the Manhattan Project, and the Apollo program combined.

In the fictional book and TV series *The Three-Body Problem*, our species is trying to prepare technologically for an alien invasion that is four hundred years away; this preparation leads to some of humanity's most expensive and ambitious intellectual projects. In reality, we have a lot less time until certain death. We also don't have a benevolent leader with sufficient power and ambition to run an anti-aging research project of the required scale.

However, some influential figures and organizations *are* trying to solve the problem of aging and have their own visions of a solution. In 2021, Crown Prince Mohammed bin Salman of Saudi Arabia pledged roughly a billion dollars a year for the nonprofit Hevolution Foundation, which supports "innovation in life sciences and medicine that focuses on the biology of aging itself, rather than disease[1246]". The Hevolution Foundation has given grants to academic groups researching aging, and it has contributed 40 million dollars to the 101-million-dollar XPRIZE Healthspan competition, which incentivizes "groundbreaking therapeutics that target biological aging to improve function and extend healthy life by ten years with a target of twenty years[1247]."

Another large program that may be relevant to our goals is the Advanced Research Projects Agency for Health (ARPA-H), which the US government modeled after the Defense Advanced Research Projects Agency (DARPA) and launched in 2022. In 2023 its budget was 1.5 billion dollars, and its funding was increased for 2024. The agency's website states that "ARPA-H advances high-potential, high-impact biomedical and health research that cannot be readily accomplished through traditional research or commercial activity. ARPA-H awardees are developing entirely new ways to tackle the hardest challenges in health[1248]."

Current proposed ARPA-H projects include cellular therapies, blood-glucose monitors, vaccines that can boost immune response against malignant cancer cells, and other health-related interventions not specifically focused on aging. However, ARPA-H is recruiting program managers who will be authorized to decide which health problems to tackle, and there is no reason why such a manager shouldn't pursue aging. A simple application form is available online.

One of the most influential nongovernmental organizations in the field of aging research is the Methuselah Foundation—named after Noah's grandfather, who, according to the Hebrew Bible, lived for 969 years. A highly unlikely number. The Methuselah Foundation is a nonprofit medical charity that was established in 2001 and gives grants, investments, and prizes to promising projects and initiatives. For example, the foundation helped fund the genome sequencing of the incredibly long-lived bowhead whale[1249].

It also established the Methuselah Mouse Prize for researchers who break the world record for the longest-living mouse or develop successful rejuvenation strategies. This prize has been given to Zelton Dave Sharp, who extended the lifespan of aged mice with rapamycin; Andrzej Bartke, who created genetically modified mice that lived for 1,819 days; and Stephen Spindler, who prolonged murine lifespan with calorie restriction. An additional prize was awarded to Huber Warner, for his founding of the National Institute on Aging's Interventions Testing Program—one of the most important projects for reproducing longevity research. In 2021, the Methuselah Foundation gave one million dollars to the Albert Einstein College of Medicine for a project led by professor of neurosciences and genetics Jean M. Hebert, who is developing ways to remove damaged brain cells and tissues and replace them with precursor brain cells to reverse brain aging[1250]. In 2023, Hebert and his colleagues artificially introduced neuronal precursors into the brain of a mouse. The cells differentiated into neurons and took part in normal brain activity by forming connections with preexisting cells[1251].

The number of organizations independently funding longevity research has been steadily growing, which is a sign of increasing interest. At the moment of writing, 34 million dollars in funding for aging science and clinical trials has been deployed through Longevity Impetus Grants[1252] from Norn Group[1253]. Some funded research is fascinating. Consider a recent study published in *Nature Aging* that found that a roundworm's lifespan can be increased by a light-activated proton pump installed via genetic engineering into the worm's mitochondria. The proton pump was used to maintain potential between the inner and outer mitochondrial membranes, which is typically reduced due to aging in these organelles[1254]. This is exactly the kind of original and technologically advanced idea that the field of longevity studies requires.

Another relatively new organization in the anti-aging field is the Amaranth Foundation. Between 2021 and 2023 Amaranth reported more than thirty million dollars in funding for aging research, education, and policy change. The foundation's website lists a number of reasonable research directions, some of which have been underappreciated in conventional academic research; the website also offers recommendations for funders and allows submission of grants and proposals[1255]. Scientific directions suggested by the Amaranth Foundation include developing new end points of aging to accelerate clinical trials; generating and transplanting brain cells; immune-system replacement; biopreservation, such as vitrification of animal organs; developing new anti-aging drugs with the help of centenarian genetics data; collecting biological samples from long-lived animals; developing extracellular-matrix-remodeling molecular tools; and gaining insights from germ line immortality (such as those described in this book's chapter entitled "The immortal cell").

Other directions suggested or funded by the Amaranth Foundation include recruiting talented young people, increasing the number of ARPA-H longevity proposals, creating new stories and movies about aging, and conducting data analysis of the economic costs and benefits of defeating aging. James Fickel, the founder of the Amaranth Foundation, also funds the TIME Initiative[1256] for accelerating talent in aging biology, which gives fellowships and grants to bright young people who are interested in ridding the world of age-related diseases. It's inspiring to see how the longevity movement is connecting and educating people while creating new communities and facilitating new collaborations. Meanwhile, other organizations, such as age1[1257] and the Longevity Fund[1258], are searching for new talent and funding innovative antiaging start-ups.

The TIME Initiative was launched under the guidance of the American Federation for Aging Research (AFAR), which has funded many important anti-aging discoveries involving cellular reprogramming[1259], the elimination of senescent cells, and the prevention of protein aggregation. AFAR is also behind the Targeting Aging with Metformin (TAME) trial, with the goal of proving that aging "can be treated, just as we treat diseases." While I'm not sure that metformin is the best solution to the problem of aging, I agree that the FDA and other regulators should evaluate anti-aging treatments just as they evaluate other medical treatments. Proof that aging can be even slightly delayed would mean a great deal for our future.

In addition to these foundations and others like them, several large companies are developing anti-aging therapies. We already mentioned Altos Labs in earlier chapters. The company was established in 2022 and has raised three billion dollars from investors such as Yuri Milner. Altos Labs has focused on researching induced pluripotent stem cells to improve human healthspan.

The other most renowned anti-aging research company, Calico Life Sciences, was founded by Google in 2013. Calico's recent advancements include developing small-molecule drugs for cancer immunotherapy and for the treatment of amyotrophic lateral sclerosis (ALS), a severe neurodegenerative disease. These drugs are in phase I clinical trials[1260–1262], being tested on human subjects.

Calico's website also mentions much basic research on molecules and molecular pathways involved in aging. For example, a recent publication provided a detailed representation of the three-dimensional structure of a metalloproteinase that cleaves insulin-like growth factor (IGF) binding proteins[1263]. As you may remember, IGF plays a vital role in the nutrient-sensing hallmark of aging. Basic research is vital for the advancement of science; however, I must admit that I had greater—perhaps unrealistic—expectations of what Calico would do for the prevention of aging given the ten years of its existence, the size of its budget, and its connections to Google and Apple (the CEO of Calico is Arthur D. Levinson, who is also chairman of Apple Inc.).

Another prominent figure who is interested in longevity research is the CEO of OpenAI, Sam Altman. He recently invested 180 million dollars in a start-up called Retro Biosciences, whose mission is to add ten years of healthy human lifespan[1264]. The advisory board of Retro Biosciences includes respected aging researchers, and the team is focused on developing therapies that involve cellular reprogramming, induction of autophagy with designed molecules, and the use of blood plasma in preclinical and clinical settings, amongst other projects. According to Retro's website, the organization is currently looking for collaborators.

While establishing well-funded organizations dedicated to tackling aging and aging-related disorders is a step in the right direction, we shouldn't bet our lives on the success of one project, company, or foundation. Neither should we wait passively while someone else solves the problem. We should consider how each of us can contribute. The more plans we have, the more likely it will be that at least one plan will succeed.

We require a scientific mobilization of our societies. As more people express the opinion that aging should be stopped, more resources will be dedicated to this cause via government spending, private and corporate investments, organized communities, and other sources. If you are a scientist or medical expert, you can contribute to the anti-aging field directly by applying your skills, ideas, and knowledge. If you are an investor or philanthropist, consider funding the anti-aging field. I have named some respectable organizations, but you might create your own.

Recently, the Longevity Biotech Fellowship (LBF) conducted a "longevity bottlenecks" survey[1265] among subjects from various sectors of longevity study (researchers, entrepreneurs, investors, and so on); this survey addressed the factors that hinder advancement in radical life extension. Overall lack of funding was one of the greatest perceived hindrances, along with lack of validated biomarkers for aging and lack of public datasets. Some of the most acutely needed solutions, according to the survey, involve public outreach and changing how our culture views aging.

The final point is that you don't need to be a scientist or investor to support the global anti-aging movement. Share a book; post on social media; speak to your friends; write a story with an anti-aging message; curate a museum exhibition; produce a movie or video game or some other work of art. Use your voice and creativity. Be more political. Organize communities, inform yourself and others, and enter debates to promote scientific progress and the value of human life.

One example of public activism you could participate in is the global Say Forever campaign for life extension. The Say Forever website[1266] contains instructions on how to run surveys that can capture public attention and raise awareness of the premise of radical life extension in your city. You can also join the Longevity Biotech Fellowship community, which brings people of different backgrounds—both scientists and nonscientists—together to share, discuss, and implement ideas on radical life extension[1267].

Other notable communities include Vitalism[1268], which promotes the ideas that 120 years of life is not enough, that aging is treatable, and that research and development in this field should receive much more funding. Russian-speaking readers may consider joining the Open Longevity online community, which has similar goals and ideals. You can also subscribe to longevity-related media, such as lifespan.io[1269], to stay informed and share news with your friends, relatives, and colleagues.

Humanity, with its population of more than eight billion, has an immense amount of resources and capabilities, but only a tiny fraction of these are used for biomedical research. Just look at the typical distribution of government spending. In 2022 the United States government spent 877 billion dollars on national defense, while the entire budget of the National Institutes of Health was about 45 billion dollars—roughly five percent of the funding defense receives with only around 4 billion dollars earmarked for the National Institute on Aging, most of which is dedicated to Alzheimer's disease, not aging in general[1270]. Such a disparity can be resolved through social and political means, especially in democracies.

Consider that defeating aging would save more lives than preventing all wars, conflicts, accidents, murders, and terrorist attacks combined. It is estimated that during the twentieth century around 230 million people were killed or allowed to die by human decision in conflict[1271], which included the disastrous world wars. However, this death toll constitutes less than five percent of the 5.5 billion[1272] people who died due to other causes, mostly disease, during the same period. Most of these other causes are directly and indirectly influenced by aging. It shocks me that we dedicate twenty times less funding to problems that cause twenty times more deaths.

Another puzzling fact is that we have nothing resembling an influential social movement for radical life extension. We have human rights activists, animal rights activists, environmental movements, passionately pro-life or pro-choice social groups, and powerful religious organizations. Most people hold and express strong opinions on topics such as vaccination, human cloning, genetically modified organisms, gun control, climate change, and gender politics. But the topic of aging is barely addressed in the political realm despite being an existential threat for every human being who is currently alive or will ever live.

In 2023 a group of leading longevity scientists proposed and signed the Dublin Longevity Declaration. It gained worldwide media attention[1273], but only about 2,500 people have subsequently signed it (you can increase that number right now). Compare this number to the 6,911 people who financially supported a joke Kickstarter campaign to make potato salad. In response to the introduction of theology on the list of approved scientific specialties in Russia, I once gathered almost a thousand signatures for a satirical petition to include Pokémon studies on that list.

For some reason, we are used to the idea of dying—to such an extent that circumventing aging and death is broadly considered less important than debates about taxes, sexual preferences, or the best anime. I am not saying that other issues should be dropped. I am merely pointing out the seemingly irrational skew of our priorities, and I have come up with several hypothetical explanations for this state of affairs.

(1) The concept of radical life extension isn't nearly radical enough. It fails to provoke strong reactions, whether support or opposition, toward developing a cure for aging. Sure, people can discuss whether life extension is desirable or possible, but the debate is usually civilized, unemotional, and well-meant.

Neither side considers the other to be evil or deeply misguided. Nobody is actively sabotaging research on aging. Compare this situation to the destruction of experimental GMO crops by activists, the burning down of 5G cell towers, or the protests against and dissemination of disinformation about vaccination. The relative calmness of discussion about aging seems like a good thing, but it has a peculiar drawback.

Nonradical ideas don't tend to draw as much public attention as radical ideas. This was explained to some extent by "Trevor's Axiom," a theory of online trolling presented in *South Park*. This simple idea is that if a person makes an excessively offensive or controversial statement, a second party might react to it with a rebuttal that a third party views as extreme, leading to a fission-like chain of overreaction, with more and more parties joining in. While seeking to offend is not the best communication strategy, heated discussion can propel opposing views into the stratosphere of public debate, drawing attention to the subject of that debate. Suddenly an issue no one cared about before becomes a topic on which humanity is extremely divided (you can probably think of a few examples of this phenomenon).

Let's imagine that 1% of the public favors an idea and 1% opposes it, while 98% has no clue that such an issue even exists. After a scandal, an Internet holy war, or the explosion of a meme, the figures might become something like 40%, 40%, and 20% respectively. This is why some of the most widely discussed ideas are extremely offensive to one group of people or another: these ideas have been selected by a quasi-Darwinian process, the survival of the most provocative and emotional statements.

If one or both opposed views are harmful for society, the consequences of Trevor's Axiom can be detrimental. Imagine an extremely offensive pro-vaccination campaign that inadvertently boosts opposition to vaccinations, leading to an increase in infection rates; or a political campaign that has gone so far to the left or right that it rallies the opposition and leads to the election of an opposing candidate (this is why we should always think about the possible consequences of our public statements, and sometimes tone them down a little).

However, it doesn't really matter how many people oppose the idea of radical life extension. What matters is whether a critical number of people support progress and contribute to research on aging.

If forty percent of humanity actively wants an anti-aging therapy and another forty percent loudly declares that this idea is moronic, that's a win for the radical-life-extension movement, because it would likely lead to increased funding of research, more economic demand for commercial solutions, and most importantly, more ideas and scientific discussions. When tangible progress is eventually made, opponents of the idea of radical life extension might just switch sides, and we should welcome them, because their lives matter. Until then, though, public indifference will be our greatest opponent.

While I would love to test Trevor's Axiom by instigating a radical movement for radical life extension, I haven't figured out a way to do so. Perhaps radicalism cannot be rationally constructed, and must come from the heart. But it does seem that some of the most heated real-life conversations involve accusations of moral violation. The environmental movement guilt-trips people for harming the planet with carbon emissions and nonbiodegradable waste; anti-GMO activists claim that scientists such as myself are paid shills for global corporations; atheists are often labeled as amoral while religion is seen as justifying war and bigotry; abortions are compared to murder; conspiracy theorists claim that vaccines are used to sterilize or disable people. The list goes on. In these debates one side often considers itself either morally superior to the other or wrongfully accused of moral inferiority.

So, let's construct an argument in favor of life extension that invokes morality. Andrew Steele's thesis about the horrors of introducing aging into an ageless society (quoted in the introduction of this book) is a good example of such an argument, but it's defensive rather than offensive. Consider another scenario. Imagine the crew of a rowboat trying to escape a storm. Some sailors accept death and stop rowing. Some lose hope. Others get bored, or tired of life, or put their faith in an afterlife, or perhaps the Kraken or Cthulhu. But still others try their hardest to steer the vessel away from harm. The problem is that this task can be accomplished only by the combined effort of most of the sailors, and they will all either live or die together. How do we judge inaction in this situation? What if elderly people are aboard? It's a personal choice to give up on your own life, but those who do not seek life extension should realize that they are in the same boat as those who do.

(2) Many people are uninterested in longevity research either because they regard aging as a distant prospect or because they have found ways to cope with the idea of death. In discussion with these uninterested people, perhaps the idea of radical life extension can be associated with things people tend to care about deeply.

One great example of such an approach is the Dog Aging Project co-directed by professors Daniel Promislow and Matt Kaeberlein, which allows ordinary people to become community scientists and enroll their favorite pets in longevity research. Dogs are short-lived compared with humans, and dog owners tend to be personally invested in keeping their animal friends alive[1274].

Some research of the Dog Aging Project has direct practical value. For example, a 2022 study published in *GeroScience* showed that feeding dogs once daily is associated with better health outcomes for them[1275]. This study utilized a large dataset of animals of different ages and breeds, which was possible thanks to the participation of tens of thousands of dog owners. There have also been several small, randomized placebo-controlled trials on the use of rapamycin on dogs[1276], although with inconclusive results. Hopefully larger studies will be conducted in the near future, and maybe your dog can participate in them.

Similarly, we can consider the interests of cats and their owners. This idea is linked to an inspirational scientific discovery. According to a recent article in *PLOS One*, approximately

60% of postmortem cats display from renal abnormalities[1277]. This problem ends up being the primary cause of death for about 13% of our feline friends. University of Tokyo professor Toru Miyazaki (yes, another talented Miyazaki) discovered which gene is responsible for this problem[1278,1279]. The gene encodes a protein called apoptosis inhibitor of macrophage (AIM). Like a fighter jet stowed on an aircraft carrier, AIM spends most of its time bound to immunoglobulin M. When kidneys get injured, AIM dissociates from immunoglobulin M and enters the proximal kidney tubules, where it stimulates clearance of dead and damaged cells. The damage is repaired.

But the cat's version of AIM has a thousand-fold greater binding affinity to immunoglobulin M than, say, its murine version. It just doesn't take off on its mission. Miyazaki even created "felinized" mice with the cat version of the gene, and they experienced the same health problems[1280]. In theory, if the cat version of the gene was somehow replaced with its human or mouse version, the animals' expected lifespan could increase dramatically. As you can guess, this finding inspired cat lovers in Japan. They donated nearly two million dollars for cat kidney research[1281].

According to the University of Tokyo's website, Miyazaki became passionate about cats after losing a friend, who was a cat lover, to terminal illness. Here is a quote from Dr. Miyazaki himself:

"I believe that perhaps my friend is one of the reasons why I was destined to save cats — I wouldn't have ever imagined myself focusing on cats otherwise. In reality, however, I have witnessed many people passing away due to incurable illnesses, so I feel strongly about eventually using AIM to treat people. That is the greatest motivation that is supporting my research right now[1282]."

Perhaps some people could be persuaded of the importance of longevity research by highlighting its potential to extend the lives of both humans and beloved animals.

Here's another perspective: many people are more invested in sports than in addressing aging. Soccer star Cristiano

Ronaldo boasts over 110 million followers on X (formerly Twitter), while leading gerontologist David Sinclair has around 450,000. But imagine if we harnessed the competitive spirit of sports to drive biomedical innovation. Picture a new kind of Olympics, where nations compete not only in athletic prowess but also in scientific breakthroughs, with each discovery benefiting athletes and the broader population alike.

Such initiatives are already in development. Recently, Dr. Aron D'Souza founded Enhanced Games, an alternative sports league that allows athletes to use enhancing therapies. D'Souza and the cocreators of Enhanced Games believe that the modern Olympics are to a certain extent wasteful; athletes should be better paid given the enormous profits of grand sports events, and they should have the right to decide what to do with their bodies. One argument in favor of anti-doping rules is that some performance-enhancing methods are dangerous; however, many athletes use performance-enhancing methods anyway. Moreover, although studies typically find that professional athletes live longer than members of the general population[1283], professional sports pose numerous health risks, such as head impacts[1284]. Perhaps it would be better if people risked their health for something more significant than mass entertainment.

A third possible cultural-scientific symbiosis is in harnessing the growing popularity of biohacking and channeling it in a more scientific direction. I believe we can learn a lot from citizen-science projects, such as microdosing. Microdosing is the practice of consuming small amounts of psychedelic substances in pursuit of improved cognitive abilities or to treat anxiety or depression. Conducting rigorous clinical trials of microdosing is complicated because of ethical, legal, and scientific reasons, such as the difficulty of blinding participants, who can guess whether they are microdosing or not. However, researchers have found some practical work-arounds.

One of the largest microdosing studies, published in 2021, implemented self-blinding citizen science[1285]. Participants were instructed to prepare some nontransparent pills with an active ingredient—if they were willing to take it—and others with a placebo. The pills were placed in envelopes, which were then sealed and marked with QR codes. For the next four weeks, participants consumed pills from a random subset of envelopes. Some participants were randomized to consume microdoses for four weeks, others were assigned four weeks of placebos, and still others were assigned two weeks of each treatment.

"Some days during the test were really, really focused and colors were more vivid. This sensation was really new to me," responded one participant after the trial. "It seems I was able to generate a powerful 'altered consciousness' experience," said another. Both comments came from the placebo group, highlighting the psychological bias introduced by personal expectations. "You put spirituality into an empty pill here," summarized a third placebo-group participant. As you have probably guessed, this study found a significant effect from expectation, but not so much from microdosing itself. The study even led to the invention of a statistical correction that takes into account that some participants break blinding and attempt to guess which group they are in[1286].

The design of the self-blinding microdosing study is suitable for testing various biohacking interventions that address physical, cognitive, and molecular biomarkers of aging. Why not give people who are already experimenting on themselves the opportunity to contribute to a more objective search for truth? While citizen science has limitations, it also has benefits: it's cheaper than normal clinical trials; and it can be performed by people from all over the world, who are living their normal day-to-day lives, thus improving the generalization of findings. It also allows more people to engage with the scientific process.

(3) A third reason for pessimism might be due to a perception that the anti-aging field is a scam. Indeed, many promises of immortality, both historical and contemporary, have been made in bad faith, or based on overconfidence or ignorance (I provided a number of examples in the "false Grail" chapter).

Lying harms science in many ways. In earlier chapters, I referenced several studies involving supercentenarians. For instance, we can compare the genetic variants of individuals with average lifespans to those with above-average lifespans. A 2024 preprint raised concerns about the credibility of extraordinary age records, citing several key observations: supercentenarian birthdates tend to cluster on days divisible by five, suggesting possible birth date forgery; many supercentenarians lack official birth certificates, and as birth registration systems have been introduced, the number of verified supercentenarian records has decreased[1287]. Specifically, the authors question whether regions previously labeled as 'blue zones,' such as Okinawa, truly possess the pro-longevity dietary[1288] or cultural factors[1289] often attributed to them, or if these claims might be influenced by factors like low literacy and high crime rates (with Okinawa having the highest crime rate in Japan[1290]).

Aside from lying, there are errors. There is talk of a reproducibility crisis in science[1291], and many longevity studies, including some mentioned in this book, have never been directly replicated. This does not mean that they can't be replicated or that their conclusions are wrong, but we must be prepared for some promising therapies and ideas to fail.

There are many legitimate reasons for the appearance of false positives in longevity studies. Positive findings might be easier to publish. Overestimates of lifespan extension due to certain interventions might be due to coincidentally shorter-than-normal lifespan in control groups[1292]. Mistakes also happen. This is why the Interventions Testing Program is so important and deserves more funding, and why other, similar programs need to be established. This is also why open data sharing should be fostered. Start-ups such as First Approval[1293] intend to help researchers publish their datasets in a way that allows anyone to use them, on the submitters' terms and without affecting subsequent publications. Some scientists don't share datasets because they aren't incentivized to do so, or because they are afraid that their data will be stolen or used without proper credit. Hopefully this problem can be solved with new policies and initiatives, and consequently more data will become available, and knowledge will be shared at a greater pace.

Public perception of the anti-aging field as a scam has unfortunately been influenced by some aspects of the biohacking movement. Association between questionable supplements and the idea of life extension is making it difficult to convince well-meaning investors and philanthropists to support genuine aging research. It's understandable that a nonscientist would be unclear on whether he is being asked to fund legitimate scientific research or snake oil, or that he would be concerned that a biotech company courting funding could suffer the same fate as Theranos—a privately held corporation whose CEO continuously lied to investors and the government about its achievements and bankability, tanking its multibillion-dollar valuation in doing so.

Theranos promised to revolutionize health care by performing complex blood tests using the tiniest amounts of blood, but the company didn't even come close to fulfilling this objective. Nobody enjoys being tricked or swindled. A friend of mine who is a wealthy business owner once told me, "I would rather lose a million dollars to gambling than give a dollar to someone who will scam me." Earning and maintaining a reputation for honesty is vital for members of the pro-longevity community, and profit-driven promotion of miraculous interventions that haven't passed and can't pass proper clinical trials simply does not accord with this.

What I find most amusing and perplexing about skepticism toward radical life extension, though, is how billions of people believe in far more questionable ideas—such as ghosts, homeopathy, and astrology. Some even think benevolent aliens will save them from an impending apocalypse, yet they doubt that scientists could ever enable humans to develop biological traits already found in species like naked mole-rats, hydras, and jellyfish. Our species is truly fascinatingly inconsistent.

(4) It might sound paradoxical, but growing public interest in pursuing healthy lifestyles, emphasizing diet and physical

activity, could have detrimental long-term consequences if such lifestyles become widely perceived as the solution to aging. They are not the solution. No diet and no amount of exercise, despite the proven benefits of some diets and exercises, will yield a great chance for an average human to reach even a hundred years of age. A healthy lifestyle should be encouraged, but we should view it as a stopgap that gives scientists more time to do their job. An illusion of a solution might prevent us from searching for an actual one.

(5) It is unclear what we fear more, death itself or the specter of death. Many people want to avoid the topic of death entirely. This is why some popular arguments against the idea of radical life extension are so easy to refute: they don't result from deep consideration, but are automatically generated to drive conversation in other directions as quickly as possible. We need to promote discussions about death. After all, the most efficacious treatment for phobia is exposure therapy[1294]. Note that I'm not talking about treating the fear of death, which is a completely rational fear and helped our ancestors survive. I'm talking about the irrational fear of contemplating death, discussing it openly, and admitting that we don't want to die. Our culture celebrates bravery, selflessness, and risking one's life for the greater good—whatever we believe that "good" to be. Maybe it's time to be more honest with one another instead of trying to impress or to signal virtue.

(6) The pro-longevity movement is not well organized, and it lacks strong political representation. Consider organizations such as Greenpeace, with its well-funded campaigns and its ability to affect government decisions. There are diverse opinions about Greenpeace, and I won't go into a detailed discussion of its current goals, effectiveness, or motives (I'll just note that I disagree with its policies on GMOs and nuclear energy). I mention Greenpeace because it was initially a small group of activists, which drew attention to itself by protesting nuclear-weapons tests, and it has since transformed into a powerful international community comprising three thousand staff members and thirty thousand volunteers, with an annual

budget of more than 100 million dollars. It is recognized all over the world, although banned in some countries. The movement for radical life extension would benefit greatly from the leadership of talented and motivated people with political ambition who were prepared to campaign for anti-aging-research funding and changing our cultural perceptions on aging and death. Perhaps this movement could eventually become as effective as Greenpeace or even more effective.

Social activism could include establishing art galleries and science museums, hosting lectures and participating in public debates on the topic of aging, conducting community polls, engaging in street epistemology, and organizing movie contests and collaborations. In 2014 the Ice Bucket Challenge went viral, promoting awareness of amyotrophic lateral sclerosis (ALS) and raising more than 100 million dollars for research on this deadly disease. Perhaps something similar can be done for aging.

In 2024 I was asked to review a wonderful art project by student Polina Pakhomova. The project was called "Apples Make You Young: a study of the social aspects of the phenomenon of rejuvenation." It defended the idea of creating quercetin-rich genetically modified apples, which could be a prototype for Iðunn's apples of youth in Norse mythology. One of the project's goals was to combat the negative stereotypes people have regarding GMOs and longevity research.

Political activism might involve convincing elected officials to support the radical-life-extension movement and campaigning for politicians who understand the importance of achieving the movement's goals. I would certainly give my vote to a proponent of scientific progress. In the United States, one relevant organization that is gaining traction is the Alliance for Longevity Initiatives, a nonprofit organization "focused on advancing legislation, policies and initiatives that aim to increase healthy human lifespan[1295]." One of their recent political achievements was the passage of a bill in Montana that expanded access to investigational (phase I) treatments for terminally ill patients who have considered existing

FDA-approved options, received recommendations from their doctors, and given informed consent.

We can also benefit by creating offline and online communities. Jellyfish DAO[1296] is a young organization dedicated to producing longevity-centered movies and games and VitaDAO is a venture fund[1297] that, among other things, organizes longevity pop-up cities. Or consider the LessWrong[1298] forum, which is dedicated to teaching, discussing, and applying rationality. We can achieve a lot by gathering educated people on a similar platform that promotes informed discussion and celebrates original ideas on the topic of aging. Perhaps a suitable name for such a platform is LessDead. Currently the Vitalism community has taken this important role[1268].

(7) Immortality is not enough. It may sound surprising, but the defeat of aging may not be sufficient motivation for certain people, even some who agree that dying is bad. After all, we do all kinds of things not for practical reasons but because we simply enjoy doing them. I probably wouldn't have written this book if I found the activity boring or unpleasant. Some people like puzzles and other challenges; others might prefer connecting with people, feeling a sense of camaraderie; still others might be motivated by the possibility of personal achievement, of leaving a mark on history. Most people are gratified by alleviating the suffering of others. We should spread the message that the battle against aging offers all of the pleasures mentioned above, whether the ultimate goal is ever achieved or not. Aging is an interesting and complex puzzle, and there are wonderful people with whom to forge relationships in the process of solving it. Furthermore, longevity research is not just about advancing science. It is a powerful tool for promoting the value of human well-being, development, and dignity.

(8) So far, the longevity field has failed to provide the kinds of tangible results that could inspire humanity to believe that defeating aging really is possible. Sure, we need to promote basic scientific research for the long run, but practical results from clinical and preclinical trials could greatly alter the

general perception of aging and death. I have already stressed the importance of performing combined anti-aging therapies, first on mice, then on humans, a path advocated by Aubrey de Grey and the SENS Research Foundation. Even preliminary success in this kind of work could significantly boost public interest and investment in longevity research.

The development of new standardized, validated[1299], and accessible biomarkers of aging can both facilitate research for anti-aging therapies and provide people with personalized information on their rate of aging[1299,1300]. Aging and death are often perceived as distant concerns, but we age in real time, and perhaps quantifying aging would promote a sense of urgency for the pursuit of viable anti-aging solutions. In this sense, I think that biohackers are onto something when they organize competitions in which one must demonstrate the lowest rate of aging, based on biomarker data. Such competitions may not be rigorous or scientific, but they might make people more interested in exploring the science of longevity.

From the perspective of basic research, I think it's vital to understand the genetics of aging. The wide range of animal lifespans indicates that genes play an important role in aging. For example, consider *Turritopsis dohrnii*, the immortal jellyfish, which is capable of repeated rejuvenation, even after sexual reproduction. This is a rather novel evolutionary invention. We know this because *T. dohrnii* has many close relatives, such as *T. rubra*, that are not "immortal". Comparison between the genomes of the two just-mentioned jellyfish species helped identify variations and even duplications of genes involved in aging, including DNA repair and telomere maintenance[277]. Similar comparative-genomics studies have been conducted on cetaceans[648], giant tortoises[1301], and long-lived mammal species such as naked mole-rats and certain bats[1302], with fascinating results.

Some of my colleagues have studied genes associated with the extraordinary regenerative abilities of some cold-blooded vertebrates. These abilities were lost in the evolution of warm-blooded birds and mammals, due to the loss of the corresponding genes[1303]. Remarkably, some lost ancient features can be restored with relatively simple genetic-engineering techniques, such as when scientists created a chicken embryo with saber-shaped teeth resembling those of an alligator; they achieved this by modifying just one gene[1304]. Perhaps some lost features of human regeneration could also be turned back on if we understood the relevant genetic mechanisms in our distant relatives. Currently a Japanese pharmaceutical startup called Toregem Biopharma is launching human clinical trials of a new drug that stimulates the growth of new teeth in ferrets[1305]. The drug is an antibody that deactivates a gene that normally prevents excessive tooth formation; mice genetically engineered to lack this gene form supernumerary teeth. Imagine the possible implications for future dentistry.

More than 2,400 aging-associated genes have been cataloged in the publicly available Open Genes database[1306], a tool that could bolster the search for aging-therapy targets; and the AnAge database[1307] contains validated lifespan data for thousands of animals, some of which have yet to be analyzed from a genetics viewpoint. Each sequenced animal genome brings us closer to understanding how aging evolves, and what genetic changes accompany decreases and increases of lifespan. The ongoing Earth BioGenome Project ambitiously aims to sequence all of Earth's eukaryotic biodiversity within the next decade[1308]. While the project plans to inform such areas as biodiversity studies and the conservation of endangered species, it will certainly also help us understand why and how we age.

The growing availability of genetic data, along with recent advances in machine learning, data analysis, and computational power, will revolutionize our understanding of the inner workings of cells and organisms. These are areas in which computational experts and IT specialists will be invaluable. Evolutionary and comparative genomics will allow us to discover new aging-related genes, some of which just might be usable for anti-aging gene therapies in the future.

The battle against aging is becoming a point of convergence for all human advancement, both cultural and scientific.

CHAPTER 14

We need artists, psychologists, sociologists, and entertainers to promote the idea of radical life extension; we need biologists to discover aging-related genes and gene-delivery systems, and doctors to apply them; we need chemists to create new drugs; we need physicists to create new optical devices and detection apparatus for cellular, genetic, and chemical analysis; we need programmers and data analysts, as well as AI researchers, to process the increasingly huge amount of biological and chemical data we are collecting; we need as many people as possible to support policy changes to accelerate the creation of our dragon-piercing weapons.

Let's fantasize for a moment about what awaits us after the epic battle for immortality has been won. I call it humanity's final battle, because if aging is ever defeated, our perception of human life and its value will undoubtedly change forever. We will never be able to say that a person who suffers an accident or falls as a casualty of war would have died sooner or later anyway. War might finally become a thing of the past.

Society will change. People will have time to pursue all possible professions, not just one or two. We will live to witness the birth of the as-yet unimaginable technologies that will be invented in a thousand years. We will experience incredible works of art and literature, such as those that are produced by artists who appear only once in a several generations; the next William Shakespeare, Vladimir Nabokov, Ray Bradbury, Edgar Allan Poe, or Stanisław Lem will emerge in our lifetime. Perhaps we will even experience artworks that take hundreds of years to create. Religions' promises of an afterlife will lose their appeal. It will become normal to learn multiple languages and visit every country—not only as a tourist, but actually to live in a new place and get a true feel for its culture. People will no longer get married "until death do us part," but rather will remain together "until we want to try something new"—or "forever," if that's what they want. We will have time—time to do whatever makes us happy, with all our unique and diverse preferences.

Life coaches and gurus like to speak of time as our most valuable resource. In this they are correct. Money can't buy time. Status can't buy time. Time should not be wasted on bad relationships, failed projects, or unsuccessful careers. Immortality will give us all the time we want—time to fail, to learn from our mistakes and try again without ever regretting wasted time. Think of the things you would do if not for the fear that doing them would mean not having time to do something else.

I have never understood people who are afraid that immortality would be boring. I ask if they have read every book, watched every movie, played every game, climbed every mountain, or gotten to know every person, but the question is rhetorical. It's impossible to do in one lifetime, or a thousand lifetimes, everything worth doing. The reason we get tired and bored is because our aging bodies can no longer support the activities that bring us joy, and the constant maintenance we require becomes a large part of our life—along with chronic pain, with physical and cognitive decline. Imagine that this will no longer be so.

When asked how long I would like to live, I say that, at the very least, I would like to live long enough to witness, from the inside of a giant spaceship, the final hours of our home planet, Earth, before it is engulfed by the expanding sun, approximately 7.5 billion years from now[1309]; and afterward I would like to return safely to humanity's new interstellar home. I know that this is an unlikely scenario, but I am looking forward to it nonetheless.

Nobody can tell what the future holds for us, but one thing is certain: change is inevitable. In a few hundred years—a rather short time—you, I, and every other person who's alive today will have met one of two fates: immortality or death. It's time to choose, and to act accordingly.

Aging must be overcome for the good of all humankind. No one should be left behind. We are all in this boat together—let's make it soar. Let's fight for our freedom to choose whether we live or die.

Acknowledgments

I would like to thank Mikhail Batin for believing in me, for his support and for inspiring me to write this book, and thereby giving something meaningful to do, during a difficult time of my life, when I thought all meaning was lost. Writing this book took more than two years, and it would have been an impossible task without the inspiration I received thanks to the illustrations made by Olga Posukh who worked alongside me. I thank my editor, Zachary Vigna, for the numerous suggestions that helped this book reach its final readable form, and Kira Severinova for designing the book cover.

Many people, including other scientists, have read this book and helped me to improve its scientific quality. I would like to thank my colleagues Alexander Tyshkovskiy, Timofey Glinin, Alexander Alexandrov, Vasily Vlassov, Anna Parfenenkova, Elena Kleshenko, Maxim Kholin, Dmitry Voronov, Nikolai Nikolayev, Aleksey Rzhevsky, and Arkadi Mazin working through the entire draft and giving me the most eye-opening comments. I would like to thank Alexander Anisimov for providing expert comments on the topic of Alzheimer's disease; Roman Litvinov for thoughtful conversations on the aging of the extracellular matrix; Sergei Pustylnikov for educating me on the subject of immunology; and Leon Peshkin for helpful and blunt criticism. I thank Mark Hamalainen, Anastasia Egorova, Yulia Shaykhutdinova, Ayur Sandanov, and Georgy Kurakin for providing additional useful comments. I thank Luca Kormiltsev for technical help. I also thank Andrei Sobolevski, who for many years created a remarkable working atmosphere in which both research and popular science could be pursued. Finally, I would like to thank both my parents Yuri, and Nadia, for raising me with a respect for life and a desire to create.

References

1. Barnett, M. D. & Helphrey, J. H. Who wants to live forever? Age cohort differences in attitudes toward life extension. *Journal of Aging Studies* **57**, 100931 (2021).

2. The billion-dollar search for immortality — UnHerd. https://unherd.com/2023/06/the-billion-dollar-search-for-immortality/.

3. Average life expectancy by country. https://www.worlddata.info/life-expectancy.php.

4. Zuo, W., Jiang, S., Guo, Z., Feldman, M. W. & Tuljapurkar, S. Advancing front of old-age human survival. *Proceedings of the National Academy of Sciences* **115**, 11209–11214 (2018).

5. Ayyadevara, S., Alla, R., Thaden, J. J. & Shmookler Reis, R. J. Remarkable longevity and stress resistance of nematode PI3K-null mutants. *Aging Cell* **7**, 13–22 (2008).

6. Wei, M. *et al.* Life span extension by calorie restriction depends on Rim15 and transcription factors downstream of Ras/PKA, Tor, and Sch9. *PLoS Genet* **4**, e13 (2008).

7. Feynman, R. P. & Robbins, J. *The Pleasure of Finding Things out: The Best Short Works of Richard P. Feynman.* (Perseus Books, Cambridge, Mass, 1999).

8. Le Bras, A. New insights into axolotl brain regeneration. *Lab Anim* **51**, 250–250 (2022).

9. Hartmann, D., Smith, J. M., Mazzotti, G., Chowdhry, R. & Booth, M. J. Controlling gene expression with light: a multidisciplinary endeavour. *Biochemical Society Transactions* **48**, 1645 (2020).

10. Donner, Y. *et al.* Great Desire for Extended Life and Health amongst the American Public. *Frontiers in Genetics* **6**, 353 (2016).

11. Barbi, E., Lagona, F., Marsili, M., Vaupel, J. W. & Wachter, K. W. The plateau of human mortality: Demography of longevity pioneers. *Science* **360**, 1459–1461 (2018).

12. Andersen, S. L., Sebastiani, P., Dworkis, D. A., Feldman, L. & Perls, T. T. Health span approximates life span among many

supercentenarians: compression of morbidity at the approximate limit of life span. *J Gerontol A Biol Sci Med Sci* **67**, 395–405 (2012).

13. Cervellati, M. & Sunde, U. Life Expectancy and Economic Growth: The Role of the Demographic Transition. *IZA Discussion Papers* (2009).

14. Gavrilov, L. A. & Gavrilova, N. S. Demographic Consequences of Defeating Aging. *Rejuvenation Research* **13**, 329 (2010).

15. Steele, A. *Ageless: The New Science of Getting Older without Getting Old*. (Doubleday, New York, 2020).

16. Arshavsky YuI, Beloozerova, I. N., Orlovsky, G. N., Panchin YuV & Pavlova, G. A. Control of locomotion in marine mollusc Clione limacina. I. Efferent activity during actual and fictitious swimming. *Exp Brain Res* **58**, 255–262 (1985).

17. Schrödinger, E. *What Is Life? The Physicist's Approach to the Subject -- with an Epilogue on Determinism and Free Will*. (Cambridge University Press, 1946).

18. Conselice, C. J., Wilkinson, A., Duncan, K. & Mortlock, A. The evolution of galaxy number density at z < 8 and its implications. *ApJ* **830**, 83 (2016).

19. Castelvecchi, D. Universe has ten times more galaxies than researchers thought. *Nature* nature.2016.20809 (2016) doi:10.1038/nature.2016.20809.

20. Lavergne, J. Commentary on: "Photosynthesis and negative entropy production" by Jennings and coworkers. *Biochimica et Biophysica Acta (BBA) - Bioenergetics* **1757**, 1453–1459 (2006).

21. Johnson, R. D. & Jasin, M. Sister chromatid gene conversion is a prominent double-strand break repair pathway in mammalian cells. *The EMBO Journal* **19**, 3398 (2000).

22. Lu, L.-Y. & Yu, X. Double-strand break repair on sex chromosomes: challenges during male meiotic prophase. *Cell Cycle* **14**, 516 (2015).

23. Fijalkowska, I. J., Schaaper, R. M. & Jonczyk, P. DNA replication fidelity in Escherichia coli: a multi-DNA polymerase affair. *FEMS Microbiol Rev* **36**, 1105–1121 (2012).

24. Lajoie, M. J. *et al.* Genomically Recoded Organisms Expand Biological Functions. *Science* **342**, 357–360 (2013).

25. Cagan, A. *et al.* Somatic mutation rates scale with lifespan across mammals. *Nature* **604**, 517–524 (2022).

26. Pal, S. & Tyler, J. K. Epigenetics and aging. *Science Advances* **2**, e1600584 (2016).

27. Meer, M. V., Podolskiy, D. I., Tyshkovskiy, A. & Gladyshev, V. N. A whole lifespan mouse multi-tissue DNA methylation clock. *eLife* **7**, e40675 (2018).

28. Lu, Y. R., Tian, X. & Sinclair, D. A. The Information Theory of Aging. *Nat Aging* **3**, 1486–1499 (2023).

29. Avanesov, A. S. *et al.* Age- and diet-associated metabolome remodeling characterizes the aging process driven by damage accumulation. *eLife* **3**, e02077 (2014).

30. Haithcock, E. *et al.* Age-related changes of nuclear architecture in *Caenorhabditis elegans*. *Proc. Natl. Acad. Sci. U.S.A.* **102**, 16690–16695 (2005).

31. van Loosdregt, I. a. E. W. *et al.* Lmna knockout mouse embryonic fibroblasts are less contractile than their wild-type counterparts. *Integr Biol (Camb)* **9**, 709–721 (2017).

32. Gainotti, G. & Marra, C. Differential Contribution of Right and Left Temporo-Occipital and Anterior Temporal Lesions to Face Recognition Disorders. *Frontiers in Human Neuroscience* **5**, 55 (2011).

33. Norup, A. & Mortensen, E. L. Prevalence and predictors of personality change after severe brain injury. *Arch Phys Med Rehabil* **96**, 56–62 (2015).

34. Van Lommel, P., Van Wees, R., Meyers, V. & Elfferich, I. Near-death experience in survivors of cardiac arrest: a prospective study in the Netherlands. *The Lancet* **358**, 2039–2045 (2001).

35. Mobbs, D. & Watt, C. There is nothing paranormal about near-death experiences: how neuroscience can explain seeing bright lights, meeting the dead, or being convinced you are one of them. *Trends in Cognitive Sciences* **15**, 447–449 (2011).

36. Blanke, O., Ortigue, S., Landis, T. & Seeck, M. Stimulating illusory own-body perceptions. *Nature* **419**, 269–270 (2002).

37. Loftus, E. F. & Pickrell, J. E. The formation of false memories. *Psychiatric Annals* **25**, 720–725 (1995).

38. Laney, C. & Loftus, E. F. Recent advances in false memory research. *South African Journal of Psychology* **43**, 137–146 (2013).

39. Hamblin, J. Cheating Death and Being Okay With God. *The Atlantic* https://www.theatlantic.com/health/archive/2013/08/cheating-death-and-being-okay-with-god/278381/ (2013).

40. Chapter 2: Views on Radical Life Extension, by Religious Affiliation, Beliefs and Practices | Pew Research Center. https://www.pewresearch.org/religion/2013/08/06/chapter-2-views-on-radical-life-extension-by-religious-affiliation-beliefs-and-practices/.

41. Sinclair, C. T. & Rosielle, D. A. Avoid Stigmatizing Language About Atheist Patients. *Journal of Pain and Symptom Management* **60**, e30 (2020).

42. Bostrom, N. Are You living in a computer simulation?

43. Kwang-Ju Lee & Byoung-Tak Zhang. Learning robot behaviors by evolving genetic programs. in *2000 26th Annual Conference of the IEEE Industrial Electronics Society. IECON 2000. 2000 IEEE International Conference on Industrial Electronics, Control and Instrumentation. 21st Century Technologies and Industrial Opportunities (Cat. No.00CH37141)* vol. 4 2867–2872 (IEEE, Nagoya, Japan, 2000).

44. Kutzner, C. *et al.* GROMACS in the cloud: A global supercomputer to speed up alchemical drug design. *J. Chem. Inf. Model.* **62**, 1691–1711 (2022).

45. Jung, J. *et al.* Scaling molecular dynamics beyond 100,000 processor cores for large-scale biophysical simulations. *J Comput Chem* **40**, 1919–1930 (2019).

46. Oba, Y. *et al.* Identifying the wide diversity of extraterrestrial purine and pyrimidine nucleobases in carbonaceous meteorites. *Nat Commun* **13**, 2008 (2022).

47. Kaiho, K. & Oshima, N. Site of asteroid impact changed the history of life on Earth: the low probability of mass extinction. *Scientific Reports* **7**, 14855 (2017).

48. Chiarenza, A. A. *et al.* Asteroid impact, not volcanism, caused the end-Cretaceous dinosaur extinction. *Proceedings of the National Academy of Sciences of the United States of America* **117**, 17084 (2020).

49. Dunk, R. D. P., Petto, A. J., Wiles, J. R. & Campbell, B. C. A multifactorial analysis of acceptance of evolution. *Evolution: Education and Outreach* **10**, 4 (2017).

50. Monette, M. Spending eternity in liquid nitrogen. *CMAJ : Canadian Medical Association Journal* **184**, 747 (2012).

51. Best, B. P. Scientific Justification of Cryonics Practice. *Rejuvenation Research* **11**, 493 (2008).

52. Shiurba, R. A. *et al.* Immunocytochemistry of formalin-fixed human brain tissues: microwave irradiation of free-floating sections. *Brain Res Brain Res Protoc* **2**, 109–119 (1998).

53. McKenzie, A. T. *et al.* Structural brain preservation: a potential bridge to future medical technologies. *Front. Med. Technol.* **6**, (2024).

54. Chian, R.-C., Wang, Y. & Li, Y.-R. Oocyte vitrification: advances, progress and future goals. *Journal of Assisted Reproduction and Genetics* **31**, 411 (2014).

55. Shah, D., Rasappan, Shila & Gunasekaran, K. A simple method of human sperm vitrification. *MethodsX* **6**, 2198 (2019).

56. Fahy, G. M. *et al.* Physical and biological aspects of renal vitrification. *Organogenesis* **5**, 167 (2009).

57. Xue, W. *et al.* Effective cryopreservation of human brain tissue and neural organoids. *Cell Reports Methods* **4**, (2024).

58. Bojic, S. *et al.* Winter is coming: the future of cryopreservation. *BMC Biology* **19**, 56 (2021).

59. Shatilovich, A. *et al.* A novel nematode species from the Siberian permafrost shares adaptive mechanisms for cryptobiotic survival with C. elegans dauer larva. *PLoS Genet* **19**, e1010798 (2023).

60. Beattie, G. M. *et al.* Trehalose: a cryoprotectant that enhances recovery and preserves function of human pancreatic islets after long-term storage. *Diabetes* **46**, 519–523 (1997).

61. Martinetti, D. *et al.* Effect of trehalose on cryopreservation of pure peripheral blood stem cells. *Biomedical Reports* **6**, 314 (2017).

62. Shmakova, L. *et al.* A living bdelloid rotifer from 24,000-year-old Arctic permafrost. *Curr Biol* **31**, R712–R713 (2021).

63. Toxopeus, J. & Sinclair, B. J. Mechanisms underlying insect freeze tolerance. *Biol Rev Camb Philos Soc* **93**, 1891–1914 (2018).

64. Storey, K. B. Life in a frozen state: adaptive strategies for natural freeze tolerance in amphibians and reptiles. *Am J Physiol* **258**, R559-568 (1990).

65. Strimbeck, G. R., Schaberg, P. G., Fossdal, C. G., Schröder, W. P. & Kjellsen, T. D. Extreme low temperature tolerance in woody plants. *Frontiers in Plant Science* **6**, 884 (2015).

66. Jiang, H. *et al.* Multi-omics Investigation of Freeze Tolerance in the Amur Sleeper, an Aquatic Ectothermic Vertebrate. *Molecular Biology and Evolution* **40**, msad040 (2023).

67. de Graaf, I. a. M. & Koster, H. J. Cryopreservation of precision-cut tissue slices for application in drug metabolism research. *Toxicol In Vitro* **17**, 1–17 (2003).

68. Iverach, L., Menzies, R. G. & Menzies, R. E. Death anxiety and its role in psychopathology: Reviewing the status of a transdiagnostic construct. *Clinical Psychology Review* **34**, 580–593 (2014).

69. Graham, D. L. *et al.* A scale for identifying 'Stockholm syndrome' reactions in young dating women: factor structure, reliability, and validity. *Violence Vict* **10**, 3–22 (1995).

70. Ketamine Reverses Neural Changes Underlying Depression-Related Behaviors in Mice - National Institute of Mental Health (NIMH).

REFERENCES

71. Yavi, M., Lee, H., Henter, I. D., Park, L. T. & Carlos A Zarate, J. Ketamine treatment for depression: a review. *Discover Mental Health* **2**, 9 (2022).

72. Kalsi, S. S., Wood, D. M. & Dargan, P. I. The epidemiology and patterns of acute and chronic toxicity associated with recreational ketamine use. *Emerging Health Threats Journal* **4**, 7107 (2011).

73. Projects - BioCurious. https://biocurious.org/projects/.

74. Gruber, K. Biohackers: A growing number of amateurs join the do-it-yourself molecular biology movement outside academic laboratories. *EMBO Reports* **20**, e48397 (2019).

75. Blueprint — Blueprint Bryan Johnson. https://blueprint.bryanjohnson.com/.

76. Boardman, C. R. & Sonnenberg, A. Magical Thinking. *Clinical and Translational Gastroenterology* **5**, e63 (2014).

77. Russian Scientist Injects Himself With 3.5-Million-Year-Old Bacteria In Quest For Immortality | HuffPost UK Tech. https://www.huffingtonpost.co.uk/2015/10/01/russian-scientist-injects-himself-with-3-5-million-year-old-bacteria-in-quest-for-immortality-_n_8224992.html.

78. Brenner, E. V. *et al.* Draft Genome Sequence of Bacillus cereus Strain F, Isolated from Ancient Permafrost. *Genome Announcements* **1**, e00561 (2013).

79. Cahoon, D. *et al.* Walnut intake, cognitive outcomes and risk factors: a systematic review and meta-analysis. *Annals of Medicine* **53**, 972 (2021).

80. Heretsch, P., Tzgkaroulaki, L. & Giannis, A. Cyclopamine and Hedgehog Signaling: Chemistry, Biology, Medical Perspectives. *Angew Chem Int Ed* **49**, 3418–3427 (2010).

81. Ernst, E. A systematic review of systematic reviews of homeopathy. *British Journal of Clinical Pharmacology* **54**, 577 (2002).

82. Hawke, K., King, D., Van Driel, M. L. & McGuire, T. M. Homeopathic medicinal products for preventing and treating acute respiratory tract infections in children. *Cochrane Database of Systematic Reviews* **2022**, (2022).

83. Pakpoor, J. Homeopathy is not an effective treatment for any health condition, report concludes. *BMJ* **350**, h1478–h1478 (2015).

84. Matute, H. *et al.* Illusions of causality: how they bias our everyday thinking and how they could be reduced. *Front. Psychol.* **6**, (2015).

85. Hróbjartsson, A. & Gøtzsche, P. C. Placebo interventions for all clinical conditions. *Cochrane Database of Systematic Reviews* (2010) doi:10.1002/14651858.CD003974.pub3.

86. Skinner, B. F. 'Superstition' in the pigeon. *Journal of Experimental Psychology* **38**, 168–172 (1948).

87. Fenton, T. R. & Huang, T. Systematic review of the association between dietary acid load, alkaline water and cancer. *BMJ Open* **6**, e010438 (2016).

88. Sun, D., Gao, W., Hu, H. & Zhou, S. Why 90% of clinical drug development fails and how to improve it? *Acta Pharmaceutica Sinica. B* **12**, 3049 (2022).

89. Baati, T. *et al.* The prolongation of the lifespan of rats by repeated oral administration of [60]fullerene. *Biomaterials* **33**, 4936–4946 (2012).

90. Shytikov, D. *et al.* Effect of Long-Term Treatment with C60 Fullerenes on the Lifespan and Health Status of CBA/Ca Mice. *Rejuvenation Res* **24**, 345–353 (2021).

91. Grohn, K. J. *et al.* C60 in olive oil causes light-dependent toxicity and does not extend lifespan in mice. *GeroScience* **43**, 579 (2020).

92. Keykhosravi, S. *et al.* [60]Fullerene for Medicinal Purposes, A Purity Criterion towards Regulatory Considerations. *Materials* **12**, 2571 (2019).

93. Campbell, E. K., Holz, M., Gerlich, D. & Maier, J. P. Laboratory confirmation of C60(+) as the carrier of two diffuse interstellar bands. *Nature* **523**, 322–323 (2015).

94. Cami, J., Bernard-Salas, J., Peeters, E. & Malek, S. E. Detection of C60 and C70 in a young planetary nebula. *Science* **329**, 1180–1182 (2010).

95. Catalgol, B., Batirel, S., Taga, Y. & Ozer, N. K. Resveratrol: French Paradox Revisited. *Frontiers in Pharmacology* **3**, 141 (2012).

96. Renaud, S. & de Lorgeril, M. Wine, alcohol, platelets, and the French paradox for coronary heart disease. *Lancet* **339**, 1523–1526 (1992).

97. Howitz, K. T. *et al.* Small molecule activators of sirtuins extend Saccharomyces cerevisiae lifespan. *Nature* **425**, 191–196 (2003).

98. Kaeberlein, M., McVey, M. & Guarente, L. The SIR2/3/4 complex and SIR2 alone promote longevity in Saccharomyces cerevisiae by two different mechanisms. *Genes & Development* **13**, 2570 (1999).

99. Tissenbaum, H. A. & Guarente, L. Increased dosage of a sir-2 gene extends lifespan in Caenorhabditis elegans. *Nature* **410**, 227–230 (2001).

REFERENCES

100. Rogina, B. & Helfand, S. L. Sir2 mediates longevity in the fly through a pathway related to calorie restriction. *Proc Natl Acad Sci U S A* **101**, 15998–16003 (2004).

101. Majeed, Y. *et al.* SIRT1 promotes lipid metabolism and mitochondrial biogenesis in adipocytes and coordinates adipogenesis by targeting key enzymatic pathways. *Sci Rep* **11**, 8177 (2021).

102. Yoshizaki, T. *et al.* SIRT1 inhibits inflammatory pathways in macrophages and modulates insulin sensitivity. *American Journal of Physiology-Endocrinology and Metabolism* **298**, E419–E428 (2010).

103. Rahman, S. & Islam, R. Mammalian Sirt1: insights on its biological functions. *Cell Communication and Signaling : CCS* **9**, 11 (2011).

104. Couzin, J. Scientific community. Aging research's family feud. *Science* **303**, 1276–1279 (2004).

105. Baur, J. A. *et al.* Resveratrol improves health and survival of mice on a high-calorie diet. *Nature* **444**, 337 (2006).

106. Nair, A. B. & Jacob, S. A simple practice guide for dose conversion between animals and human. *Journal of Basic and Clinical Pharmacy* **7**, 27 (2016).

107. Shin, J.-W., Seol, I.-C. & Son, C.-G. Interpretation of Animal Dose and Human Equivalent Dose for Drug Development.

108. Valenzano, D. R. *et al.* Resveratrol prolongs lifespan and retards the onset of age-related markers in a short-lived vertebrate. *Curr Biol* **16**, 296–300 (2006).

109. Beher, D. *et al.* Resveratrol is not a direct activator of SIRT1 enzyme activity. *Chem Biol Drug Des* **74**, 619–624 (2009).

110. Pacholec, M. *et al.* SRT1720, SRT2183, SRT1460, and Resveratrol Are Not Direct Activators of SIRT1. *The Journal of Biological Chemistry* **285**, 8340 (2010).

111. Macchiarini, F., Miller, R. A., Strong, R., Rosenthal, N. & Harrison, D. E. Chapter 10 - NIA Interventions Testing Program: A collaborative approach for investigating interventions to promote healthy aging. in *Handbook of the Biology of Aging (Ninth Edition)* (eds. Musi, N. & Hornsby, P. J.) 219–235 (Academic Press, 2021). doi:10.1016/B978-0-12-815962-0.00010-X.

112. Strong, R. *et al.* Evaluation of resveratrol, green tea extract, curcumin, oxaloacetic acid, and medium-chain triglyceride oil on life span of genetically heterogeneous mice. *J Gerontol A Biol Sci Med Sci* **68**, 6–16 (2013).

113. Miller, R. A. *et al.* Rapamycin, But Not Resveratrol or Simvastatin, Extends Life Span of Genetically Heterogeneous Mice. *The Journals of Gerontology: Series A* **66A**, 191–201 (2011).

114. Burnett, C. *et al.* Absence of effects of Sir2 over-expression on lifespan in C. elegans and Drosophila. *Nature* **477**, 482 (2011).

115. Mercken, E. M. *et al.* SIRT1 but not its increased expression is essential for lifespan extension in caloric-restricted mice. *Aging Cell* **13**, 193–196 (2014).

116. Herranz, D. *et al.* Sirt1 improves healthy ageing and protects from metabolic syndrome-associated cancer syndrome. *Nature communications* **1**, 3 (2010).

117. Herranz, D. & Serrano, M. Sirt1: recent lessons from mouse models. *Nature reviews. Cancer* **10**, 819 (2010).

118. Satoh, A. *et al.* Sirt1 extends life span and delays aging in mice through the regulation of Nk2 homeobox 1 in the DMH and LH. *Cell metabolism* **18**, 416 (2013).

119. Staats, S. *et al.* Dietary Resveratrol Does Not Affect Life Span, Body Composition, Stress Response, and Longevity-Related Gene Expression in Drosophila melanogaster. *IJMS* **19**, 223 (2018).

120. Curry, A. M., White, D. S., Donu, D. & Cen, Y. Human Sirtuin Regulators: The "Success" Stories. *Front. Physiol.* **12**, (2021).

121. Kjær, T. N. *et al.* No Beneficial Effects of Resveratrol on the Metabolic Syndrome: A Randomized Placebo-Controlled Clinical Trial. *J Clin Endocrinol Metab* **102**, 1642–1651 (2017).

122. Avalos, J. L., Bever, K. M. & Wolberger, C. Mechanism of Sirtuin Inhibition by Nicotinamide: Altering the NAD+ Cosubstrate Specificity of a Sir2 Enzyme. *Molecular Cell* **17**, 855–868 (2005).

123. Camacho-Pereira, J. *et al.* CD38 dictates age-related NAD decline and mitochondrial dysfunction through a SIRT3-dependent mechanism. *Cell metabolism* **23**, 1127 (2016).

124. Yoshida, M. *et al.* Extracellular vesicle-contained eNAMPT delays aging and extends lifespan in mice. *Cell metabolism* **30**, 329 (2019).

125. Tarragó, M. G. *et al.* A potent and specific CD38 inhibitor ameliorates age-related metabolic dysfunction by reversing tissue NAD+ decline. *Cell metabolism* **27**, 1081 (2018).

126. Global Nicotinamide Riboside (NR) Market Research Report 2023-Competitive Analysis, Status and Outlook by Type, Downstream Industry, and Geography, Forecast to 2029. https://www.

marketresearch.com/Maia-Research-v4212/Global-Nicotinamide-Riboside-NR-Research-33920597/.

127. FDA Purges NMN From the List of Supplements | Lifespan. io. https://www.lifespan.io/news/fda-purges-nmn-from-the-list-of-supplements/.

128. Zhang, H. et al. NAD⁺ repletion improves mitochondrial and stem cell function and enhances life span in mice. *Science* **352**, 1436–1443 (2016).

129. Mitchell, S. J. et al. Nicotinamide Improves Aspects of Healthspan, but Not Lifespan, in Mice. *Cell Metab* **27**, 667-676.e4 (2018).

130. Harrison, D. E. et al. 17-a-estradiol late in life extends lifespan in aging UM-HET3 male mice; nicotinamide riboside and three other drugs do not affect lifespan in either sex. *Aging Cell* **20**, e13328 (2021).

131. Bonkowski, M. S. & Sinclair, D. A. Slowing ageing by design: the rise of NAD+ and sirtuin-activating compounds. *Nature reviews. Molecular cell biology* **17**, 679 (2016).

132. Nadeeshani, H., Li, J., Ying, T., Zhang, B. & Lu, J. Nicotinamide mononucleotide (NMN) as an anti-aging health product — Promises and safety concerns. *Journal of Advanced Research* **37**, 267–278 (2022).

133. Picciotto, N. E. de et al. Nicotinamide mononucleotide supplementation reverses vascular dysfunction and oxidative stress with aging in mice. *Aging Cell* **15**, 522 (2016).

134. Das, A. et al. Impairment of an endothelial NAD+-H2S signaling network is a reversible cause of vascular aging. *Cell* **173**, 74 (2018).

135. Mills, K. F. et al. Long-term administration of nicotinamide mononucleotide mitigates age-associated physiological decline in mice. *Cell metabolism* **24**, 795 (2016).

136. Fu, R. & Zhang, Q. Use of beta-nicotinamide mononucleotide in preparation of anti-aging drugs or health-care products. (2017).

137. A, T. et al. Distinct longevity mechanisms across and within species and their association with aging. *Cell* **186**, (2023).

138. Madsen, K. S., Chi, Y., Metzendorf, M.-I., Richter, B. & Hemmingsen, B. Metformin for prevention or delay of type 2 diabetes mellitus and its associated complications in persons at increased risk for the development of type 2 diabetes mellitus. *Cochrane Database of Systematic Reviews* **2019**, (2019).

139. Mohammed, I., Hollenberg, M. D., Ding, H. & Triggle, C. R. A Critical Review of the Evidence That Metformin Is a Putative

Anti-Aging Drug That Enhances Healthspan and Extends Lifespan. *Front. Endocrinol.* **12**, 718942 (2021).

140. Xie, D. *et al.* Let-7 underlies metformin-induced inhibition of hepatic glucose production. *Proc. Natl. Acad. Sci. U.S.A.* **119**, e2122217119 (2022).

141. Martin-Montalvo, A. *et al.* Metformin improves healthspan and lifespan in mice. *Nat Commun* **4**, 2192 (2013).

142. Bannister, C. A. *et al.* Can people with type 2 diabetes live longer than those without? A comparison of mortality in people initiated with metformin or sulphonylurea monotherapy and matched, non-diabetic controls. *Diabetes Obes Metab* **16**, 1165–1173 (2014).

143. Strong, R. *et al.* Longer lifespan in male mice treated with a weakly estrogenic agonist, an antioxidant, an α-glucosidase inhibitor or a Nrf2-inducer. *Aging Cell* **15**, 872–884 (2016).

144. Keys, M. T. *et al.* Reassessing the evidence of a survival advantage in Type 2 diabetes treated with metformin compared with controls without diabetes: a retrospective cohort study. *Int J Epidemiol* **51**, 1886–1898 (2022).

145. H, N., A, G., T, T. & M, N. Cancer risk in diabetic patients treated with metformin: a systematic review and meta-analysis. *PloS one* **7**, (2012).

146. Gandini, S. *et al.* Metformin and cancer risk and mortality: a systematic review and meta-analysis taking into account biases and confounders. *Cancer Prev Res (Phila)* **7**, 867–885 (2014).

147. Lee, C. G. *et al.* Effect of Metformin and Lifestyle Interventions on Mortality in the Diabetes Prevention Program and Diabetes Prevention Program Outcomes Study. *Diabetes Care* **44**, 2775 (2021).

148. Li, X. *et al.* Metformin and health outcomes: An umbrella review of systematic reviews with meta-analyses. *Eur J Clin Investigation* **51**, e13536 (2021).

149. Infante, M., Leoni, M., Caprio, M. & Fabbri, A. Long-term metformin therapy and vitamin B12 deficiency: An association to bear in mind. *World Journal of Diabetes* **12**, 916 (2021).

150. DeFronzo, R., Fleming, G. A., Chen, K. & Bicsak, T. A. Metformin-associated lactic acidosis: Current perspectives on causes and risk. *Metabolism* **65**, 20–29 (2016).

151. Blough, B., Moreland, A. & Adan Mora, J. Metformin-induced lactic acidosis with emphasis on the anion gap. *Proceedings (Baylor University. Medical Center)* **28**, 31 (2015).

REFERENCES

152. Vecchio, S. & Protti, A. Metformin-induced lactic acidosis: no one left behind. *Crit Care* **15**, 107 (2011).

153. Roussel, R. Metformin Use and Mortality Among Patients With Diabetes and Atherothrombosis. *Arch Intern Med* **170**, 1892 (2010).

154. Li, T. *et al.* Association of Metformin with the Mortality and Incidence of Cardiovascular Events in Patients with Pre-existing Cardiovascular Diseases. *Drugs* **82**, 311–322 (2022).

155. Han, Y. *et al.* Effect of metformin on all-cause and cardiovascular mortality in patients with coronary artery diseases: a systematic review and an updated meta-analysis. *Cardiovasc Diabetol* **18**, 96 (2019).

156. Padki, M. M. & Stambler, I. Targeting Aging with Metformin (TAME). in *Encyclopedia of Gerontology and Population Aging* (eds. Gu, D. & Dupre, M. E.) 4908–4910 (Springer International Publishing, Cham, 2021). doi:10.1007/978-3-030-22009-9_400.

157. Mullard, A. Cancer reproducibility project yields first results. *Nat Rev Drug Discov* **16**, 77–77 (2017).

158. Wen, H., Wang, H.-Y., He, X. & Wu, C.-I. On the low reproducibility of cancer studies. *National science review* **5**, 619 (2018).

159. Vj, N. *et al.* Liver injury from herbal and dietary supplements. *Hepatology (Baltimore, Md.)* **65**, (2017).

160. Geller, A. I. *et al.* Emergency Department Visits for Adverse Events Related to Dietary Supplements. *The New England journal of medicine* **373**, 1531 (2015).

161. Ronis, M. J. J., Pedersen, K. B. & Watt, J. Adverse Effects of Nutraceuticals and Dietary Supplements. *Annu. Rev. Pharmacol. Toxicol.* **58**, 583–601 (2018).

162. White, C. M. Continued Risk of Dietary Supplements Adulterated With Approved and Unapproved Drugs: Assessment of the US Food and Drug Administration's Tainted Supplements Database 2007 Through 2021. *The Journal of Clinical Pharma* **62**, 928–934 (2022).

163. Affairs, O. of R. Health Fraud Product Database. *FDA* (2024).

164. Dunnick', J. K. & Hailey, J. R. Phenolphthalein Exposure Causes Multiple Carcinogenic Effects in Experimental Model Systems.

165. Program, H. F. DMAA in Products Marketed as Dietary Supplements. *FDA* (2024).

166. Dolan, S. B. & Gatch, M. B. Abuse liability of the dietary supplement dimethylamylamine. *Drug and Alcohol Dependence* **146**, 97–102 (2015).

167. Karnatovskaia, L. V., Leoni, J. C. & Freeman, M. L. Cardiac arrest in a 21-year-old man after ingestion of 1,3-DMAA-containing workout supplement. *Clin J Sport Med* **25**, e23-25 (2015).

168. Eliason, M. J. *et al.* Case Reports: Death of Active Duty Soldiers Following Ingestion of Dietary Supplements Containing 1,3-Dimethylamylamine (DMAA). *Military Medicine* **177**, 1455–1459 (2012).

169. Scheen, A. J. Sibutramine on Cardiovascular Outcome. *Diabetes Care* **34**, S114 (2011).

170. Bjelakovic, G., Nikolova, D., Gluud, L. L., Simonetti, R. G. & Gluud, C. Antioxidant supplements for prevention of mortality in healthy participants and patients with various diseases. *Cochrane Database of Systematic Reviews* **2012**, (2012).

171. Spindler, S. R., Mote, P. L., Flegal, J. M. & Teter, B. Influence on longevity of blueberry, cinnamon, green and black tea, pomegranate, sesame, curcumin, morin, pycnogenol, quercetin, and taxifolin fed iso-calorically to long-lived, F1 hybrid mice. *Rejuvenation Res* **16**, 143–151 (2013).

172. Gladyshev, V. N. The Free Radical Theory of Aging Is Dead. Long Live the Damage Theory! *Antioxidants & Redox Signaling* **20**, 727 (2014).

173. De-Regil, L. M., Peña-Rosas, J. P., Fernández-Gaxiola, A. C. & Rayco-Solon, P. Effects and safety of periconceptional oral folate supplementation for preventing birth defects. *Cochrane Database of Systematic Reviews* **2015**, (2015).

174. Bird, J. K., Murphy, R. A., Ciappio, E. D. & McBurney, M. I. Risk of Deficiency in Multiple Concurrent Micronutrients in Children and Adults in the United States. *Nutrients* **9**, 655 (2017).

175. Knapik, J. J., Farina, E. K., Victor L Fulgoni, I. I. I. & Lieberman, H. R. Clinically-diagnosed vitamin deficiencies and disorders in the entire United States military population, 1997–2015. *Nutrition Journal* **20**, 55 (2021).

176. Miller, E. R. *et al.* Meta-analysis: high-dosage vitamin E supplementation may increase all-cause mortality. *Ann Intern Med* **142**, 37–46 (2005).

177. Abner, E. L., Schmitt, F. A., Mendiondo, M. S., Marcum, J. L. & Kryscio, R. J. Vitamin E and all-cause mortality: A meta-analysis. *Current aging science* **4**, 158 (2011).

REFERENCES

178. Kryscio, R. J. *et al.* Association of Antioxidant Supplement Use and Dementia in the Prevention of Alzheimer's Disease by Vitamin E and Selenium Trial (PREADViSE). *JAMA Neurol* **74**, 567 (2017).

179. Ernst, I. M. A. *et al.* Vitamin E supplementation and lifespan in model organisms. *Ageing Research Reviews* **12**, 365–375 (2013).

180. Selman, C., McLaren, J. S., Collins, A. R., Duthie, G. G. & Speakman, J. R. Deleterious consequences of antioxidant supplementation on lifespan in a wild-derived mammal. *Biol. Lett.* **9**, 20130432 (2013).

181. The Effect of Vitamin E and Beta Carotene on the Incidence of Lung Cancer and Other Cancers in Male Smokers. *N Engl J Med* **330**, 1029–1035 (1994).

182. Cortés-Jofré, M., Rueda, J.-R., Asenjo-Lobos, C., Madrid, E. & Bonfill Cosp, X. Drugs for preventing lung cancer in healthy people. *Cochrane Database of Systematic Reviews* **2020**, (2020).

183. Zhang, Y. *et al.* Association between vitamin D supplementation and mortality: systematic review and meta-analysis. *BMJ* l4673 (2019) doi:10.1136/bmj.l4673.

184. Manson, J. E., Bassuk, S. S., Buring, J. E., & VITAL Research Group. Principal results of the VITamin D and OmegA-3 TriaL (VITAL) and updated meta-analyses of relevant vitamin D trials. *J Steroid Biochem Mol Biol* **198**, 105522 (2020).

185. Vitamin D and Omega-3 Trial - American College of Cardiology. https://www.acc.org/Latest-in-Cardiology/Clinical-Trials/2018/11/08/22/42/VITAL.

186. O'Connor, E. A. *et al.* Vitamin and Mineral Supplements for the Primary Prevention of Cardiovascular Disease and Cancer: Updated Evidence Report and Systematic Review for the US Preventive Services Task Force. *JAMA* **327**, 2334 (2022).

187. Ferrell, M. *et al.* A terminal metabolite of niacin promotes vascular inflammation and contributes to cardiovascular disease risk. *Nat Med* **30**, 424–434 (2024).

188. Schandelmaier, S. *et al.* Niacin for primary and secondary prevention of cardiovascular events. *The Cochrane Database of Systematic Reviews* **2017**, CD009744 (2017).

189. D'Andrea, E., Hey, S. P., Ramirez, C. L. & Kesselheim, A. S. Assessment of the Role of Niacin in Managing Cardiovascular Disease Outcomes: A Systematic Review and Meta-analysis. *JAMA Netw Open* **2**, e192224 (2019).

190. Jenkins, D. J. A. *et al.* Supplemental Vitamins and Minerals for Cardiovascular Disease Prevention and Treatment: JACC Focus Seminar. *Journal of the American College of Cardiology* **77**, 423–436 (2021).

191. Leosdottir, M., Nilsson, P., Nilsson, J.-A., Månsson, H. & Berglund, G. The association between total energy intake and early mortality: data from the Malmö Diet and Cancer Study. *J Intern Med* **256**, 499–509 (2004).

192. Lee, C.-L., Liu, W.-J. & Wang, J.-S. Association of diurnal calorie trajectory with all-cause mortality: Findings from the National Health and Nutrition Examination Survey. *Clinical Nutrition* **40**, 1920–1925 (2021).

193. Seidelmann, S. B. *et al.* Dietary carbohydrate intake and mortality: a prospective cohort study and meta-analysis. *The Lancet. Public health* **3**, e419 (2018).

194. Mazidi, M., Katsiki, N., Mikhailidis, D. P., Sattar, N. & Banach, M. Lower carbohydrate diets and all-cause and cause-specific mortality: a population-based cohort study and pooling of prospective studies. *European Heart Journal* **40**, 2870–2879 (2019).

195. English, L. K. *et al.* Evaluation of Dietary Patterns and All-Cause Mortality: A Systematic Review. *JAMA Netw Open* **4**, e2122277 (2021).

196. Finicelli, M., Salle, A. D., Galderisi, U. & Peluso, G. The Mediterranean Diet: An Update of the Clinical Trials. *Nutrients* **14**, 2956 (2022).

197. Papadaki, A., Nolen-Doerr, E. & Mantzoros, C. S. The Effect of the Mediterranean Diet on Metabolic Health: A Systematic Review and Meta-Analysis of Controlled Trials in Adults. *Nutrients* **12**, 3342 (2020).

198. Guasch-Ferré, M. & Willett, W. C. The Mediterranean diet and health: a comprehensive overview. *J Intern Med* **290**, 549–566 (2021).

199. Farvid, M. S. *et al.* Consumption of red meat and processed meat and cancer incidence: a systematic review and meta-analysis of prospective studies. *Eur J Epidemiol* **36**, 937–951 (2021).

200. Lescinsky, H. *et al.* Health effects associated with consumption of unprocessed red meat: a Burden of Proof study. *Nat Med* **28**, 2075–2082 (2022).

201. Iqbal, R. *et al.* Associations of unprocessed and processed meat intake with mortality and cardiovascular disease in 21 countries [Prospective Urban Rural Epidemiology (PURE) Study]: a prospective

cohort study. *The American Journal of Clinical Nutrition* **114**, 1049–1058 (2021).

202. Khaw, K.-T. *et al.* Combined impact of health behaviours and mortality in men and women: the EPIC-Norfolk prospective population study. *PLoS Med* **5**, e12 (2008).

203. Show & Tell Projects Archive. *Quantified Self* https://quantifiedself.com/show-and-tell/.

204. Ruegsegger, G. N. *et al.* A meal enriched in saturated fat acutely impairs cognitive performance in obese men. *Physiol Behav* **244**, 113664 (2022).

205. Cao, G.-Y. *et al.* Dietary Fat Intake and Cognitive Function among Older Populations: A Systematic Review and Meta-Analysis. *J Prev Alzheimers Dis* **6**, 204–211 (2019).

206. Hooper, L. *et al.* Reduction in saturated fat intake for cardiovascular disease. *Cochrane Database of Systematic Reviews* (2020) doi:10.1002/14651858.CD011737.pub2.

207. REYNOLDS, A. N. *Saturated Fat and Trans-Fat Intakes and Their Replacement with Other Macronutrients: A Systematic Review and Meta-Analysis of Prospective Observational Studies.* (World Health Organization, Geneva, 2023).

208. Healthy diet. https://www.who.int/news-room/fact-sheets/detail/healthy-diet.

209. Welcome to my colon: Tech pioneer guides his surgeon with VR. https://www.statnews.com/2017/03/09/colon-virtual-reality/.

210. Kale, M. S. & Korenstein, D. Overdiagnosis in primary care: framing the problem and finding solutions. *The BMJ* **362**, k2820 (2018).

211. Albarqouni, L. *et al.* Overdiagnosis and overuse of diagnostic and screening tests in low-income and middle-income countries: a scoping review. *BMJ Global Health* **7**, e008696 (2022).

212. Sharon, T. & Zandbergen, D. From data fetishism to quantifying selves: Self-tracking practices and the other values of data. *New Media & Society* **19**, (2017).

213. Chiodo, S. Quantified Self as Epistemological Anarchism. *Philosophia* **50**, 1665–1685 (2022).

214. Holgersson, J., Ceric, A., Sethi, N., Nielsen, N. & Jakobsen, J. C. Fever therapy in febrile adults: systematic review with meta-analyses and trial sequential analyses. *The BMJ* **378**, e069620 (2022).

215. Glycated haemoglobin (HbA1c) for the diagnosis of diabetes. in *Use of Glycated Haemoglobin (HbA1c) in the Diagnosis of Diabetes*

Mellitus: Abbreviated Report of a WHO Consultation (World Health Organization, 2011).

216. Levine, M. E. *et al.* An epigenetic biomarker of aging for lifespan and healthspan. *Aging (Albany NY)* **10**, 573 (2018).

217. Föhr, T. *et al.* Does the epigenetic clock GrimAge predict mortality independent of genetic influences: an 18 year follow-up study in older female twin pairs. *Clin Epigenet* **13**, 128 (2021).

218. Lu, A. T. *et al.* DNA methylation GrimAge strongly predicts lifespan and healthspan. *Aging (Albany NY)* **11**, 303–327 (2019).

219. Ying, K. *et al.* Causality-Enriched Epigenetic Age Uncouples Damage and Adaptation. Preprint at https://doi.org/10.1101/2022.10.07.511382 (2022).

220. Bankhead, C. Benefits of acarbose and metformin in type 2 diabetes: *Inpharma Weekly* **NA;**, 13–14 (1998).

221. DiNicolantonio, J. J., Bhutani, J. & O'Keefe, J. H. Acarbose: safe and effective for lowering postprandial hyperglycaemia and improving cardiovascular outcomes. *Open Heart* **2**, e000327 (2015).

222. Harrison, D. E. *et al.* Acarbose improves health and lifespan in aging HET3 mice. *Aging Cell* **18**, e12898 (2019).

223. Harrison, D. E. *et al.* Acarbose, 17-α-estradiol, and nordihydroguaiaretic acid extend mouse lifespan preferentially in males. *Aging Cell* **13**, 273–282 (2014).

224. Miller, R. A. *et al.* Glycine supplementation extends lifespan of male and female mice. *Aging Cell* **18**, e12953 (2019).

225. Brind, J. *et al.* Dietary glycine supplementation mimics lifespan extension by dietary methionine restriction in Fisher 344 rats. *The FASEB Journal* **25**, (2011).

226. Johnson, A. A. & Cuellar, T. L. Glycine and aging: Evidence and mechanisms. *Ageing Res Rev* **87**, 101922 (2023).

227. Paz-Lugo, P. de, Lupiáñez, J. A. & Meléndez-Hevia, E. High glycine concentration increases collagen synthesis by articular chondrocytes in vitro: acute glycine deficiency could be an important cause of osteoarthritis. *Amino Acids* **50**, 1357 (2018).

228. Ripps, H. & Shen, W. Review: Taurine: A "very essential" amino acid. *Molecular Vision* **18**, 2673 (2012).

229. Singh, P. *et al.* Taurine deficiency as a driver of aging. *Science* **380**, eabn9257 (2023).

230. Qaradakhi, T. *et al.* The Anti-Inflammatory Effect of Taurine on Cardiovascular Disease. *Nutrients* **12**, 2847 (2020).

231. Ansar, M. *et al.* Taurine treatment of retinal degeneration and cardiomyopathy in a consanguineous family with SLC6A6 taurine transporter deficiency. *Hum Mol Genet* **29**, 618–623 (2020).

232. Tofalo, R., Cocchi, S. & Suzzi, G. Polyamines and Gut Microbiota. *Front. Nutr.* **6**, 16 (2019).

233. Ali, M. A., Poortvliet, E., Strömberg, R. & Yngve, A. Polyamines in foods: development of a food database. *Food & Nutrition Research* **55**, 10.3402/fnr.v55i0.5572 (2011).

234. Kiechl, S. *et al.* Higher spermidine intake is linked to lower mortality: a prospective population-based study. *Am J Clin Nutr* **108**, 371–380 (2018).

235. Wu, H. *et al.* The association of dietary spermidine with all-cause mortality and CVD mortality: The U.S. National Health and Nutrition Examination Survey, 2003 to 2014. *Front Public Health* **10**, 949170 (2022).

236. Cassidy, L. D. *et al.* Temporal inhibition of autophagy reveals segmental reversal of ageing with increased cancer risk. *Nat Commun* **11**, 307 (2020).

237. Eisenberg, T. *et al.* Cardioprotection and lifespan extension by the natural polyamine spermidine. *Nature medicine* **22**, 1428 (2016).

238. Filfan, M. *et al.* Long-term treatment with spermidine increases health span of middle-aged Sprague-Dawley male rats. *GeroScience* **42**, 937 (2020).

239. Hofer, S. J. *et al.* Mechanisms of spermidine-induced autophagy and geroprotection. *Nat Aging* **2**, 1112–1129 (2022).

240. Kim, Y., Je, Y. & Giovannucci, E. Coffee consumption and all-cause and cause-specific mortality: a meta-analysis by potential modifiers. *Eur J Epidemiol* **34**, 731–752 (2019).

241. Yamada-Fowler, N. & Söderkvist, P. Coffee, Genetic Variants, and Parkinson's Disease: Gene–Environment Interactions. *Journal of Caffeine Research* **5**, 3 (2015).

242. Yang, A., Palmer, A. A. & Wit, H. de. Genetics of caffeine consumption and responses to caffeine. *Psychopharmacology* **211**, 245 (2010).

243. Chung, M. *et al.* Dose-Response Relation between Tea Consumption and Risk of Cardiovascular Disease and All-Cause Mortality: A Systematic Review and Meta-Analysis of Population-Based Studies. *Adv Nutr* **11**, 790–814 (2020).

244. Cornelis, M. C., El-Sohemy, A., Kabagambe, E. K. & Campos, H. Coffee, CYP1A2 genotype, and risk of myocardial infarction. *JAMA* **295**, 1135–1141 (2006).

245. Palatini, P. *et al.* CYP1A2 genotype modifies the association between coffee intake and the risk of hypertension. *J Hypertens* **27**, 1594–1601 (2009).

246. Opinion | My Medical Choice by Angelina Jolie — The New York Times. https://www.nytimes.com/2013/05/14/opinion/my-medical-choice.html.

247. Kotsopoulos, J. BRCA Mutations and Breast Cancer Prevention. *Cancers* **10**, 524 (2018).

248. Desai, S. & Jena, A. B. Do celebrity endorsements matter? Observational study of BRCA gene testing and mastectomy rates after Angelina Jolies New York Times editorial. *BMJ* i6357 (2016) doi:10.1136/bmj.i6357.

249. Basu, N. N. *et al.* The Angelina Jolie effect: Contralateral risk-reducing mastectomy trends in patients at increased risk of breast cancer. *Sci Rep* **11**, 2847 (2021).

250. Friedenson, B. The BRCA1/2 pathway prevents hematologic cancers in addition to breast and ovarian cancers. *BMC Cancer* **7**, 152 (2007).

251. Castro, E. & Eeles, R. The role of BRCA1 and BRCA2 in prostate cancer. *Asian Journal of Andrology* **14**, 409 (2012).

252. Mai, P. L. *et al.* Potential Excess Mortality in BRCA1/2 Mutation Carriers beyond Breast, Ovarian, Prostate, and Pancreatic Cancers, and Melanoma. *PLoS ONE* **4**, e4812 (2009).

253. O'Sullivan, B. P. & Freedman, S. D. Cystic fibrosis. *The Lancet* **373**, 1891–1904 (2009).

254. Martin, C. & Burgel, P.-R. Carriers of a single *CFTR* mutation are asymptomatic: an evolving dogma? *Eur Respir J* **56**, 2002645 (2020).

255. CBIPM Genomics Team *et al.* Exome sequencing reveals a high prevalence of BRCA1 and BRCA2 founder variants in a diverse population-based biobank. *Genome Med* **12**, 2 (2020).

256. Smith, K. R., Hanson, H. A., Mineau, G. P. & Buys, S. S. Effects of BRCA1 and BRCA2 mutations on female fertility. *Proceedings of the Royal Society B: Biological Sciences* **279**, 1389 (2011).

257. Kwiatkowski, F. *et al.* BRCA Mutations Increase Fertility in Families at Hereditary Breast/Ovarian Cancer Risk. *PLoS One* **10**, e0127363 (2015).

REFERENCES

258. Eriksson, N. *et al.* Genetic variants associated with breast size also influence breast cancer risk. *BMC Med Genet* **13**, 53 (2012).

259. Daum, H., Peretz, T. & Laufer, N. BRCA mutations and reproduction. *Fertil Steril* **109**, 33–38 (2018).

260. Solberg, O. K., Filkuková, P., Frich, J. C. & Feragen, K. J. B. Age at Death and Causes of Death in Patients with Huntington Disease in Norway in 1986–2015. *JHD* **7**, 77–86 (2018).

261. Mastromauro, C. A. *et al.* Estimation of fertility and fitness in Huntington disease in New England. *Am. J. Med. Genet.* **33**, 248–254 (1989).

262. Pridmore, S. A. & Adams, G. C. The fertility of HD-affected individuals in Tasmania. *Aust N Z J Psychiatry* **25**, 262–264 (1991).

263. Walker, D. A., Harper, P. S., Newcombe, R. G. & Davies, K. Huntington's chorea in South Wales: mutation, fertility, and genetic fitness. *Journal of Medical Genetics* **20**, 12 (1983).

264. Shokeir, M. H. Investigation on Huntington's disease in the Canadian Prairies. II. Fecundity and fitness. *Clin Genet* **7**, 349–353 (1975).

265. Carter, A. J. & Nguyen, A. Q. Antagonistic pleiotropy as a widespread mechanism for the maintenance of polymorphic disease alleles. *BMC Medical Genetics* **12**, 160 (2011).

266. Morton, A. J., Skillings, E. A., Wood, N. I. & Zheng, Z. Antagonistic pleiotropy in mice carrying a CAG repeat expansion in the range causing Huntington's disease. *Sci Rep* **9**, 37 (2019).

267. Sørensen, S. A., Fenger, K. & Olsen, J. H. Significantly lower incidence of cancer among patients with Huntington disease: An apoptotic effect of an expanded polyglutamine tract? *Cancer* **86**, 1342–1346 (1999).

268. McNulty, P. *et al.* Reduced Cancer Incidence in Huntington's Disease: Analysis in the Registry Study. *J Huntingtons Dis* **7**, 209–222 (2018).

269. Li, S.-H. Intranuclear huntingtin increases the expression of caspase-1 and induces apoptosis. *Human Molecular Genetics* **9**, 2859–2867 (2000).

270. Saudou, F., Finkbeiner, S., Devys, D. & Greenberg, M. E. Huntingtin acts in the nucleus to induce apoptosis but death does not correlate with the formation of intranuclear inclusions. *Cell* **95**, 55–66 (1998).

271. Dawkins, R. *The Selfish Gene*. (Oxford University Press, 1976).

272. Williams, G. C. Pleiotropy, Natural Selection, and the Evolution of Senescence. *Evolution* **11**, 398–411 (1957).

273. Medawar, P. B. *An Unsolved Problem of Biology: An Inaugural Lecture Delivered at University College, London, 6 December, 1951.* (H.K. Lewis and Company, 1952).

274. Turan, Z. G. *et al.* Molecular footprint of Medawar's mutation accumulation process in mammalian aging. *Aging Cell* **18**, e12965 (2019).

275. Dobzhansky, T. Nothing in Biology Makes Sense except in the Light of Evolution. *The American Biology Teacher* **35**, 125–129 (1973).

276. Rose, M. R. LABORATORY EVOLUTION OF POSTPONED SENESCENCE IN DROSOPHILA MELANOGASTER. *Evolution* **38**, 1004–1010 (1984).

277. Pascual-Torner, M. *et al.* Comparative genomics of mortal and immortal cnidarians unveils novel keys behind rejuvenation. *Proceedings of the National Academy of Sciences* **119**, e2118763119 (2022).

278. Gatti, S. *The Role of Sponges in High-Antarctic Carbon and Silicon Cycling- a Modelling Approach (Die Rolle Der Schwämme Im Hochantarktischen Kohlenstoff- Und Silikatkreislauf - Ein Modellierungsansatz). Berichte zur Polar- und Meeresforschung = Reports on Polar and Marine Research* vol. 434 1–124 https://www.tib.eu/suchen/id/awi:2d537bfca8893aab9e166d5a30772efd250a0eb6 (2002).

279. Durkin, A., Fisher, C. R. & Cordes, E. E. Extreme longevity in a deep-sea vestimentiferan tubeworm and its implications for the evolution of life history strategies. *Naturwissenschaften* **104**, 63 (2017).

280. Nielsen, J. *et al.* Eye lens radiocarbon reveals centuries of longevity in the Greenland shark (*Somniosus microcephalus*). *Science* **353**, 702–704 (2016).

281. Nielsen, J. *et al.* Eye lens radiocarbon reveals centuries of longevity in the Greenland shark (*Somniosus microcephalus*). *Science* **353**, 702–704 (2016).

282. Sweeney, B. W. & Vannote, R. L. *Population Synchrony in Mayflies: A Predator Satiation Hypothesis.* (Evolution 36:810-821, 1982).

283. Depczynski, M. & Bellwood, D. R. Shortest recorded vertebrate lifespan found in a coral reef fish. *Current Biology* **15**, R288–R289 (2005).

284. Sun, Y. *et al.* Aging Studies in Drosophila melanogaster. *Methods in molecular biology (Clifton, N.J.)* **1048**, 77 (2013).

285. Healy, K. *et al.* Ecology and mode-of-life explain lifespan variation in birds and mammals. *Proc. R. Soc. B.* **281**, 20140298 (2014).

286. Stark, G. & Meiri, S. Cold and dark captivity: Drivers of amphibian longevity. *Global Ecol Biogeogr* **27**, 1384–1397 (2018).

287. Hossie, T. J., Hassall, C., Knee, W. & Sherratt, T. N. Species with a chemical defence, but not chemical offence, live longer. *J Evol Biol* **26**, 1598–1602 (2013).

288. Zhu, P. *et al.* Correlated evolution of social organization and lifespan in mammals. *Nat Commun* **14**, 372 (2023).

289. Stearns, S. C., Ackermann, M., Doebeli, M. & Kaiser, M. Experimental evolution of aging, growth, and reproduction in fruitflies. *Proc. Natl. Acad. Sci. U.S.A.* **97**, 3309–3313 (2000).

290. Shokhirev, M. N. & Johnson, A. A. Effects of Extrinsic Mortality on the Evolution of Aging: A Stochastic Modeling Approach. *PLoS ONE* **9**, e86602 (2014).

291. Chen, H.-Y. & Maklakov, A. A. Longer life span evolves under high rates of condition-dependent mortality. *Curr Biol* **22**, 2140–2143 (2012).

292. Anderson, J. L., Reynolds, R. M., Morran, L. T., Tolman-Thompson, J. & Phillips, P. C. Experimental Evolution Reveals Antagonistic Pleiotropy in Reproductive Timing but Not Life Span in Caenorhabditis elegans. *The Journals of Gerontology: Series A* **66A**, 1300–1308 (2011).

293. Platt, O. S. *et al.* Mortality In Sickle Cell Disease -- Life Expectancy and Risk Factors for Early Death. *N Engl J Med* **330**, 1639–1644 (1994).

294. Eridani, S. Sickle cell protection from malaria. *Hematology Reports* **3**, e24 (2011).

295. Werner, H., Sarfstein, R., Nagaraj, K. & Laron, Z. Laron Syndrome Research Paves the Way for New Insights in Oncological Investigation. *Cells* **9**, 2446 (2020).

296. Zhu, F. *et al.* Symbiotic polydnavirus and venom reveal parasitoid to its hyperparasitoids. *Proceedings of the National Academy of Sciences* **115**, 5205–5210 (2018).

297. Kenyon, C., Chang, J., Gensch, E., Rudner, A. & Tabtiang, R. A C. elegans mutant that lives twice as long as wild type. *Nature* **366**, 461–464 (1993).

298. Garsin, D. A. *et al.* Long-Lived *C. elegans daf-2* Mutants Are Resistant to Bacterial Pathogens. *Science* **300**, 1921–1921 (2003).

299. Jenkins, N. L., McColl, G. & Lithgow, G. J. Fitness cost of extended lifespan in Caenorhabditis elegans. *Proceedings. Biological sciences / The Royal Society* **271**, 2523 (2004).

300. Henderson, S. T. & Johnson, T. E. daf-16 integrates developmental and environmental inputs to mediate aging in the nematode Caenorhabditis elegans. *Curr Biol* **11**, 1975–1980 (2001).

301. Savory, F. R., Benton, T. G., Varma, V., Hope, I. A. & Sait, S. M. Stressful environments can indirectly select for increased longevity. *Ecol Evol* **4**, 1176–1185 (2014).

302. Lithgow, G. J., White, T. M., Hinerfeld, D. A. & Johnson, T. E. Thermotolerance of a long-lived mutant of Caenorhabditis elegans. *J Gerontol* **49**, B270-276 (1994).

303. Maklakov, A. A. *et al.* Antagonistically pleiotropic allele increases lifespan and late-life reproduction at the cost of early-life reproduction and individual fitness. *Proc Biol Sci* **284**, 20170376 (2017).

304. Kelemen, D. Why are rocks pointy? Children's preference for teleological explanations of the natural world. *Dev Psychol* **35**, 1440–1452 (1999).

305. Wang, Z. Y. & Ragsdale, C. W. Multiple optic gland signaling pathways implicated in octopus maternal behaviors and death. *Journal of Experimental Biology* **221**, jeb185751 (2018).

306. Wang, Z. Y., Pergande, M. R., Ragsdale, C. W. & Cologna, S. M. Steroid hormones of the octopus self-destruct system. *Current Biology* **32**, 2572-2579.e4 (2022).

307. Wells, M. J. & Wells, J. Hormonal Control of Sexual Maturity in Octopus. *Journal of Experimental Biology* **36**, 1–33 (1959).

308. Wodinsky, J. Hormonal inhibition of feeding and death in octopus: control by optic gland secretion. *Science* **198**, 948–951 (1977).

309. Huang, X. *et al.* Pancreatic cancer cell-derived IGFBP-3 contributes to muscle wasting. *J Exp Clin Cancer Res* **35**, 46 (2016).

310. Kwon, Y. *et al.* Systemic organ wasting induced by localized expression of the secreted insulin/IGF antagonist ImpL2. *Dev Cell* **33**, 36–46 (2015).

311. Figueroa-Clarevega, A. & Bilder, D. Malignant Drosophila tumors interrupt insulin signaling to induce cachexia-like wasting. *Dev Cell* **33**, 47–55 (2015).

312. Argilés, J. M., Busquets, S., Stemmler, B. & López-Soriano, F. J. Cancer cachexia: understanding the molecular basis. *Nat Rev Cancer* **14**, 754–762 (2014).

REFERENCES

313. Robertson, O. H. & Wexler, B. C. HISTOLOGICAL CHANGES IN THE ORGANS AND TISSUES OF MIGRATING AND SPAWNING PACIFIC SALMON (*GENUS ONCORHYNCHUS*)[1]. *Endocrinology* **66**, 222–239 (1960).

314. Carruth, L. L. Cortisol and Pacific Salmon: A New Look at the Role of Stress Hormones in Olfaction and Home-stream Migration. *Integrative and Comparative Biology* **42**, 574–581 (2002).

315. Limumpornpetch, P. et al. The Effect of Endogenous Cushing Syndrome on All-cause and Cause-specific Mortality. *J Clin Endocrinol Metab* **107**, 2377–2388 (2022).

316. Robertson, O. H. Prolongation of the Life Span of Kokanee Salmon (Oncorhynchus Nerka Kennerlyi) by Castration before Beginning of Gonad Development. *Proceedings of the National Academy of Sciences of the United States of America* **47**, 609–621 (1961).

317. Dickhoff, W. SALMONIDS AND ANNUAL FISHES: DEATH AFTER SEX. (1989).

318. McBride, J. R., Fagerlund, U. H. M., Smith, M. & Tomlinson, N. Resumption of Feeding by and Survival of Adult Sockeye Salmon (*Oncorhynchus nerka*) Following Advanced Gonad Development. *J. Fish. Res. Bd. Can.* **20**, 95–100 (1963).

319. Coburn, C. et al. Anthranilate Fluorescence Marks a Calcium-Propagated Necrotic Wave That Promotes Organismal Death in C. elegans. *PLoS Biology* **11**, e1001613 (2013).

320. Bargiotas, P. et al. Connexin 36 promotes cortical spreading depolarization and ischemic brain damage. *Brain Res* **1479**, 80 85 (2012).

321. Chen, B., Sun, T., Wu, X. & Ma, J. Correlation between connexin and traumatic brain injury in patients. *Brain Behav* **7**, e00770 (2017).

322. García-Dorado, D., Rodríguez-Sinovas, A. & Ruiz-Meana, M. Gap junction-mediated spread of cell injury and death during myocardial ischemia-reperfusion. *Cardiovasc Res* **61**, 386–401 (2004).

323. Schulz, R. et al. Connexin 43 is an emerging therapeutic target in ischemia/reperfusion injury, cardioprotection and neuroprotection. *Pharmacology & therapeutics* **153**, 90 (2015).

324. Prinzinger, R. Programmed ageing: the theory of maximal metabolic scope. *EMBO Reports* **6**, S14 (2005).

325. Skulachev, V. P. Aging is a specific biological function rather than the result of a disorder in complex living systems: biochemical

evidence in support of Weismann's hypothesis. *Biochemistry (Mosc)* **62**, 1191–1195 (1997).

326. Foellmer, M. W. & Fairbairn, D. J. Spontaneous male death during copulation in an orb-weaving spider. *Proceedings of the Royal Society B: Biological Sciences* **270**, S183 (2003).

327. Finocchiaro, G. *et al.* Association of Sexual Intercourse With Sudden Cardiac Death in Young Individuals in the United Kingdom. *JAMA Cardiol* **7**, 358 (2022).

328. Hammock, E. A. & Young, L. J. Oxytocin, vasopressin and pair bonding: implications for autism. *Philosophical Transactions of the Royal Society B: Biological Sciences* **361**, 2187 (2006).

329. Kim, J., Chang, C. & Tucker, M. L. To grow old: regulatory role of ethylene and jasmonic acid in senescence. *Frontiers in Plant Science* **6**, 20 (2015).

330. Xiao, Y. *et al.* Banana ethylene response factors are involved in fruit ripening through their interactions with ethylene biosynthesis genes. *Journal of Experimental Botany* **64**, 2499 (2013).

331. Young, T. P. & Augspurger, C. K. Ecology and evolution of long-lived semelparous plants. *Trends in Ecology & Evolution* **6**, 285–289 (1991).

332. Kindsvater, H. K., Braun, D. C., Otto, S. P. & Reynolds, J. D. Costs of reproduction can explain the correlated evolution of semelparity and egg size: theory and a test with salmon. *Ecology Letters* **19**, 687–696 (2016).

333. Bradley, A. J., McDonald, I. R. & Lee, A. K. Stress and mortality in a small marsupial (*Antechinus stuartii*, Macleay). *General and Comparative Endocrinology* **40**, 188–200 (1980).

334. Fisher, D. O., Dickman, C. R., Jones, M. E. & Blomberg, S. P. Sperm competition drives the evolution of suicidal reproduction in mammals. *Proc. Natl. Acad. Sci. U.S.A.* **110**, 17910–17914 (2013).

335. Lazenby-Cohen, K. A. & Cockburn, A. Lek Promiscuity in a Semelparous Mammal, Antechinus stuartii (Marsupialia: Dasyuridae)? *Behavioral Ecology and Sociobiology* **22**, 195–202 (1988).

336. McCulloch, D. & Gems, D. Evolution of male longevity bias in nematodes. *Aging Cell* **2**, 165–173 (2003).

337. Hawkes, K., O'Connell, J. F., Jones, N. G. B., Alvarez, H. & Charnov, E. L. Grandmothering, menopause, and the evolution of human life histories. *Proceedings of the National Academy of Sciences* **95**, 1336–1339 (1998).

REFERENCES

338. Thompson, M. E. *et al*. Aging and fertility patterns in wild chimpanzees provide insights into the evolution of menopause. *Current biology : CB* **17**, 2150 (2007).

339. Hawkes, K., O'Connell, J. F. & Blurton Jones, N. G. Hadza Women's Time Allocation, Offspring Provisioning, and the Evolution of Long Postmenopausal Life Spans. *Current Anthropology* **38**, 551–577 (1997).

340. Lieberman, D. E., Kistner, T. M., Richard, D., Lee, I.-M. & Baggish, A. L. The active grandparent hypothesis: Physical activity and the evolution of extended human healthspans and lifespans. *Proceedings of the National Academy of Sciences* **118**, e2107621118 (2021).

341. Dy, R. L., Przybilski, R., Semeijn, K., Salmond, G. P. C. & Fineran, P. C. A widespread bacteriophage abortive infection system functions through a Type IV toxin–antitoxin mechanism. *Nucleic Acids Research* **42**, 4590–4605 (2014).

342. Weismann, A. *Essays Upon Heredity and Kindred Biological Problems*. vol. 1 (Clarendon Press, Oxford, 1891).

343. Fedorova, L. *et al*. Analysis of Common SNPs across Continents Reveals Major Genomic Differences between Human Populations. *Genes* **13**, 1472 (2022).

344. Kong, A. *et al*. Rate of de novo mutations, father's age, and disease risk. *Nature* **488**, 471 (2012).

345. Walker, R. F. *et al*. A case study of 'disorganized development' and its possible relevance to genetic determinants of aging. *Mech Ageing Dev* **130**, 350–356 (2009).

346. Walker, R. F. *et al*. Epigenetic age analysis of children who seem to evade aging. *Aging (Albany NY)* **7**, 334 (2015).

347. Walker, R. F. Cessation of Somatic Growth Aging Theory. in *Encyclopedia of Gerontology and Population Aging* (eds. Gu, D. & Dupre, M. E.) 914–923 (Springer International Publishing, Cham, 2021). doi:10.1007/978-3-030-22009-9_39.

348. Bidder, G. P. SENESCENCE. *British Medical Journal* **2**, 583 (1932).

349. Kirkwood, T. B. L. Evolution of ageing. *Nature* **270**, 301–304 (1977).

350. Stoldt, M. *et al*. Parasite Presence Induces Gene Expression Changes in an Ant Host Related to Immunity and Longevity. *Genes (Basel)* **12**, 95 (2021).

351. Feldmeyer, B. *et al.* Gene expression patterns underlying parasite-induced alterations in host behaviour and life history. *Mol Ecol* **25**, 648–660 (2016).

352. Hurd, H., Warr, E. & Polwart, A. A parasite that increases host lifespan. *Proceedings of the Royal Society B: Biological Sciences* **268**, 1749 (2001).

353. Dianne, L. *et al.* Protection first then facilitation: a manipulative parasite modulates the vulnerability to predation of its intermediate host according to its own developmental stage. *Evolution* **65**, 2692–2698 (2011).

354. Ledberg, A. Exponential increase in mortality with age is a generic property of a simple model system of damage accumulation and death. *PLoS ONE* **15**, e0233384 (2020).

355. Olshansky, S. J., Hayflick, L. & Carnes, B. A. Position Statement on Human Aging. *The Journals of Gerontology Series A: Biological Sciences and Medical Sciences* **57**, B292–B297 (2002).

356. Bhaskaran, K., dos-Santos-Silva, I., Leon, D. A., Douglas, I. J. & Smeeth, L. Association of BMI with overall and cause-specific mortality: a population-based cohort study of 3·6 million adults in the UK. *The Lancet Diabetes & Endocrinology* **6**, 944–953 (2018).

357. Flegal, D. K. M., Kit, D. B. K., Orpana, D. H. & Graubard, D. B. I. Association of All-Cause Mortality With Overweight and Obesity Using Standard Body Mass Index Categories: A Systematic Review and Meta-analysis. *JAMA* **309**, 71 (2013).

358. Weindruch, R., Walford, R. L., Fligiel, S. & Guthrie, D. The retardation of aging in mice by dietary restriction: longevity, cancer, immunity and lifetime energy intake. *J Nutr* **116**, 641–654 (1986).

359. Brown-Borg, H. M. & Bartke, A. G. and IGF1: Roles in Energy Metabolism of Long-Living GH Mutant Mice. *The Journals of Gerontology Series A: Biological Sciences and Medical Sciences* **67A**, 652 (2012).

360. Colman, R. J. *et al.* Caloric restriction reduces age-related and all-cause mortality in rhesus monkeys. *Nat Commun* **5**, 3557 (2014).

361. Mattison, J. A. *et al.* Impact of caloric restriction on health and survival in rhesus monkeys from the NIA study. *Nature* **489**, 318–321 (2012).

362. Mattison, J. A. *et al.* Caloric restriction improves health and survival of rhesus monkeys. *Nat Commun* **8**, 14063 (2017).

363. Pifferi, F. *et al.* Caloric restriction increases lifespan but affects brain integrity in grey mouse lemur primates. *Commun Biol* **1**, 1–8 (2018).

364. Szafranski, K. & Mekhail, K. The fine line between lifespan extension and shortening in response to caloric restriction. *Nucleus* **5**, 56 (2014).

365. Vatner, S. F., Vatner, D. E. & Yan, L. Models of longevity (Calorie Restriction and AC5 KO): Result of three bad hypotheses. *Aging (Albany NY)* **4**, 662 (2012).

366. Yan, L. *et al.* Common Mechanisms for Calorie Restriction and AC5 Knockout Models of Longevity. *Aging cell* **11**, 1110 (2012).

367. Keys, A., Brožek, J., Henschel, A., Mickelsen, O. & Taylor, H. L. *The Biology of Human Starvation: Volume I*. (University of Minnesota Press, 1950). doi:10.5749/j.ctv9b2tqv.

368. Most, J. & Redman, L. M. Impact of calorie restriction on energy metabolism in humans. *Experimental gerontology* **133**, 110875 (2020).

369. Romashkan, S. V. *et al.* Safety of two-year caloric restriction in non-obese healthy individuals. *Oncotarget* **7**, 19124–19133 (2016).

370. Ravussin, E. *et al.* A 2-Year Randomized Controlled Trial of Human Caloric Restriction: Feasibility and Effects on Predictors of Health Span and Longevity. *J Gerontol A Biol Sci Med Sci* **70**, 1097–1104 (2015).

371. Longo, V. D., Tano, M. D., Mattson, M. P. & Guidi, N. Intermittent and periodic fasting, longevity and disease. *Nature aging* **1**, 47 (2021).

372. Spadaro, O. *et al.* Caloric restriction in humans reveals immunometabolic regulators of health span. *Science* **375**, 671–677 (2022).

373. McCrory, C. *et al.* GrimAge Outperforms Other Epigenetic Clocks in the Prediction of Age-Related Clinical Phenotypes and All-Cause Mortality. *J Gerontol A Biol Sci Med Sci* **76**, 741–749 (2021).

374. Waziry, R. *et al.* Effect of long-term caloric restriction on DNA methylation measures of biological aging in healthy adults from the CALERIE trial. *Nat Aging* (2023) doi:10.1038/s43587-022-00357-y.

375. Nagai, M. *et al.* Association of Total Energy Intake with 29-Year Mortality in the Japanese: NIPPON DATA80. *J Atheroscler Thromb* **23**, 339–354 (2016).

376. Lassale, C. *et al.* Energy Balance and Risk of Mortality in Spanish Older Adults. *Nutrients* **13**, 1545 (2021).

377. Di Francesco, A. *et al.* Dietary restriction impacts health and lifespan of genetically diverse mice. *Nature* **634**, 684–692 (2024).

378. Acosta-Rodríguez, V. *et al.* Circadian alignment of early onset caloric restriction promotes longevity in male C57BL/6J mice. *Science* **376**, 1192–1202 (2022).

379. Lopez-Minguez, J., Gómez-Abellán, P. & Garaulet, M. Timing of Breakfast, Lunch, and Dinner. Effects on Obesity and Metabolic Risk. *Nutrients* **11**, 2624 (2019).

380. Rong, S. *et al.* Association of Skipping Breakfast With Cardiovascular and All-Cause Mortality. *J Am Coll Cardiol* **73**, 2025–2032 (2019).

381. Han, T. *et al.* The Association of Energy and Macronutrient Intake at Dinner Versus Breakfast With Disease-Specific and All-Cause Mortality Among People With Diabetes: The U.S. National Health and Nutrition Examination Survey, 2003-2014. *Diabetes Care* **43**, 1442–1448 (2020).

382. Bekker, A. *et al.* Dating the rise of atmospheric oxygen. *Nature* **427**, 117–120 (2004).

383. Hodgskiss, M. S. W., Crockford, P. W., Peng, Y., Wing, B. A. & Horner, T. J. A productivity collapse to end Earth's Great Oxidation. *Proceedings of the National Academy of Sciences of the United States of America* **116**, 17207 (2019).

384. Lee, S.-J., Hwang, A. B. & Kenyon, C. Inhibition of respiration extends C. elegans' lifespan via reactive oxygen species that increase HIF-1 activity. *Current biology : CB* **20**, 2131 (2010).

385. Conti, B. *et al.* Transgenic mice with a reduced core body temperature have an increased life span. *Science* **314**, 825–828 (2006).

386. Mołoń, M. *et al.* Effects of Temperature on Lifespan of Drosophila melanogaster from Different Genetic Backgrounds: Links between Metabolic Rate and Longevity. *Insects* **11**, 470 (2020).

387. Hsu, C. & Chiu, Y. Ambient temperature influences aging in an annual fish (*Nothobranchius rachovii*). *Aging Cell* **8**, 726–737 (2009).

388. Dampc, J., Kula-Maximenko, M., Molon, M. & Durak, R. Enzymatic Defense Response of Apple Aphid Aphis pomi to Increased Temperature. *Insects* **11**, 436 (2020).

389. Molon, M. & Zadrag-Tecza, R. Effect of temperature on replicative aging of the budding yeast Saccharomyces cerevisiae. *Biogerontology* **17**, 347–357 (2016).

390. Van Voorhies, W. A. & Ward, S. Genetic and environmental conditions that increase longevity in Caenorhabditis elegans decrease

metabolic rate. *Proceedings of the National Academy of Sciences* **96**, 11399–11403 (1999).

391. Xiao, R. et al. A genetic program promotes C. elegans longevity at cold temperatures via a thermosensitive TRP channel. *Cell* **152**, 806–817 (2013).

392. Sergi, G. et al. Body composition and resting energy expenditure in elderly male patients with chronic obstructive pulmonary disease. *Respir Med* **100**, 1918–1924 (2006).

393. Schrack, J. A., Knuth, N. D., Simonsick, E. M. & Ferrucci, L. "IDEAL" Aging is Associated with Lower Resting Metabolic Rate: The Baltimore Longitudinal Study of Aging. *Journal of the American Geriatrics Society* **62**, 667 (2014).

394. Austad, S. N. & Fischer, K. E. Mammalian aging, metabolism, and ecology: evidence from the bats and marsupials. *J Gerontol* **46**, B47-53 (1991).

395. Magalhães, J. P. de, Costa, J. & Church, G. M. An Analysis of the Relationship Between Metabolism, Developmental Schedules, and Longevity Using Phylogenetic Independent Contrasts. *The journals of gerontology. Series A, Biological sciences and medical sciences* **62**, 149 (2007).

396. Stark, G., Pincheira-Donoso, D. & Meiri, S. No evidence for the 'rate-of-living' theory across the tetrapod tree of life. *Global Ecol Biogeogr* **29**, 857–884 (2020).

397. Shilovsky, G. A., Putyatina, T. S. & Markov, A. V. Evolution of Longevity in Tetrapods: Safety Is More Important than Metabolism Level. *Biochemistry Moscow* **89**, 322–340 (2024).

398. Zhao, Z. et al. Body temperature is a more important modulator of lifespan than metabolic rate in two small mammals. *Nat Metab* **4**, 320–326 (2022).

399. Kalyani, R. R. & Egan, J. M. Diabetes and Altered Glucose Metabolism with Aging. *Endocrinology and metabolism clinics of North America* **42**, 333 (2013).

400. Holzenberger, M. et al. IGF-1 receptor regulates lifespan and resistance to oxidative stress in mice. *Nature* **421**, 182–187 (2003).

401. Shimokawa, I. et al. The life-extending effect of dietary restriction requires Foxo3 in mice. *Aging Cell* **14**, 707–709 (2015).

402. Domazet-Lošo, T. et al. Naturally occurring tumours in the basal metazoan Hydra. *Nat Commun* **5**, 4222 (2014).

403. Yoshida, K., Fujisawa, T., Hwang, J. S., Ikeo, K. & Gojobori, T. Degeneration after sexual differentiation in hydra and its relevance to the evolution of aging. *Gene* **385**, 64–70 (2006).

404. Schaible, R. *et al.* Constant mortality and fertility over age in Hydra. *Proceedings of the National Academy of Sciences* **112**, 15701–15706 (2015).

405. Boehm, A.-M. *et al.* FoxO is a critical regulator of stem cell maintenance in immortal Hydra. *Proceedings of the National Academy of Sciences* **109**, 19697–19702 (2012).

406. Hannenhalli, S. & Kaestner, K. H. The evolution of Fox genes and their role in development and disease. *Nature reviews. Genetics* **10**, 233 (2009).

407. Weigel, D., Jürgens, G., Küttner, F., Seifert, E. & Jäckle, H. The homeotic gene fork head encodes a nuclear protein and is expressed in the terminal regions of the Drosophila embryo. *Cell* **57**, 645–658 (1989).

408. Seim, I. *et al.* Genome analysis reveals insights into physiology and longevity of the Brandt's bat Myotis brandtii. *Nat Commun* **4**, 2212 (2013).

409. Chionh, Y. T. *et al.* High basal heat-shock protein expression in bats confers resistance to cellular heat/oxidative stress. *Cell Stress Chaperones* **24**, 835–849 (2019).

410. Donovan, M. R. & Michael T Marr, I. I. dFOXO Activates Large and Small Heat Shock Protein Genes in Response to Oxidative Stress to Maintain Proteostasis in Drosophila. *The Journal of Biological Chemistry* **291**, 19042 (2016).

411. Murphy, C. T., Lee, S.-J. & Kenyon, C. Tissue entrainment by feedback regulation of insulin gene expression in the endoderm of *Caenorhabditis elegans*. *Proc. Natl. Acad. Sci. U.S.A.* **104**, 19046–19050 (2007).

412. Giannakou, M. E., Goss, M. & Partridge, L. Role of dFOXO in lifespan extension by dietary restriction in Drosophila melanogaster: not required, but its activity modulates the response. *Aging Cell* **7**, 187–198 (2008).

413. Hwangbo, D. S., Gershman, B., Tu, M.-P., Palmer, M. & Tatar, M. Drosophila dFOXO controls lifespan and regulates insulin signalling in brain and fat body. *Nature* **429**, 562–566 (2004).

414. Blüher, M., Kahn, B. B. & Kahn, C. R. Extended longevity in mice lacking the insulin receptor in adipose tissue. *Science* **299**, 572–574 (2003).

415. Essers, M. A. G. *et al.* FOXO transcription factor activation by oxidative stress mediated by the small GTPase Ral and JNK. *EMBO J* **23**, 4802–4812 (2004).

REFERENCES

416. Akasaki, Y. *et al.* FOXO transcription factors support oxidative stress resistance in human chondrocytes. *Arthritis & rheumatology (Hoboken, N.J.)* **66**, 3349 (2014).

417. Kuningas, M. *et al.* Haplotypes in the human Foxo1a and Foxo3a genes; impact on disease and mortality at old age. *Eur J Hum Genet* **15**, 294–301 (2007).

418. Willcox, B. J. *et al.* FOXO3A genotype is strongly associated with human longevity. *Proceedings of the National Academy of Sciences* **105**, 13987–13992 (2008).

419. Li, Y. *et al.* Genetic association of FOXO1A and FOXO3A with longevity trait in Han Chinese populations. *Human Molecular Genetics* **18**, 4897–4904 (2009).

420. Flachsbart, F. *et al.* Association of FOXO3A variation with human longevity confirmed in German centenarians. *Proceedings of the National Academy of Sciences* **106**, 2700–2705 (2009).

421. Davies, G. *et al.* Study of 300,486 individuals identifies 148 independent genetic loci influencing general cognitive function. *Nat Commun* **9**, 2098 (2018).

422. Sniekers, S. *et al.* Genome-wide association meta-analysis of 78,308 individuals identifies new loci and genes influencing human intelligence. *Nat Genet* **49**, 1107–1112 (2017).

423. Torigoe, T. H. *et al.* Novel protective effect of the FOXO3 longevity genotype on mechanisms of cellular aging in Okinawans. *npj Aging* **10**, 18 (2024).

424. Zhang, W. *et al.* A single-cell transcriptomic landscape of primate arterial aging. *Nat Commun* **11**, 2202 (2020).

425. De Candia, P., Blekhman, R., Chabot, A. E., Oshlack, A. & Gilad, Y. A Combination of Genomic Approaches Reveals the Role of FOXO1a in Regulating an Oxidative Stress Response Pathway. *PLoS ONE* **3**, e1670 (2008).

426. Glick, D., Barth, S. & Macleod, K. F. Autophagy: cellular and molecular mechanisms. *The Journal of pathology* **221**, 3 (2010).

427. Arensman, M. D. & Eng, C. H. Self-Digestion for Lifespan Extension: Enhanced Autophagy Delays Aging. *Molecular Cell* **71**, 485–486 (2018).

428. Tanida, I., Ueno, T. & Kominami, E. LC3 and Autophagy. *Methods Mol Biol* **445**, 77–88 (2008).

429. Chung, K. W. & Chung, H. Y. The Effects of Calorie Restriction on Autophagy: Role on Aging Intervention. *Nutrients* **11**, 2923 (2019).

430. Noda, T. & Ohsumi, Y. Tor, a phosphatidylinositol kinase homologue, controls autophagy in yeast. *J Biol Chem* **273**, 3963–3966 (1998).

431. Cutler, N. S., Heitman, J. & Cardenas, M. E. TOR kinase homologs function in a signal transduction pathway that is conserved from yeast to mammals. *Mol Cell Endocrinol* **155**, 135–142 (1999).

432. Solon-Biet, S. M. *et al.* The Ratio of Macronutrients, Not Caloric Intake, Dictates Cardiometabolic Health, Aging, and Longevity in Ad Libitum-Fed Mice. *Cell Metabolism* **19**, 418–430 (2014).

433. Kitada, M., Xu, J., Ogura, Y., Monno, I. & Koya, D. Mechanism of Activation of Mechanistic Target of Rapamycin Complex 1 by Methionine. *Front. Cell Dev. Biol.* **8**, (2020).

434. Richie, J. P. *et al.* Methionine restriction increases blood glutathione and longevity in F344 rats. *FASEB J* **8**, 1302–1307 (1994).

435. Bárcena, C. *et al.* Methionine Restriction Extends Lifespan in Progeroid Mice and Alters Lipid and Bile Acid Metabolism. *Cell Reports* **24**, 2392 (2018).

436. Miller, R. A. *et al.* Methionine-deficient diet extends mouse lifespan, slows immune and lens aging, alters glucose, T4, IGF-I and insulin levels, and increases hepatocyte MIF levels and stress resistance. *Aging Cell* **4**, 119–125 (2005).

437. Harrison, D. E. *et al.* Rapamycin fed late in life extends lifespan in genetically heterogeneous mice. *Nature* **460**, 392–395 (2009).

438. Rapamycin — the Easter Island drug that extends lifespan of old mice. *Science* https://www.nationalgeographic.com/science/article/rapamycin-the-easter-island-drug-that-extends-lifespan-of-old-mice (2009).

439. Bitto, A. *et al.* Transient rapamycin treatment can increase lifespan and healthspan in middle-aged mice. *eLife* **5**, e16351 (2016).

440. Strong, R. *et al.* Rapamycin-mediated mouse lifespan extension: Late-life dosage regimes with sex-specific effects. *Aging Cell* **19**, e13269 (2020).

441. Miller, R. A. *et al.* Rapamycin-mediated lifespan increase in mice is dose and sex dependent and metabolically distinct from dietary restriction. *Aging Cell* **13**, 468–477 (2014).

442. Shindyapina, A. V. *et al.* Rapamycin treatment during development extends life span and health span of male mice and *Daphnia magna*. *Sci. Adv.* **8**, eabo5482 (2022).

443. Singh, A. K. *et al.* Rapamycin Confers Neuroprotection Against Aging-Induced Oxidative Stress, Mitochondrial Dysfunction,

REFERENCES

and Neurodegeneration in Old Rats Through Activation of Autophagy. *Rejuvenation Research* **22**, 60–70 (2019).

444. Van Skike, C. E. *et al.* mTOR drives cerebrovascular, synaptic, and cognitive dysfunction in normative aging. *Aging Cell* **19**, e13057 (2020).

445. Altschuler, R. A. *et al.* Rapamycin Added to Diet in Late Mid-Life Delays Age-Related Hearing Loss in UMHET4 Mice. *Front. Cell. Neurosci.* **15**, 658972 (2021).

446. Seto, B. Rapamycin and mTOR: a serendipitous discovery and implications for breast cancer. *Clinical and Translational Medicine* **1**, 29 (2012).

447. Ferrer, I. R., Araki, K. & Ford, M. L. Paradoxical Aspects of Rapamycin Immunobiology in Transplantation. *American journal of transplantation : official journal of the American Society of Transplantation and the American Society of Transplant Surgeons* **11**, 654 (2011).

448. Lamming, D. W. *et al.* Rapamycin-induced insulin resistance is mediated by mTORC2 loss and uncoupled from longevity. *Science (New York, N.y.)* **335**, 1638 (2012).

449. Phillips, E. J. & Simons, M. J. P. Rapamycin not dietary restriction improves resilience against pathogens: a meta-analysis. *GeroScience* **45**, 1263–1270 (2023).

450. Mannick, J. B. *et al.* TORC1 inhibition enhances immune function and reduces infections in the elderly. *Sci Transl Med* **10**, eaaq1564 (2018).

451. Baghdadi, M. *et al.* Intermittent rapamycin feeding recapitulates some effects of continuous treatment while maintaining lifespan extension. *Molecular Metabolism* **81**, 101902 (2024).

452. Arriola Apelo, S. I., Pumper, C. P., Baar, E. L., Cummings, N. E. & Lamming, D. W. Intermittent Administration of Rapamycin Extends the Life Span of Female C57BL/6J Mice. *J Gerontol A Biol Sci Med Sci* **71**, 876–881 (2016).

453. Strong, R. *et al.* Lifespan benefits for the combination of rapamycin plus acarbose and for captopril in genetically heterogeneous mice. *Aging Cell* **21**, e13724 (2022).

454. Chung, C. L. *et al.* Topical rapamycin reduces markers of senescence and aging in human skin: an exploratory, prospective, randomized trial. *Geroscience* **41**, 861–869 (2019).

455. AgelessRx. *Participatory Evaluation (of) Aging (With) Rapamycin (for) Longevity Study (PEARL): A Prospective, Double-Blind,*

Placebo-Controlled Trial for Rapamycin in Healthy Individuals Assessing Safety and Efficacy in Reducing Aging Effects. https://clinicaltrials.gov/study/NCT04488601 (2024).

456. Stanfield, B. *et al.* A single-center, double-blind, randomized, placebo-controlled, two-arm study to evaluate the safety and efficacy of once-weekly sirolimus (rapamycin) on muscle strength and endurance in older adults following a 13-week exercise program. *Trials* **25**, 642 (2024).

457. Viribay, A., Burgos, J., Fernández-Landa, J., Seco-Calvo, J. & Mielgo-Ayuso, J. Effects of Arginine Supplementation on Athletic Performance Based on Energy Metabolism: A Systematic Review and Meta-Analysis. *Nutrients* **12**, 1300 (2020).

458. Chin, R. M. *et al.* The metabolite alpha-ketoglutarate extends lifespan by inhibiting the ATP synthase and TOR. *Nature* **510**, 397 (2014).

459. Su, Y. *et al.* Alpha-ketoglutarate extends Drosophila lifespan by inhibiting mTOR and activating AMPK. *Aging (Albany NY)* **11**, 4183 (2019).

460. Shahmirzadi, A. A. *et al.* Alpha-Ketoglutarate, an Endogenous Metabolite, Extends Lifespan and Compresses Morbidity in Aging Mice. *Cell metabolism* **32**, 447 (2020).

461. Tyshkovskiy, A. *et al.* Identification and Application of Gene Expression Signatures Associated with Lifespan Extension. *Cell Metab* **30**, 573-593.e8 (2019).

462. DrugAge: Database of Ageing-Related Drugs. https://genomics.senescence.info/drugs/.

463. Lublin, A. *et al.* FDA-approved drugs that protect mammalian neurons from glucose toxicity slow aging dependent on cbp and protect against proteotoxicity. *PLoS One* **6**, e27762 (2011).

464. Saiki, S. *et al.* Caffeine induces apoptosis by enhancement of autophagy via PI3K/Akt/mTOR/p70S6K inhibition. *Autophagy* **7**, 176 (2011).

465. Takahashi, K., Yanai, S., Shimokado, K. & Ishigami, A. Coffee consumption in aged mice increases energy production and decreases hepatic mTOR levels. *Nutrition* **38**, 1–8 (2017).

466. Reinke, A., Chen, J. C.-Y., Aronova, S. & Powers, T. Caffeine Targets TOR Complex I and Provides Evidence for a Regulatory Link between the FRB and Kinase Domains of Tor1p *. *Journal of Biological Chemistry* **281**, 31616–31626 (2006).

REFERENCES

467. Pietrocola, F. *et al.* Coffee induces autophagy in vivo. *Cell Cycle* **13**, 1987 (2014).

468. Wax, B., Kavazis, A. N., Webb, H. E. & Brown, S. P. Acute L-arginine alpha ketoglutarate supplementation fails to improve muscular performance in resistance trained and untrained men. *Journal of the International Society of Sports Nutrition* **9**, 17 (2012).

469. Kalender, A. *et al.* Metformin, Independent of AMPK, Inhibits mTORC1 In a Rag GTPase-Dependent Manner. *Cell metabolism* **11**, 390 (2010).

470. Rabheru, K., Byles, J. E. & Kalache, A. How "old age" was withdrawn as a diagnosis from ICD-11. *The Lancet Healthy Longevity* **3**, e457–e459 (2022).

471. Lynch, A. J. J., Barnes, R. W., Cambecedes, J. & Vaillancourt, R. Genetic evidence that Lomatia tasmanica (Proteaceae) is an ancient clone. (1998).

472. Vasek, F. C. Creosote Bush: Long-Lived Clones in the Mojave Desert. *American Journal of Botany* **67**, 246–255 (1980).

473. Witte, L. C. de & Stöcklin, J. Longevity of clonal plants: why it matters and how to measure it. *Annals of Botany* **106**, 859 (2010).

474. Anderson, J. B. *et al.* Clonal evolution and genome stability in a 2500-year-old fungal individual. *Proc. R. Soc. B.* **285**, 20182233 (2018).

475. Hodnett, B. & Anderson, J. B. Genomic stability of two individuals of *Armillaria gallica*. *Mycologia* **92**, 894–899 (2000).

476. Garcia-Cisneros, A. *et al.* Long telomeres are associated with clonality in wild populations of the fissiparous starfish Coscinasterias tenuispina. *Heredity* **115**, 437–443 (2015).

477. Hayflick, L. & Moorhead, P. S. The serial cultivation of human diploid cell strains. *Experimental Cell Research* **25**, 585–621 (1961).

478. Hayflick, L. The limited *in vitro* lifetime of human diploid cell strains. *Experimental Cell Research* **37**, 614–636 (1965).

479. Bianconi, E. *et al.* An estimation of the number of cells in the human body. *Ann Hum Biol* **40**, 463–471 (2013).

480. Gilbert, S. F. The Cell Death Pathways. in *Developmental Biology. 6th edition* (Sinauer Associates, 2000).

481. Olovnikov, A. M. A theory of marginotomy. The incomplete copying of template margin in enzymic synthesis of polynucleotides and biological significance of the phenomenon. *J Theor Biol* **41**, 181–190 (1973).

REFERENCES

482. Greider, C. W. & Blackburn, E. H. Identification of a specific telomere terminal transferase activity in Tetrahymena extracts. *Cell* **43**, 405–413 (1985).

483. Greider, C. W. & Blackburn, E. H. A telomeric sequence in the RNA of Tetrahymena telomerase required for telomere repeat synthesis. *Nature* **337**, 331–337 (1989).

484. Morin, G. B. The human telomere terminal transferase enzyme is a ribonucleoprotein that synthesizes TTAGGG repeats. *Cell* **59**, 521–529 (1989).

485. Hiyama, E. & Hiyama, K. Telomere and telomerase in stem cells. *British Journal of Cancer* **96**, 1020 (2007).

486. Ozturk, S., Sozen, B. & Demir, N. Telomere length and telomerase activity during oocyte maturation and early embryo development in mammalian species. *Mol Hum Reprod* **20**, 15–30 (2014).

487. Chaïlakhian, L. M., Veprintsev, B. N., Sviridova, T. A. & Nikitin, V. A. [Electrostimulated cell fusion in cell engineering]. *Biofizika* **32**, 874–887 (1987).

488. The story of Dolly the sheep. *National Museums Scotland* https://www.nms.ac.uk/discover-catalogue/the-story-of-dolly-the-sheep.

489. *Ashworth, D., Bishop, M., Campbell, K. et al. DNA Microsatellite Analysis of Dolly. Nature 394, 329 (1998).*

490. *Signer, E., Dubrova, Y., Jeffreys, A. et al. DNA Fingerprinting Dolly. Nature 394, 329–330 (1998).*

491. Sinclair, K. D. *et al.* Healthy ageing of cloned sheep. *Nature Communications* **7**, 12359 (2016).

492. Xu, J. & Yang, X. Will cloned animals suffer premature aging — The story at the end of clones' chromosomes. *Reprod Biol Endocrinol* **1**, 105 (2003).

493. Wakayama, S. *et al.* Successful Serial Recloning in the Mouse over Multiple Generations. *Cell Stem Cell* **12**, 293–297 (2013).

494. Liu, Z. *et al.* Cloning of a gene-edited macaque monkey by somatic cell nuclear transfer. *National Science Review* **6**, 101–108 (2019).

495. Schaetzlein, S. *et al.* Telomere length is reset during early mammalian embryogenesis. *Proc Natl Acad Sci U S A* **101**, 8034–8038 (2004).

496. Sarek, G., Marzec, P., Margalef, P. & Boulton, S. J. Molecular basis of telomere dysfunction in human genetic diseases. *Nat Struct Mol Biol* **22**, 867–874 (2015).

497. Mitchell, J. R., Wood, E. & Collins, K. A telomerase component is defective in the human disease dyskeratosis congenita. *Nature* **402**, 551–555 (1999).

498. Garus, A. & Autexier, C. Dyskerin: an essential pseudouridine synthase with multifaceted roles in ribosome biogenesis, splicing, and telomere maintenance. *RNA* **27**, 1441 (2021).

499. Mangaonkar, A. A. & Patnaik, M. M. Short Telomere Syndromes in Clinical Practice: Bridging Bench and Bedside. *Mayo Clin Proc* **93**, 904–916 (2018).

500. Whittemore, K., Vera, E., Martínez-Nevado, E., Sanpera, C. & Blasco, M. A. Telomere shortening rate predicts species life span. *Proceedings of the National Academy of Sciences* **116**, 15122–15127 (2019).

501. Telomeres Mendelian Randomization Collaboration *et al.* Association Between Telomere Length and Risk of Cancer and Non-Neoplastic Diseases: A Mendelian Randomization Study. *JAMA Oncol* **3**, 636–651 (2017).

502. Blasco, M. A. *et al.* Telomere shortening and tumor formation by mouse cells lacking telomerase RNA. *Cell* **91**, 25–34 (1997).

503. Tomás-Loba, A. *et al.* Telomerase Reverse Transcriptase Delays Aging in Cancer-Resistant Mice. *Cell* **135**, 609–622 (2008).

504. Muñoz-Lorente, M. A., Cano-Martin, A. C. & Blasco, M. A. Mice with hyper-long telomeres show less metabolic aging and longer lifespans. *Nat Commun* **10**, 4723 (2019).

505. Jesus, B. B. de *et al.* Telomerase gene therapy in adult and old mice delays aging and increases longevity without increasing cancer. *EMBO Molecular Medicine* **4**, 691 (2012).

506. Jaijyan, D. K. *et al.* New intranasal and injectable gene therapy for healthy life extension. *Proc. Natl. Acad. Sci. U.S.A.* **119**, e2121499119 (2022).

507. Graber, T. G., Kim, J.-H., Grange, R. W., McLoon, L. K. & Thompson, L. V. C57BL/6 life span study: age-related declines in muscle power production and contractile velocity. *Age* **37**, 36 (2015).

508. Wang, H. *et al.* Cytomegalovirus Infection and Relative Risk of Cardiovascular Disease (Ischemic Heart Disease, Stroke, and Cardiovascular Death): A Meta-Analysis of Prospective Studies Up to 2016. *Journal of the American Heart Association: Cardiovascular and Cerebrovascular Disease* **6**, e005025 (2017).

509. Kallemeijn, M. J. *et al.* Ageing and latent CMV infection impact on maturation, differentiation and exhaustion profiles of T-cell receptor gammadelta T-cells. *Sci Rep* **7**, 5509 (2017).

510. DeFrancesco, L. Church to de-extinct woolly mammoths. *Nat Biotechnol* **39**, 1171–1171 (2021).

REFERENCES

511. Church, G. M. & Gilbert, W. Genomic sequencing. *Proceedings of the National Academy of Sciences of the United States of America* **81**, 1991 (1984).

512. Gibney, E. R. & Nolan, C. M. Epigenetics and gene expression. *Heredity* **105**, 4–13 (2010).

513. López-Otín, C., Blasco, M. A., Partridge, L., Serrano, M. & Kroemer, G. The Hallmarks of Aging. *Cell* **153**, 1194–1217 (2013).

514. Sen, P., Shah, P. P., Nativio, R. & Berger, S. L. Epigenetic Mechanisms of Longevity and Aging. *Cell* **166**, 822–839 (2016).

515. Niemann, H. *et al.* DNA methylation patterns reflect epigenetic reprogramming in bovine embryos. *Cell Reprogram* **12**, 33–42 (2010).

516. Reik, W., Dean, W. & Walter, J. Epigenetic reprogramming in mammalian development. *Science* **293**, 1089–1093 (2001).

517. Kerepesi, C., Zhang, B., Lee, S.-G., Trapp, A. & Gladyshev, V. N. Epigenetic clocks reveal a rejuvenation event during embryogenesis followed by aging. *Sci Adv* **7**, eabg6082 (2021).

518. Zeng, Y. & Chen, T. DNA Methylation Reprogramming during Mammalian Development. *Genes* **10**, 257 (2019).

519. Mitalipov, S. & Wolf, D. Totipotency, Pluripotency and Nuclear Reprogramming. *Advances in biochemical engineering/biotechnology* **114**, 185 (2009).

520. Zakrzewski, W., Dobrzyński, M., Szymonowicz, M. & Rybak, Z. Stem cells: past, present, and future. *Stem Cell Research & Therapy* **10**, 68 (2019).

521. Takahashi, K. *et al.* Induction of pluripotent stem cells from adult human fibroblasts by defined factors. *Cell* **131**, 861–872 (2007).

522. Petkovich, D. A. *et al.* Using DNA Methylation Profiling to Evaluate Biological Age and Longevity Interventions. *Cell Metab* **25**, 954-960.e6 (2017).

523. Murakami, K. *et al.* Generation of functional oocytes from male mice in vitro. *Nature* **615**, 900–906 (2023).

524. Lu, Y. R., Tian, X. & Sinclair, D. A. The Information Theory of Aging. *Nat Aging* **3**, 1486–1499 (2023).

525. Miller, D. M., Thomas, S. D., Islam, A., Muench, D. & Sedoris, K. c-Myc and Cancer Metabolism. *Clinical cancer research : an official journal of the American Association for Cancer Research* **18**, 5546 (2012).

526. Yang, J.-H. *et al.* Loss of epigenetic information as a cause of mammalian aging. *Cell* **186**, 305-326.e27 (2023).

REFERENCES

527. Browder, K. C. *et al.* In vivo partial reprogramming alters age-associated molecular changes during physiological aging in mice. *Nat Aging* **2**, 243–253 (2022).

528. Chondronasiou, D. *et al.* Multi-omic rejuvenation of naturally aged tissues by a single cycle of transient reprogramming. *Aging Cell* **21**, e13578 (2022).

529. Wang, C. *et al.* In vivo partial reprogramming of myofibers promotes muscle regeneration by remodeling the stem cell niche. *Nat Commun* **12**, 3094 (2021).

530. Lu, Y. *et al.* Reprogramming to recover youthful epigenetic information and restore vision. *Nature* **588**, 124–129 (2020).

531. Macip, C. C. *et al.* Gene Therapy-Mediated Partial Reprogramming Extends Lifespan and Reverses Age-Related Changes in Aged Mice. *Cell Reprogram* **26**, 24–32 (2024).

532. Sarkar, T. J. *et al.* Transient non-integrative expression of nuclear reprogramming factors promotes multifaceted amelioration of aging in human cells. *Nature Communications* **11**, 1545 (2020).

533. Yang, J.-H. *et al.* Chemically induced reprogramming to reverse cellular aging. *Aging* **15**, 5966–5989 (2023).

534. Takebe, T. *et al.* Vascularized and functional human liver from an iPSC-derived organ bud transplant. *Nature* **499**, 481–484 (2013).

535. Eisenstein, M. Rejuvenation by controlled reprogramming is the latest gambit in anti-aging. *Nature Biotechnology* **40**, 144–146 (2022).

536. Bohnert, K. A. & Kenyon, C. A lysosomal switch triggers proteostasis renewal in the immortal C. elegans germ lineage. *Nature* **551**, 629–633 (2017).

537. Samaddar, M. *et al.* A genetic screen identifies new steps in oocyte maturation that enhance proteostasis in the immortal germ lineage. *eLife* **10**, e62653 (2021).

538. Bard, J. A. M. *et al.* Structure and Function of the 26S Proteasome. *Annu. Rev. Biochem.* **87**, 697–724 (2018).

539. Hoeller, D. & Dikic, I. How the proteasome is degraded. *Proc. Natl. Acad. Sci. U.S.A.* **113**, 13266–13268 (2016).

540. Vilchez, D. *et al.* Increased proteasome activity in human embryonic stem cells is regulated by PSMD11. *Nature* **489**, 304–308 (2012).

541. Cole, L. W. The Evolution of Per-cell Organelle Number. *Front Cell Dev Biol* **4**, 85 (2016).

542. Wolf, D. P., Mitalipov, N. & Mitalipov, S. Mitochondrial Replacement Therapy in Reproductive Medicine. *Trends Mol Med* **21**, 68–76 (2015).

543. Fan, W. *et al.* A mouse model of mitochondrial disease reveals germline selection against severe mtDNA mutations. *Science* **319**, 958–962 (2008).

544. Luo, S. *et al.* Biparental Inheritance of Mitochondrial DNA in Humans. *Proceedings of the National Academy of Sciences* **115**, 13039–13044 (2018).

545. Trifunovic, A. & Larsson, N.-G. Mitochondrial dysfunction as a cause of ageing. *J Intern Med* **263**, 167–178 (2008).

546. Scheibye-Knudsen, M., Fang, E. F., Croteau, D. L., Wilson, D. M. & Bohr, V. A. Protecting the Mitochondrial Powerhouse. *Trends Cell Biol* **25**, 158–170 (2015).

547. Jin, S. M. & Youle, R. J. PINK1- and Parkin-mediated mitophagy at a glance. *Journal of Cell Science* **125**, 795–799 (2012).

548. Miller, S. & Muqit, M. M. K. Therapeutic approaches to enhance PINK1/Parkin mediated mitophagy for the treatment of Parkinson's disease. *Neuroscience Letters* **705**, 7–13 (2019).

549. Manders, F., van Boxtel, R. & Middelkamp, S. The Dynamics of Somatic Mutagenesis During Life in Humans. *Front. Aging* **2**, (2021).

550. Milholland, B. *et al.* Differences between germline and somatic mutation rates in humans and mice. *Nat Commun* **8**, 1–8 (2017).

551. Moore, L. *et al.* The mutational landscape of human somatic and germline cells. *Nature* **597**, 381–386 (2021).

552. Vermezovic, J., Stergiou, L., Hengartner, M. O. & d'Adda di Fagagna, F. Differential regulation of DNA damage response activation between somatic and germline cells in Caenorhabditis elegans. *Cell Death Differ* **19**, 1847–1855 (2012).

553. Reiman, A. *et al.* Lymphoid tumours and breast cancer in ataxia telangiectasia; substantial protective effect of residual ATM kinase activity against childhood tumours. *Br J Cancer* **105**, 586–591 (2011).

554. Tissue expression of ATM - Summary - The Human Protein Atlas. https://www.proteinatlas.org/ENSG00000149311-ATM/tissue.

555. Titus, S. *et al.* Impairment of BRCA1-related DNA Double Strand Break Repair Leads to Ovarian Aging in Mice and Humans. *Sci Transl Med* **5**, 172ra21 (2013).

556. Clark-Hachtel, C. M. *et al.* The tardigrade Hypsibius exemplaris dramatically upregulates DNA repair pathway genes in response to ionizing radiation. *Curr Biol* **34**, 1819-1830.e6 (2024).

557. Bloom, J. C., Loehr, A. R., Schimenti, J. C. & Weiss, R. S. Germline genome protection: implications for gamete quality and germ cell tumorigenesis. *Andrology* **7**, 516–526 (2019).

558. Liu, X. *et al.* Resurrection of endogenous retroviruses during aging reinforces senescence. *Cell* **186**, 287-304.e26 (2023).

559. Campisi, J. & d'Adda di Fagagna, F. Cellular senescence: when bad things happen to good cells. *Nat Rev Mol Cell Biol* **8**, 729–740 (2007).

560. Mijit, M., Caracciolo, V., Melillo, A., Amicarelli, F. & Giordano, A. Role of p53 in the Regulation of Cellular Senescence. *Biomolecules* **10**, 420 (2020).

561. Childs, B. G., Baker, D. J., Kirkland, J. L., Campisi, J. & van Deursen, J. M. Senescence and apoptosis: dueling or complementary cell fates? *EMBO Rep* **15**, 1139–1153 (2014).

562. Seluanov, A. *et al.* Distinct tumor suppressor mechanisms evolve in rodent species that differ in size and lifespan. *Aging Cell* **7**, 813–823 (2008).

563. Medrano, E. E., Im, S., Yang, F. & Abdel-Malek, Z. A. Ultraviolet B light induces G1 arrest in human melanocytes by prolonged inhibition of retinoblastoma protein phosphorylation associated with long-term expression of the p21Waf-1/SDI-1/Cip-1 protein. *Cancer Res* **55**, 4047–4052 (1995).

564. Suzuki, M. & Boothman, D. A. Stress-induced premature senescence (SIPS)--influence of SIPS on radiotherapy. *J Radiat Res* **49**, 105–112 (2008).

565. Benkafadar, N. *et al.* ROS-Induced Activation of DNA Damage Responses Drives Senescence-Like State in Postmitotic Cochlear Cells: Implication for Hearing Preservation. *Mol Neurobiol* **56**, 5950–5969 (2019).

566. Chen, Q., Fischer, A., Reagan, J. D., Yan, L. J. & Ames, B. N. Oxidative DNA damage and senescence of human diploid fibroblast cells. *Proc Natl Acad Sci U S A* **92**, 4337–4341 (1995).

567. Barnes, R. P., Fouquerel, E. & Opresko, P. L. The impact of oxidative DNA damage and stress on telomere homeostasis. *Mech Ageing Dev* **177**, 37–45 (2019).

568. Liu, X., Ding, J. & Meng, L. Oncogene-induced senescence: a double edged sword in cancer. *Acta Pharmacol Sin* **39**, 1553–1558 (2018).

569. Joselow, A., Lynn, D., Terzian, T. & Box, N. F. Senescence-like Phenotypes in Human Nevi. *Methods Mol Biol* **1534**, 175–184 (2017).

REFERENCES

570. Rhinn, M., Ritschka, B. & Keyes, W. M. Cellular senescence in development, regeneration and disease. *Development* **146**, dev151837 (2019).

571. Storer, M. *et al.* Senescence is a developmental mechanism that contributes to embryonic growth and patterning. *Cell* **155**, 1119–1130 (2013).

572. Muñoz-Espín, D. *et al.* Programmed Cell Senescence during Mammalian Embryonic Development. *Cell* **155**, 1104–1118 (2013).

573. Choy, E. & Rose-John, S. Interleukin-6 as a Multifunctional Regulator: Inflammation, Immune Response, and Fibrosis. *Journal of Scleroderma and Related Disorders* **2**, S1–S5 (2017).

574. Di, G. *et al.* IL-6 Secreted from Senescent Mesenchymal Stem Cells Promotes Proliferation and Migration of Breast Cancer Cells. *PLoS One* **9**, e113572 (2014).

575. Furman, D. *et al.* Chronic inflammation in the etiology of disease across the life span. *Nat Med* **25**, 1822–1832 (2019).

576. Ferrucci, L. & Fabbri, E. Inflammageing: chronic inflammation in ageing, cardiovascular disease, and frailty. *Nat Rev Cardiol* **15**, 505–522 (2018).

577. Arai, Y. *et al.* Inflammation, But Not Telomere Length, Predicts Successful Ageing at Extreme Old Age: A Longitudinal Study of Semi-supercentenarians. *EBioMedicine* **2**, 1549–1558 (2015).

578. Krtolica, A., Parrinello, S., Lockett, S., Desprez, P.-Y. & Campisi, J. Senescent fibroblasts promote epithelial cell growth and tumorigenesis: A link between cancer and aging. *Proc Natl Acad Sci U S A* **98**, 12072–12077 (2001).

579. Kim, S. R. *et al.* Transplanted senescent renal scattered tubular-like cells induce injury in the mouse kidney. *American Journal of Physiology - Renal Physiology* **318**, F1167 (2020).

580. Xu, M. *et al.* Senolytics Improve Physical Function and Increase Lifespan in Old Age. *Nat Med* **24**, 1246–1256 (2018).

581. Sharpless, N. E., Ramsey, M. R., Balasubramanian, P., Castrillon, D. H. & DePinho, R. A. The differential impact of p16(INK4a) or p19(ARF) deficiency on cell growth and tumorigenesis. *Oncogene* **23**, 379–385 (2004).

582. Kirkland, J. L. & Tchkonia, T. Senolytic drugs: from discovery to translation. *J Intern Med* **288**, 518–536 (2020).

583. Jun, J.-I. & Lau, L. F. The matricellular protein CCN1 induces fibroblast senescence and restricts fibrosis in cutaneous wound healing. *Nat Cell Biol* **12**, 676–685 (2010).

REFERENCES

584. Saito, Y. & Chikenji, T. S. Diverse Roles of Cellular Senescence in Skeletal Muscle Inflammation, Regeneration, and Therapeutics. *Front Pharmacol* **12**, 739510 (2021).

585. Krizhanovsky, V. *et al.* Senescence of activated stellate cells limits liver fibrosis. *Cell* **134**, 657–667 (2008).

586. Helman, A. *et al.* p16Ink4a-induced senescence of pancreatic beta cells enhances insulin secretion. *Nat Med* **22**, 412–420 (2016).

587. Salinas-Saavedra, M. *et al.* Senescence-induced cellular reprogramming drives cnidarian whole-body regeneration. *Cell Reports* **42**, 112687 (2023).

588. Zhao, Y. *et al.* Naked mole rats can undergo developmental, oncogene-induced and DNA damage-induced cellular senescence. *Proc. Natl. Acad. Sci. U.S.A.* **115**, 1801–1806 (2018).

589. VanHook, A. M. Naked mole-rats slay zombies. *Sci. Signal.* **16**, eadj8555 (2023).

590. Kawamura, Y. *et al.* Cellular senescence induction leads to progressive cell death via the INK4a-RB pathway in naked mole-rats. *The EMBO Journal* **42**, e111133 (2023).

591. Baker, D. J. *et al.* Clearance of p16Ink4a-positive senescent cells delays ageing-associated disorders. *Nature* **479**, 232–236 (2011).

592. Liu, J.-Y. *et al.* Cells exhibiting strong p16INK4a promoter activation in vivo display features of senescence. *Proceedings of the National Academy of Sciences* **116**, 2603–2611 (2019).

593. Buj, R., Leon, K. E., Anguelov, M. A. & Aird, K. M. Suppression of p16 alleviates the senescence-associated secretory phenotype. *Aging (Albany NY)* **13**, 3290–3312 (2021)

594. Baker, D. J. *et al.* Naturally occurring p16Ink4a-positive cells shorten healthy lifespan. *Nature* **530**, 184–189 (2016).

595. Schafer, M. J. *et al.* Cellular senescence mediates fibrotic pulmonary disease. *Nat Commun* **8**, 14532 (2017).

596. Hickson, L. J. *et al.* Senolytics decrease senescent cells in humans: Preliminary report from a clinical trial of Dasatinib plus Quercetin in individuals with diabetic kidney disease. *EBioMedicine* **47**, 446–456 (2019).

597. Raffaele, M. & Vinciguerra, M. The costs and benefits of senotherapeutics for human health. *The Lancet Healthy Longevity* **3**, e67–e77 (2022).

598. Spirito, P. *et al.* Magnitude of Left Ventricular Hypertrophy and Risk of Sudden Death in Hypertrophic Cardiomyopathy. *N Engl J Med* **342**, 1778–1785 (2000).

599. Anderson, R. *et al.* Length-independent telomere damage drives post-mitotic cardiomyocyte senescence. *EMBO J* **38**, e100492 (2019).

600. Dookun, E. *et al.* Clearance of senescent cells during cardiac ischemia–reperfusion injury improves recovery. *Aging Cell* **19**, e13249 (2020).

601. Walaszczyk, A. *et al.* Pharmacological clearance of senescent cells improves survival and recovery in aged mice following acute myocardial infarction. *Aging Cell* **18**, e12945 (2019).

602. Deierborg, T., Roybon, L., Inacio, A. R., Pesic, J. & Brundin, P. Brain injury activates microglia that induce neural stem cell proliferation ex vivo and promote differentiation of neurosphere-derived cells into neurons and oligodendrocytes. *Neuroscience* **171**, 1386–1396 (2010).

603. Cornell, J., Salinas, S., Huang, H.-Y. & Zhou, M. Microglia regulation of synaptic plasticity and learning and memory. *Neural Regen Res* **17**, 705–716 (2021).

604. Hansen, D. V., Hanson, J. E. & Sheng, M. Microglia in Alzheimer's disease. *J Cell Biol* **217**, 459–472 (2018).

605. Jurk, D. *et al.* Postmitotic neurons develop a p21-dependent senescence-like phenotype driven by a DNA damage response. *Aging Cell* **11**, 996–1004 (2012).

606. Martinelli, C., Sartori, P., Ledda, M. & Pannese, E. A study of mitochondria in spinal ganglion neurons during life: quantitative changes from youth to extremely advanced age. *Tissue Cell* **38**, 93–98 (2006).

607. Simmnacher, K. *et al.* Unique signatures of stress-induced senescent human astrocytes. *Experimental Neurology* **334**, 113466 (2020).

608. Han, M. J. *et al.* Inhibition of neural stem cell aging through the transient induction of reprogramming factors. *J Comp Neurol* **529**, 595–604 (2021).

609. Campbell, I. L. *et al.* Neurologic disease induced in transgenic mice by cerebral overexpression of interleukin 6. *Proc Natl Acad Sci U S A* **90**, 10061–10065 (1993).

610. Blum-Degen, D. *et al.* Interleukin-1 beta and interleukin-6 are elevated in the cerebrospinal fluid of Alzheimer's and de novo Parkinson's disease patients. *Neurosci Lett* **202**, 17–20 (1995).

611. Tan, F. C. C., Hutchison, E. R., Eitan, E. & Mattson, M. P. Are There Roles for Brain Cell Senescence in Aging and Neurodegenerative Disorders? *Biogerontology* **15**, 643–660 (2014).

REFERENCES

612. Zhang, P. *et al*. Senolytic therapy alleviates Aβ-associated oligodendrocyte progenitor cell senescence and cognitive deficits in an Alzheimer's disease model. *Nat Neurosci* **22**, 719–728 (2019).

613. Ogrodnik, M. *et al*. Whole-body senescent cell clearance alleviates age-related brain inflammation and cognitive impairment in mice. *Aging Cell* **20**, e13296 (2021).

614. Bussian, T. J. *et al*. Clearance of senescent glial cells prevents tau-dependent pathology and cognitive decline. *Nature* **562**, 578–582 (2018).

615. Sikora, E. *et al*. Cellular Senescence in Brain Aging. *Front. Aging Neurosci.* **13**, (2021).

616. Acklin, S. *et al*. Depletion of senescent-like neuronal cells alleviates cisplatin-induced peripheral neuropathy in mice. *Sci Rep* **10**, 14170 (2020).

617. Musi, N. *et al*. Tau protein aggregation is associated with cellular senescence in the brain. *Aging Cell* **17**, e12840 (2018).

618. Dasatinib on Fast Track for Approval by FDA for Gleevec-Resistant CML. *Cancer Biology & Therapy* **5**, 704–704 (2006).

619. Dasatinib (oral route) - Mayo Clinic. https://www.mayoclinic.org/drugs-supplements/dasatinib-oral-route/description/drg-20070797.

620. TruDiagnostic. *The Safety and Effectivness of Quercetin and Dasatinib on the Epigenetic Aging Rates in Healthy Individuals*. https://clinicaltrials.gov/study/NCT04946383 (2021).

621. Navitoclax. https://go.drugbank.com/drugs/DB12340.

622. Yousefzadeh, M. J. *et al*. Fisetin is a senotherapeutic that extends health and lifespan. *EBioMedicine* **36**, 18–28 (2018).

623. Harrison, D. E. *et al*. Astaxanthin and meclizine extend lifespan in UM-HET3 male mice; fisetin, SG1002 (hydrogen sulfide donor), dimethyl fumarate, mycophenolic acid, and 4-phenylbutyrate do not significantly affect lifespan in either sex at the doses and schedules used. *GeroScience* **46**, 795–816 (2023).

624. Hall, B. M. *et al*. Aging of mice is associated with p16(Ink4a)- and β-galactosidase-positive macrophage accumulation that can be induced in young mice by senescent cells. *Aging (Albany NY)* **8**, 1294–1315 (2016).

625. Hall, B. M. *et al*. p16(Ink4a) and senescence-associated β-galactosidase can be induced in macrophages as part of a reversible response to physiological stimuli. *Aging (Albany NY)* **9**, 1867–1884 (2017).

REFERENCES

626. Frescas, D. *et al.* Murine mesenchymal cells that express elevated levels of the CDK inhibitor p16(Ink4a) in vivo are not necessarily senescent. *Cell Cycle* **16**, 1526–1533 (2017).

627. Wissler Gerdes, E. O., Misra, A., Netto, J. M. E., Tchkonia, T. & Kirkland, J. L. Strategies for late phase preclinical and early clinical trials of senolytics. *Mechanisms of Ageing and Development* **200**, 111591 (2021).

628. Rainey, P. B. & Rainey, K. Evolution of cooperation and conflict in experimental bacterial populations. *Nature* **425**, 72–74 (2003).

629. Hardin, G. The Tragedy of the Commons: The population problem has no technical solution; it requires a fundamental extension in morality. *Science* **162**, 1243–1248 (1968).

630. Pye, R. J. *et al.* A second transmissible cancer in Tasmanian devils. *Proceedings of the National Academy of Sciences* **113**, 374–379 (2016).

631. Strakova, A. & Murchison, E. P. The cancer which survived: insights from the genome of an 11000 year-old cancer. *Current Opinion in Genetics & Development* **30**, 49–55 (2015).

632. Metzger, M. J. *et al.* Widespread transmission of independent cancer lineages within multiple bivalve species. *Nature* **534**, 705–709 (2016).

633. Scherer, W. F., Syverton, J. T. & Gey, G. O. STUDIES ON THE PROPAGATION IN VITRO OF POLIOMYELITIS VIRUSES. *J Exp Med* **97**, 695–710 (1953).

634. Landry, J. J. M. *et al.* The Genomic and Transcriptomic Landscape of a HeLa Cell Line. *G3 (Bethesda)* **3**, 1213–1224 (2013).

635. Muehlenbachs, A. *et al.* Malignant Transformation of *Hymenolepis nana* in a Human Host. *N Engl J Med* **373**, 1845–1852 (2015).

636. D'Alterio, C., Scala, S., Sozzi, G., Roz, L. & Bertolini, G. Paradoxical effects of chemotherapy on tumor relapse and metastasis promotion. *Semin Cancer Biol* **60**, 351–361 (2020).

637. Ozaki, T. & Nakagawara, A. Role of p53 in Cell Death and Human Cancers. *Cancers (Basel)* **3**, 994–1013 (2011).

638. Teicher, B. A. *Tumor Models in Cancer Research*. (Humana Press, Totowa, 2002).

639. Donehower, L. A. The p53-deficient mouse: a model for basic and applied cancer studies. *Semin Cancer Biol* **7**, 269–278 (1996).

640. Cheng, Q., Chen, L., Li, Z., Lane, W. S. & Chen, J. ATM activates p53 by regulating MDM2 oligomerization and E3 processivity. *EMBO J* **28**, 3857–3867 (2009).

REFERENCES

641. Aubrey, B. J., Kelly, G. L., Janic, A., Herold, M. J. & Strasser, A. How does p53 induce apoptosis and how does this relate to p53-mediated tumour suppression? *Cell Death Differ* **25**, 104–113 (2018).

642. Biaoxue, R., Hui, P., Wenlong, G. & Shuanying, Y. Evaluation of efficacy and safety for recombinant human adenovirus-p53 in the control of the malignant pleural effusions via thoracic perfusion. *Sci Rep* **6**, 39355 (2016).

643. Li, Y. *et al.* Selective intra-arterial infusion of rAd-p53 with chemotherapy for advanced oral cancer: a randomized clinical trial. *BMC Med* **12**, 16 (2014).

644. Zhang, W.-W. *et al.* The First Approved Gene Therapy Product for Cancer Ad-p53 (Gendicine): 12 Years in the Clinic. *Hum Gene Ther* **29**, 160–179 (2018).

645. Lee, C.-L., Blum, J. M. & Kirsch, D. G. Role of p53 in regulating tissue response to radiation by mechanisms independent of apoptosis. *Transl Cancer Res* **2**, 412–421 (2013).

646. Frebourg, T., Bajalica Lagercrantz, S., Oliveira, C., Magenheim, R. & Evans, D. G. Guidelines for the Li–Fraumeni and heritable TP53-related cancer syndromes. *Eur J Hum Genet* **28**, 1379–1386 (2020).

647. Sulak, M. *et al.* TP53 copy number expansion is associated with the evolution of increased body size and an enhanced DNA damage response in elephants. *eLife* **5**, e11994 (2016).

648. Tejada-Martinez, D., de Magalhães, J. P. & Opazo, J. C. Positive selection and gene duplications in tumour suppressor genes reveal clues about how cetaceans resist cancer. *Proceedings of the Royal Society B: Biological Sciences* **288**, 20202592 (2021).

649. Abegglen, L. M. *et al.* Potential Mechanisms for Cancer Resistance in Elephants and Comparative Cellular Response to DNA Damage in Humans. *JAMA* **314**, 1850–1860 (2015).

650. Butler, G., Baker, J., Amend, S. R., Pienta, K. J. & Venditti, C. No evidence for Peto's paradox in terrestrial vertebrates. *Proc Natl Acad Sci U S A* **122**, e2422861122 (2025).

651. Nagy, J. D., Victor, E. M. & Cropper, J. H. Why don't all whales have cancer? A novel hypothesis resolving Peto's paradox. *Integrative and Comparative Biology* **47**, 317–328 (2007).

652. Galimov, E. R., Lohr, J. N. & Gems, D. When and How Can Death Be an Adaptation? *Biochemistry (Mosc)* **84**, 1433–1437 (2019).

653. Severin, F. F. & Hyman, A. A. Pheromone Induces Programmed Cell Death in S. cerevisiae. *Current Biology* **12**, R233–R235 (2002).

654. Teng, X. & Hardwick, J. M. Reliable Method for Detection of Programmed Cell Death in Yeast. *Methods Mol Biol* **559**, 335–342 (2009).

655. Tower, J. Programmed cell death in aging. *Ageing Res Rev* **23**, 90–100 (2015).

656. Wang, Y., Shnyra, A., Africa, C., Warholic, C. & McArthur, C. Activation of the extrinsic apoptotic pathway by TNF-alpha in human salivary gland (HSG) cells in vitro, suggests a role for the TNF receptor (TNF-R) and intercellular adhesion molecule-1 (ICAM-1) in Sjögren's syndrome-associated autoimmune sialadenitis. *Arch Oral Biol* **54**, 986–996 (2009).

657. Wang, C. & Youle, R. J. The Role of Mitochondria in Apoptosis. *Annu Rev Genet* **43**, 95–118 (2009).

658. Harris, M. H. & Thompson, C. B. The role of the Bcl-2 family in the regulation of outer mitochondrial membrane permeability. *Cell Death Differ* **7**, 1182–1191 (2000).

659. Oberst, A., Bender, C. & Green, D. R. Living with death: The evolution of the mitochondrial pathway of apoptosis in animals. *Cell Death Differ* **15**, 1139–1146 (2008).

660. Böttger, A. & Alexandrova, O. Programmed cell death in Hydra. *Semin Cancer Biol* **17**, 134–146 (2007).

661. Koonin, E. V. Origin of eukaryotes from within archaea, archaeal eukaryome and bursts of gene gain: eukaryogenesis just made easier? *Philos Trans R Soc Lond B Biol Sci* **370**, 20140333 (2015).

662. Zachar, I. & Boza, G. Endosymbiosis before eukaryotes: mitochondrial establishment in protoeukaryotes. *Cell Mol Life Sci* **77**, 3503–3523 (2020).

663. Yan, G., Elbadawi, M. & Efferth, T. Multiple cell death modalities and their key features (Review). *World Acad Sci J* (2020) doi:10.3892/wasj.2020.40.

664. Chaabane, W. *et al.* Autophagy, Apoptosis, Mitoptosis and Necrosis: Interdependence Between Those Pathways and Effects on Cancer. *Arch. Immunol. Ther. Exp.* **61**, 43–58 (2013).

665. Mijaljica, D., Prescott, M. & Devenish, R. J. Mitophagy and mitoptosis in disease processes. *Methods Mol Biol* **648**, 93–106 (2010).

666. Sapunar, D., Vilović, K., England, M. & Saraga-Babić, M. Morphological diversity of dying cells during regression of the human tail. *Ann Anat* **183**, 217–222 (2001).

REFERENCES

667. Chimal-Monroy, J. *et al.* Molecular control of cell differentiation and programmed cell death during digit development. *IUBMB Life* **63**, 922–929 (2011).

668. Božič, B. & Rozman, B. Apoptosis and Autoimmunity. *EJIFCC* **17**, 69–74 (2006).

669. Kuida, K. *et al.* Reduced Apoptosis and Cytochrome c–Mediated Caspase Activation in Mice Lacking Caspase 9. *Cell* **94**, 325–337 (1998).

670. Ke, F. F. S., Brinkmann, K., Voss, A. K. & Strasser, A. Some mice lacking intrinsic, as well as death receptor induced apoptosis and necroptosis, can survive to adulthood. *Cell Death Dis* **13**, 317 (2022).

671. Collado, J. A., Guitart, C., Ciudad, M. T., Alvarez, I. & Jaraquemada, D. The Repertoires of Peptides Presented by MHC-II in the Thymus and in Peripheral Tissue: A Clue for Autoimmunity? *Front. Immunol.* **4**, (2013).

672. Westera, L. *et al.* Closing the gap between T-cell life span estimates from stable isotope-labeling studies in mice and humans. *Blood* **122**, 2205–2212 (2013).

673. Hammarlund, E. *et al.* Duration of antiviral immunity after smallpox vaccination. *Nat Med* **9**, 1131–1137 (2003).

674. Gotuzzo, E., Yactayo, S. & Córdova, E. Efficacy and Duration of Immunity after Yellow Fever Vaccination: Systematic Review on the Need for a Booster Every 10 Years. *Am J Trop Med Hyg* **89**, 434–444 (2013).

675. Macallan, D. C., Borghans, J. A. M. & Asquith, B. Human T Cell Memory: A Dynamic View. *Vaccines (Basel)* **5**, 5 (2017).

676. Giltiay, N. V., Chappell, C. P. & Clark, E. A. B-cell selection and the development of autoantibodies. *Arthritis Res Ther* **14**, S1 (2012).

677. Nemazee, D. Mechanisms of central tolerance for B cells. *Nat Rev Immunol* **17**, 281–294 (2017).

678. Charles A Janeway, J., Travers, P., Walport, M. & Shlomchik, M. J. T cell-mediated cytotoxicity. in *Immunobiology: The Immune System in Health and Disease. 5th edition* (Garland Science, 2001).

679. Hewitt, E. W. The MHC class I antigen presentation pathway: strategies for viral immune evasion. *Immunology* **110**, 163–169 (2003).

680. Goldstein, J. R. & Lee, R. D. Demographic perspectives on the mortality of COVID-19 and other epidemics. *Proceedings of the National Academy of Sciences* **117**, 22035–22041 (2020).

REFERENCES

681. Chalan, P., van den Berg, A., Kroesen, B.-J., Brouwer, L. & Boots, A. Rheumatoid Arthritis, Immunosenescence and the Hallmarks of Aging. *Curr Aging Sci* **8**, 131–146 (2015).

682. Mittelbrunn, M. & Kroemer, G. Hallmarks of T cell aging. *Nat Immunol* **22**, 687–698 (2021).

683. Flajnik, M. F. & Kasahara, M. Origin and evolution of the adaptive immune system: genetic events and selective pressures. *Nat Rev Genet* **11**, 47–59 (2010).

684. Ovadya, Y. *et al.* Impaired immune surveillance accelerates accumulation of senescent cells and aging. *Nat Commun* **9**, 5435 (2018).

685. Desdín-Micó, G. *et al.* T cells with dysfunctional mitochondria induce multimorbidity and premature senescence. *Science* **368**, 1371–1376 (2020).

686. Fajgenbaum, D. C. & June, C. H. Cytokine Storm. *N Engl J Med* **383**, 2255–2273 (2020).

687. Jang, D. *et al.* The Role of Tumor Necrosis Factor Alpha (TNF-α) in Autoimmune Disease and Current TNF-α Inhibitors in Therapeutics. *Int J Mol Sci* **22**, 2719 (2021).

688. Tang, Y. *et al.* Cytokine Storm in COVID-19: The Current Evidence and Treatment Strategies. *Front Immunol* **11**, 1708 (2020).

689. Palmer, S., Albergante, L., Blackburn, C. C. & Newman, T. J. Thymic involution and rising disease incidence with age. *Proc. Natl. Acad. Sci. U.S.A.* **115**, 1883–1888 (2018).

690. Steinmann, G. G., Klaus, B. & Müller-Hermelink, H. K. The involution of the ageing human thymic epithelium is independent of puberty. A morphometric study. *Scand J Immunol* **22**, 563–575 (1985).

691. Murray, J. M. *et al.* Naive T cells are maintained by thymic output in early ages but by proliferation without phenotypic change after age twenty. *Immunol Cell Biol* **81**, 487–495 (2003).

692. Shanley, D. P., Aw, D., Manley, N. R. & Palmer, D. B. An evolutionary perspective on the mechanisms of immunosenescence. *Trends Immunol* **30**, 374–381 (2009).

693. Turke, P. W. Microbial parasites versus developing T cells: an evolutionary 'arms race' with implications for the timing of thymic involution and HIV pathogenesis. *Thymus* **24**, 29–40 (1994).

694. Emmrich, S. *et al.* Ectopic cervical thymi and no thymic involution until midlife in naked mole rats. *Aging Cell* **20**, e13477 (2021).

695. Baran-Gale, J. *et al.* Ageing compromises mouse thymus function and remodels epithelial cell differentiation. *eLife* **9**, e56221 (2020).

696. Pawelec, G. Multiple thymi and no thymic involution in naked mole rats? *Immun Ageing* **18**, 41, s12979-021-00253-w (2021).

697. Delaney, M. A. et al. Initial Case Reports of Cancer in Naked Mole-rats (Heterocephalus glaber). *Vet Pathol* **53**, 691–696 (2016).

698. Seluanov, A. et al. Hypersensitivity to contact inhibition provides a clue to cancer resistance of naked mole-rat. *Proc Natl Acad Sci U S A* **106**, 19352–19357 (2009).

699. Deuker, M. M. et al. Unprovoked Stabilization and Nuclear Accumulation of the Naked Mole-Rat p53 Protein. *Sci Rep* **10**, 6966 (2020).

700. Cavalcanti, N. V., Palmeira, P., Jatene, M. B., de Barros Dorna, M. & Carneiro-Sampaio, M. Early Thymectomy Is Associated With Long-Term Impairment of the Immune System: A Systematic Review. *Front Immunol* **12**, 774780 (2021).

701. Gudmundsdottir, J. et al. Long-term clinical effects of early thymectomy: Associations with autoimmune diseases, cancer, infections, and atopic diseases. *Journal of Allergy and Clinical Immunology* **141**, 2294-2297.e8 (2018).

702. Poulin, J.-F. et al. Direct Evidence for Thymic Function in Adult Humans. *J Exp Med* **190**, 479–486 (1999).

703. Feinstein, L. et al. Population Distributions of Thymic Function in Adults: Variation by Sociodemographic Characteristics and Health Status. *Biodemography Soc Biol* **62**, 208–221 (2016).

704. Douek, D. C. & Koup, R. A. Evidence for thymic function in the elderly. *Vaccine* **18**, 1638 1641 (2000).

705. Rossi, S. W. et al. Keratinocyte growth factor (KGF) enhances postnatal T-cell development via enhancements in proliferation and function of thymic epithelial cells. *Blood* **109**, 3803–3811 (2007).

706. Finch, P. W. & Rubin, J. S. Keratinocyte growth factor/fibroblast growth factor 7, a homeostatic factor with therapeutic potential for epithelial protection and repair. *Adv Cancer Res* **91**, 69–136 (2004).

707. Zhang, Y. et al. The starvation hormone, fibroblast growth factor-21, extends lifespan in mice. *Elife* **1**, e00065 (2012).

708. Youm, Y.-H., Horvath, T. L., Mangelsdorf, D. J., Kliewer, S. A. & Dixit, V. D. Prolongevity hormone FGF21 protects against immune senescence by delaying age-related thymic involution. *Proc Natl Acad Sci U S A* **113**, 1026–1031 (2016).

709. Dixit, V. D. et al. Ghrelin promotes thymopoiesis during aging. *J Clin Invest* **117**, 2778–2790 (2007).

REFERENCES

710. Howard, J. K. *et al.* Leptin protects mice from starvation-induced lymphoid atrophy and increases thymic cellularity in ob/ob mice. *J Clin Invest* **104**, 1051–1059 (1999).

711. Dudakov, J. A. *et al.* Interleukin-22 drives endogenous thymic regeneration in mice. *Science* **336**, 91–95 (2012).

712. Sutherland, J. S. *et al.* Activation of Thymic Regeneration in Mice and Humans following Androgen Blockade. *The Journal of Immunology* **175**, 2741–2753 (2005).

713. Pido-Lopez, J., Imami, N. & Aspinall, R. Both age and gender affect thymic output: more recent thymic migrants in females than males as they age. *Clinical and Experimental Immunology* **125**, 409 (2001).

714. Heng, T. S. P. *et al.* Effects of castration on thymocyte development in two different models of thymic involution. *J Immunol* **175**, 2982–2993 (2005).

715. Min, K.-J., Lee, C.-K. & Park, H.-N. The lifespan of Korean eunuchs. *Current Biology* **22**, R792–R793 (2012).

716. Rode, I. *et al.* Foxn1 Protein Expression in the Developing, Aging, and Regenerating Thymus. *J Immunol* **195**, 5678–5687 (2015).

717. Amorosi, S. *et al.* FOXN1 homozygous mutation associated with anencephaly and severe neural tube defect in human athymic Nude/SCID fetus. *Clin Genet* **73**, 380–384 (2008).

718. Sun, L. *et al.* Declining expression of a single epithelial cell-autonomous gene accelerates age-related thymic involution. *Aging Cell* **9**, 347–357 (2010).

719. Kim, M.-J., Miller, C. M., Shadrach, J. L., Wagers, A. J. & Serwold, T. Young, proliferative thymic epithelial cells engraft and function in aging thymuses. *J Immunol* **194**, 4784–4795 (2015).

720. Oh, J., Wang, W., Thomas, R. & Su, D.-M. Thymic rejuvenation via FOXN1-reprogrammed embryonic fibroblasts (FREFs) to counteract age-related inflammation. *JCI Insight* **5**, e140313 (2020).

721. Quinn, K. M., Palchaudhuri, R., Palmer, C. S. & La Gruta, N. L. The clock is ticking: the impact of ageing on T cell metabolism. *Clin Transl Immunology* **8**, e01091 (2019).

722. Bharath, L. P. *et al.* Metformin Enhances Autophagy and Normalizes Mitochondrial Function to Alleviate Aging-Associated Inflammation. *Cell Metab* **32**, 44-55.e6 (2020).

723. Eikawa, S. *et al.* Immune-mediated antitumor effect by type 2 diabetes drug, metformin. *Proceedings of the National Academy of Sciences* **112**, 1809–1814 (2015).

REFERENCES

724. Wagner, C. L. et al. Short telomere syndromes cause a primary T cell immunodeficiency. *J Clin Invest* **128**, 5222–5234 (2018).

725. Tedone, E. et al. Telomere length and telomerase activity in T cells are biomarkers of high-performing centenarians. *Aging Cell* **18**, e12859 (2019).

726. Jogalekar, M. P. et al. CAR T-Cell-Based gene therapy for cancers: new perspectives, challenges, and clinical developments. *Front Immunol* **13**, 925985 (2022).

727. Zhang, C., Liu, J., Zhong, J. F. & Zhang, X. Engineering CAR-T cells. *Biomark Res* **5**, 22 (2017).

728. Quintarelli, C. et al. Co-expression of cytokine and suicide genes to enhance the activity and safety of tumor-specific cytotoxic T lymphocytes. *Blood* **110**, 2793–2802 (2007).

729. Wu, C.-Y., Roybal, K. T., Puchner, E. M., Onuffer, J. & Lim, W. A. Remote control of therapeutic T cells through a small molecule-gated chimeric receptor. *Science* **350**, aab4077 (2015).

730. Sterner, R. C. & Sterner, R. M. CAR-T cell therapy: current limitations and potential strategies. *Blood Cancer J* **11**, 69 (2021).

731. Amor, C. et al. Prophylactic and long-lasting efficacy of senolytic CAR T cells against age-related metabolic dysfunction. *Nat Aging* **4**, 336–349 (2024).

732. Amor, C. et al. Senolytic CAR T cells reverse senescence-associated pathologies. *Nature* **583**, 127–132 (2020).

733. Han, Y., Liu, D. & Li, L. PD-1/PD-L1 pathway: current researches in cancer. *Am J Cancer Res* **10**, 727–742 (2020).

734. Ostrand-Rosenberg, S., Horn, L. A. & Haile, S. T. The Programmed Death-1 Immune Suppressive Pathway: Barrier to Anti-Tumor Immunity. *J Immunol* **193**, 3835–3841 (2014).

735. Qin, W. et al. The Diverse Function of PD-1/PD-L Pathway Beyond Cancer. *Front. Immunol.* **10**, (2019).

736. Robert, C. A decade of immune-checkpoint inhibitors in cancer therapy. *Nat Commun* **11**, 3801 (2020).

737. Johnson, D. B., Nebhan, C. A., Moslehi, J. J. & Balko, J. M. Immune-checkpoint inhibitors: long-term implications of toxicity. *Nat Rev Clin Oncol* **19**, 254–267 (2022).

738. Chapman, P. B., D'Angelo, S. P. & Wolchok, J. D. Rapid Eradication of a Bulky Melanoma Mass with One Dose of Immunotherapy. *N Engl J Med* **372**, 2073–2074 (2015).

739. Ardeljan, D., Taylor, M. S., Ting, D. T. & Burns, K. H. The human LINE-1 retrotransposon: an emerging biomarker of neoplasia. *Clin Chem* **63**, 816–822 (2017).

740. Platt, R. N., Vandewege, M. W. & Ray, D. A. Mammalian transposable elements and their impacts on genome evolution. *Chromosome Res* **26**, 25–43 (2018).

741. McKerrow, W. *et al.* LINE-1 expression in cancer correlates with p53 mutation, copy number alteration, and S phase checkpoint. *Proc Natl Acad Sci U S A* **119**, e2115999119 (2022).

742. Vylegzhanina, A. V. *et al.* Cancer Relevance of Circulating Antibodies Against LINE-1 Antigens in Humans. *Cancer Res Commun* **3**, 2256–2267 (2023).

743. Rajurkar, M. *et al.* Reverse Transcriptase Inhibition Disrupts Repeat Element Life Cycle in Colorectal Cancer. *Cancer Discov* **12**, 1462–1481 (2022).

744. Novototskaya-Vlasova, K. A. *et al.* Inflammatory response to retrotransposons drives tumor drug resistance that can be prevented by reverse transcriptase inhibitors. *Proceedings of the National Academy of Sciences* **119**, e2213146119 (2022).

745. Ndhlovu, L. C. *et al.* Retro-age: A unique epigenetic biomarker of aging captured by DNA methylation states of retroelements. *Aging Cell* **23**, e14288 (2024).

746. Wahl, D. *et al.* The reverse transcriptase inhibitor 3TC protects against age-related cognitive dysfunction. *Aging Cell* **22**, e13798 (2023).

747. Riddle, M. *Repurposing Nucleoside Reverse Transcriptase Inhibitors for Treatment of AD*. https://clinicaltrials.gov/study/NCT04500847 (2024).

748. Bodea, G. O., McKelvey, E. G. Z. & Faulkner, G. J. Retrotransposon-induced mosaicism in the neural genome. *Open Biol* **8**, 180074 (2018).

749. Bushman, D. M. *et al.* Genomic mosaicism with increased amyloid precursor protein (APP) gene copy number in single neurons from sporadic Alzheimer's disease brains. *eLife* **4**, e05116.

750. Spalding, K. L., Bhardwaj, R. D., Buchholz, B. A., Druid, H. & Frisén, J. Retrospective Birth Dating of Cells in Humans. *Cell* **122**, 133–143 (2005).

751. Bergmann, O. *et al.* Evidence for cardiomyocyte renewal in humans. *Science* **324**, 98–102 (2009).

752. Bergmann, O. *et al.* Dynamics of Cell Generation and Turnover in the Human Heart. *Cell* **161**, 1566–1575 (2015).

753. Choudhury, S. *et al.* Somatic mutations in single human cardiomyocytes reveal age-associated DNA damage and widespread oxidative genotoxicity. *Nat Aging* **2**, 714–725 (2022).

754. Saad, A. M. *et al.* Characteristics, survival and incidence rates and trends of primary cardiac malignancies in the United States. *Cardiovasc Pathol* **33**, 27–31 (2018).

755. Altman, J. Are new neurons formed in the brains of adult mammals? *Science* **135**, 1127–1128 (1962).

756. Lim, D. A. & Alvarez-Buylla, A. The Adult Ventricular–Subventricular Zone (V-SVZ) and Olfactory Bulb (OB) Neurogenesis. *Cold Spring Harb Perspect Biol* **8**, a018820 (2016).

757. Kaplan, M. S. & Hinds, J. W. Neurogenesis in the adult rat: electron microscopic analysis of light radioautographs. *Science* **197**, 1092–1094 (1977).

758. BARNEA, A. & PRAVOSUDOV, V. BIRDS AS A MODEL TO STUDY ADULT NEUROGENESIS: BRIDGING EVOLUTIONARY, COMPARATIVE AND NEUROETHOLOGICAL APPROCHES. *Eur J Neurosci* **34**, 884–907 (2011).

759. Goldman, S. A. & Nottebohm, F. Neuronal production, migration, and differentiation in a vocal control nucleus of the adult female canary brain. *Proc Natl Acad Sci U S A* **80**, 2390–2394 (1983).

760. Richards, L. J., Kilpatrick, T. J. & Bartlett, P. F. De novo generation of neuronal cells from the adult mouse brain. *Proc Natl Acad Sci U S A* **89**, 8591–8595 (1992).

761. Yang, S. M., Alvarez, D. D. & Schinder, A. F. Reliable Genetic Labeling of Adult-Born Dentate Granule Cells Using Ascl1CreERT2 and GlastCreERT2 Murine Lines. *J Neurosci* **35**, 15379–15390 (2015).

762. Dolbeare, F. Bromodeoxyuridine: a diagnostic tool in biology and medicine, Part I: Historical perspectives, histochemical methods and cell kinetics. *Histochem J* **27**, 339–369 (1995).

763. Eriksson, P. S. *et al.* Neurogenesis in the adult human hippocampus. *Nat Med* **4**, 1313–1317 (1998).

764. Kuhn, H. G., Dickinson-Anson, H. & Gage, F. H. Neurogenesis in the dentate gyrus of the adult rat: age-related decrease of neuronal progenitor proliferation. *J Neurosci* **16**, 2027–2033 (1996).

765. Spalding, K. L. *et al.* Dynamics of hippocampal neurogenesis in adult humans. *Cell* **153**, 1219–1227 (2013).

766. Sorrells, S. F. *et al.* Human hippocampal neurogenesis drops sharply in children to undetectable levels in adults. *Nature* **555**, 377–381 (2018).

767. Brown, J. P. *et al.* Transient expression of doublecortin during adult neurogenesis. *J Comp Neurol* **467**, 1–10 (2003).

768. Moreno-Jiménez, E. P., Terreros-Roncal, J., Flor-García, M., Rábano, A. & Llorens-Martín, M. Evidences for Adult Hippocampal Neurogenesis in Humans. *J. Neurosci.* **41**, 2541–2553 (2021).

769. Belle, A., Tanay, A., Bitincka, L., Shamir, R. & O'Shea, E. K. Quantification of protein half-lives in the budding yeast proteome. *Proc. Natl. Acad. Sci. U.S.A.* **103**, 13004–13009 (2006).

770. Chen, W., Smeekens, J. M. & Wu, R. Systematic study of the dynamics and half-lives of newly synthesized proteins in human cells. *Chem. Sci.* **7**, 1393–1400 (2016).

771. Li, J. *et al.* Proteome-wide mapping of short-lived proteins in human cells. *Molecular Cell* **81**, 4722-4735.e5 (2021).

772. Cambridge, S. B. *et al.* Systems-wide proteomic analysis in mammalian cells reveals conserved, functional protein turnover. *J Proteome Res* **10**, 5275–5284 (2011).

773. Yen, H.-C. S., Xu, Q., Chou, D. M., Zhao, Z. & Elledge, S. J. Global protein stability profiling in mammalian cells. *Science* **322**, 918–923 (2008).

774. Swovick, K. *et al.* Interspecies Differences in Proteome Turnover Kinetics Are Correlated With Life Spans and Energetic Demands. *Mol Cell Proteomics* **20**, 100041 (2021).

775. Koyuncu, S. *et al.* Rewiring of the ubiquitinated proteome determines ageing in C. elegans. *Nature* **596**, 285–290 (2021).

776. Mills-Henry, I. A., Thol, S. L., Kosinski-Collins, M. S., Serebryany, E. & King, J. A. Kinetic Stability of Long-Lived Human Lens γ-Crystallins and Their Isolated Double Greek Key Domains. *Biophys J* **117**, 269–280 (2019).

777. Hughes, J. R. *et al.* No turnover in lens lipids for the entire human lifespan. *eLife* **4**, e06003.

778. Lynnerup, N., Kjeldsen, H., Heegaard, S., Jacobsen, C. & Heinemeier, J. Radiocarbon Dating of the Human Eye Lens Crystallines Reveal Proteins without Carbon Turnover throughout Life. *PLoS ONE* **3**, e1529 (2008).

779. Söderberg, P. G., Talebizadeh, N., Yu, Z. & Galichanin, K. Does infrared or ultraviolet light damage the lens? *Eye (Lond)* **30**, 241–246 (2016).

780. Cetinel, S. *et al.* UV-B induced fibrillization of crystallin protein mixtures. *PLoS One* **12**, e0177991 (2017).

781. Moshirfar, M., Milner, D. & Patel, B. C. Cataract Surgery. in *StatPearls* (StatPearls Publishing, Treasure Island (FL), 2024).

782. Hollick, E., Spalton, D., Ursell, P. & Pande, M. Lens epithelial cell regression on the posterior capsule with different intraocular lens materials. *Br J Ophthalmol* **82**, 1182–1188 (1998).

783. Lin, H. *et al.* Lens regeneration using endogenous stem cells with gain of visual function. *Nature* **531**, 323–328 (2016).

784. Shapiro, S. D., Endicott, S. K., Province, M. A., Pierce, J. A. & Campbell, E. J. Marked longevity of human lung parenchymal elastic fibers deduced from prevalence of D-aspartate and nuclear weapons-related radiocarbon. *J Clin Invest* **87**, 1828–1834 (1991).

785. Sivan, S.-S. *et al.* Collagen turnover in normal and degenerate human intervertebral discs as determined by the racemization of aspartic acid. *J Biol Chem* **283**, 8796–8801 (2008).

786. Verzijl, N. *et al.* Effect of collagen turnover on the accumulation of advanced glycation end products. *J Biol Chem* **275**, 39027–39031 (2000).

787. Rucklidge, G. J., Milne, G., McGaw, B. A., Milne, E. & Robins, S. P. Turnover rates of different collagen types measured by isotope ratio mass spectrometry. *Biochimica et Biophysica Acta (BBA) - General Subjects* **1156**, 57–61 (1992).

788. Wang, L. *et al.* Differences between Mice and Humans in Regulation and the Molecular Network of Collagen, Type III, Alpha-1 at the Gene Expression Level: Obstacles that Translational Research Must Overcome. *Int J Mol Sci* **16**, 15031–15056 (2015).

789. Jackson, B. C., Nebert, D. W. & Vasiliou, V. Update of human and mouse matrix metalloproteinase families. *Hum Genomics* **4**, 194–201 (2010).

790. Ricard-Blum, S. The Collagen Family. *Cold Spring Harb Perspect Biol* **3**, a004978 (2011).

791. Kohn, J. C., Lampi, M. C. & Reinhart-King, C. A. Age-related vascular stiffening: causes and consequences. *Front Genet* **6**, 112 (2015).

792. Leitinger, B. & Hohenester, E. Mammalian collagen receptors. *Matrix Biol* **26**, 146–155 (2007).

793. Mora Huertas, A. C., Schmelzer, C. E. H., Hoehenwarter, W., Heyroth, F. & Heinz, A. Molecular-level insights into aging processes of skin elastin. *Biochimie* **128–129**, 163–173 (2016).

794. Wagenseil, J. E. & Mecham, R. P. Elastin in large artery stiffness and hypertension. *J Cardiovasc Transl Res* **5**, 264–273 (2012).

795. McEniery, C. M., Wilkinson, I. B. & Avolio, A. P. AGE, HYPERTENSION AND ARTERIAL FUNCTION. *Clin Exp Pharma Physio* **34**, 665–671 (2007).

796. Li, D. Y. *et al.* Elastin is an essential determinant of arterial morphogenesis. *Nature* **393**, 276–280 (1998).

797. Hudson, D. M., Archer, M., King, K. B. & Eyre, D. R. Glycation of type I collagen selectively targets the same helical domain lysine sites as lysyl oxidase–mediated cross-linking. *J Biol Chem* **293**, 15620–15627 (2018).

798. Hennet, T. Collagen glycosylation. *Current Opinion in Structural Biology* **56**, 131–138 (2019).

799. Ansari, N. A. & Dash, D. Amadori Glycated Proteins: Role in Production of Autoantibodies in Diabetes Mellitus and Effect of Inhibitors on Non-Enzymatic Glycation. *Aging Dis* **4**, 50–56 (2012).

800. Nash, A. *et al.* Glucosepane is associated with changes to structural and physical properties of collagen fibrils. *Matrix Biol Plus* **4**, 100013 (2019).

801. Lin, L., Park, S. & Lakatta, E. G. RAGE signaling in inflammation and arterial aging. *Front Biosci* **14**, 1403–1413 (2009).

802. Monnier, V. M. *et al.* Cross-linking of the extracellular matrix by the maillard reaction in aging and diabetes: an update on 'a puzzle nearing resolution'. *Ann N Y Acad Sci* **1043**, 533–544 (2005).

803. Sell, D. R. *et al.* Glucosepane Is a Major Protein Cross-link of the Senescent Human Extracellular Matrix: RELATIONSHIP WITH DIABETES *. *Journal of Biological Chemistry* **280**, 12310–12315 (2005).

804. Fedintsev, A. & Moskalev, A. Stochastic non-enzymatic modification of long-lived macromolecules - A missing hallmark of aging. *Ageing Research Reviews* **62**, 101097 (2020).

805. Choi, W. H., Kim, S., Park, S. & Lee, M. J. Concept and application of circulating proteasomes. *Exp Mol Med* **53**, 1539–1546 (2021).

806. Van Doren, S. R. Matrix metalloproteinase interactions with collagen and elastin. *Matrix Biol* **0**, 224–231 (2015).

807. DeLeon-Pennell, K. Y., Meschiari, C. A., Jung, M. & Lindsey, M. L. Matrix Metalloproteinases in Myocardial Infarction and Heart Failure. *Prog Mol Biol Transl Sci* **147**, 75–100 (2017).

808. Vafaie, F. *et al.* Collagenase-resistant collagen promotes mouse aging and vascular cell senescence. *Aging Cell* **13**, 121–130 (2014).

809. Beare, A. H. M., O'Kane, S., Ferguson, M. W. J. & Krane, S. M. Severely Impaired Wound Healing in the Collagenase-Resistant Mouse. *Journal of Investigative Dermatology* **120**, 153–163 (2003).

810. Hamlin, C. R. & Kohn, R. R. Determination of human chronological age by study of a collagen sample. *Experimental Gerontology* **7**, 377–379 (1972).

811. Collier, T. A., Nash, A., Birch, H. L. & de Leeuw, N. H. Preferential sites for intramolecular glucosepane cross-link formation in type I collagen: A thermodynamic study. *Matrix Biol* **48**, 78–88 (2015).

812. Bourne, J. W., Lippell, J. M. & Torzilli, P. A. Glycation Cross-Linking Induced Mechanical-Enzymatic Cleavage of Microscale Tendon Fibers. *Matrix Biol* **34**, 179–184 (2014).

813. Nosaka, T., Tanaka, H., Watanabe, I., Sato, M. & Matsuda, M. Influence of Regular Exercise on Age-Related Changes in Arterial Elasticity: Mechanistic Insights From Wall Compositions in Rat Aorta. *Can. J. Appl. Physiol.* **28**, 204–212 (2003).

814. Gu, Q. *et al.* Contribution of receptor for advanced glycation end products to vasculature-protecting effects of exercise training in aged rats. *Eur J Pharmacol* **741**, 186–194 (2014).

815. Choi, S.-Y. *et al.* Long-Term Exercise Training Attenuates Age-Related Diastolic Dysfunction: Association of Myocardial Collagen Cross-Linking. *J Korean Med Sci* **24**, 32–39 (2009).

816. Fukami, K. *et al.* Ramipril inhibits AGE-RAGE-induced matrix metalloproteinase-2 activation in experimental diabetic nephropathy. *Diabetol Metab Syndr* **6**, 86 (2014).

817. Raeeszadeh-Sarmazdeh, M., Do, L. D. & Hritz, B. G. Metalloproteinases and Their Inhibitors: Potential for the Development of New Therapeutics. *Cells* **9**, 1313 (2020).

818. Vasan, S. *et al.* An agent cleaving glucose-derived protein crosslinks in vitro and in vivo. *Nature* **382**, 275–278 (1996).

819. Cooper, M. E. *et al.* The cross-link breaker, N-phenacylthiazolium bromide prevents vascular advanced glycation end-product accumulation. *Diabetologia* **43**, 660–664 (2000).

820. Vasan, S., Foiles, P. & Founds, H. Therapeutic potential of breakers of advanced glycation end product-protein crosslinks. *Arch Biochem Biophys* **419**, 89–96 (2003).

821. Zhang, B. *et al.* Alagebrium (ALT-711) improves the anti-hypertensive efficacy of nifedipine in diabetic-hypertensive rats. *Hypertens Res* **37**, 901–907 (2014).

822. Mentink, C. J. A. L., Hendriks, M., Levels, A. A. G. & Wolffenbuttel, B. H. R. Glucose-mediated cross-linking of collagen in rat tendon and skin. *Clin Chim Acta* **321**, 69–76 (2002).

REFERENCES

823. Yang, S., Litchfield, J. E. & Baynes, J. W. AGE-breakers cleave model compounds, but do not break Maillard crosslinks in skin and tail collagen from diabetic rats. *Arch Biochem Biophys* **412**, 42–46 (2003).

824. Nagai, R., Murray, D. B., Metz, T. O. & Baynes, J. W. Chelation: a fundamental mechanism of action of AGE inhibitors, AGE breakers, and other inhibitors of diabetes complications. *Diabetes* **61**, 549–559 (2012).

825. Toprak, C. & Yigitaslan, S. Alagebrium and Complications of Diabetes Mellitus. *Eurasian J Med* **51**, 285–292 (2019).

826. Perez-Sanchez, A. C. *et al.* Skin, Hair, and Nail Supplements: Marketing and Labeling Concerns. *Cureus* **12**, e12062.

827. Rustad, A. M., Nickles, M. A., McKenney, J. E., Bilimoria, S. N. & Lio, P. A. Myths and media in oral collagen supplementation for the skin, nails, and hair: A review. *J of Cosmetic Dermatology* **21**, 438–443 (2022).

828. Choi, F. D., Sung, C. T., Juhasz, M. L. W. & Mesinkovsk, N. A. Oral Collagen Supplementation: A Systematic Review of Dermatological Applications. *J Drugs Dermatol* **18**, 9–16 (2019).

829. de Miranda, R. B., Weimer, P. & Rossi, R. C. Effects of hydrolyzed collagen supplementation on skin aging: a systematic review and meta-analysis. *Int J Dermatol* **60**, 1449–1461 (2021).

830. Ioannidis, J. P. A. Why Most Published Research Findings Are False. *PLoS Med* **2**, e124 (2005).

831. Hoenig, L. J. Chicken soup for the skin! *Clin Dermatol* **40**, 764–767 (2022).

832. Head, M. L., Holman, L., Lanfear, R., Kahn, A. T. & Jennions, M. D. The Extent and Consequences of P-Hacking in Science. *PLoS Biol* **13**, e1002106 (2015).

833. Miner-Williams, W. M., Stevens, B. R. & Moughan, P. J. Are intact peptides absorbed from the healthy gut in the adult human? *Nutr. Res. Rev.* **27**, 308–329 (2014).

834. Pu, S.-Y. *et al.* Effects of Oral Collagen for Skin Anti-Aging: A Systematic Review and Meta-Analysis. *Nutrients* **15**, 2080 (2023).

835. Smith, D. W., Azadi, A., Lee, C.-J. & Gardiner, B. S. Spatial composition and turnover of the main molecules in the adult glomerular basement membrane. *Tissue Barriers* **11**, 2110798 (2023).

836. Taimen, P. *et al.* A progeria mutation reveals functions for lamin A in nuclear assembly, architecture, and chromosome organization. *Proceedings of the National Academy of Sciences* **106**, 20788–20793 (2009).

REFERENCES

837. Yao, Y. Laminin: loss-of-function studies. *Cell Mol Life Sci* **74**, 1095–1115 (2017).

838. Hirsch, T. *et al.* Regeneration of the entire human epidermis by transgenic stem cells. *Nature* **551**, 327–332 (2017).

839. Iriyama, S. *et al.* Decrease of laminin-511 in the basement membrane due to photoaging reduces epidermal stem/progenitor cells. *Sci Rep* **10**, 12592 (2020).

840. Sun, Y. *et al.* Rescuing replication and osteogenesis of aged mesenchymal stem cells by exposure to a young extracellular matrix. *FASEB J* **25**, 1474–1485 (2011).

841. Lacraz, G. *et al.* Increased Stiffness in Aged Skeletal Muscle Impairs Muscle Progenitor Cell Proliferative Activity. *PLoS ONE* **10**, e0136217 (2015).

842. Romero-Ortuno, R., Kenny, R. A. & McManus, R. Collagens and elastin genetic variations and their potential role in aging-related diseases and longevity in humans. *Experimental Gerontology* **129**, 110781 (2020).

843. Moreno-Borrallo, A. *et al.* Variation in albumin glycation rates in birds suggests resistance to relative hyperglycaemia rather than conformity to the pace of life syndrome hypothesis. Preprint at https://doi.org/10.7554/eLife.103205.2 (2025).

844. Tsunosue, M. *et al.* An alpha-glucosidase inhibitor, acarbose treatment decreases serum levels of glyceraldehyde-derived advanced glycation end products (AGEs) in patients with type 2 diabetes. *Clin Exp Med* **10**, 139–141 (2010).

845. He, K., Shi, J.-C. & Mao, X.-M. Safety and efficacy of acarbose in the treatment of diabetes in Chinese patients. *Ther Clin Risk Manag* **10**, 505–511 (2014).

846. Moelands, S. V., Lucassen, P. L., Akkermans, R. P., De Grauw, W. J. & Van De Laar, F. A. Alpha-glucosidase inhibitors for prevention or delay of type 2 diabetes mellitus and its associated complications in people at increased risk of developing type 2 diabetes mellitus. *Cochrane Database of Systematic Reviews* **2018**, (2018).

847. Reddy, G. K. Cross-linking in collagen by nonenzymatic glycation increases the matrix stiffness in rabbit achilles tendon. *Exp Diabesity Res* **5**, 143–153 (2004).

848. Harrison, D. E. Mouse erythropoietic stem cell lines function normally 100 months: loss related to number of transplantations. *Mech Ageing Dev* **9**, 427–433 (1979).

849. Soerens, A. G. *et al.* Functional T cells are capable of supernumerary cell division and longevity. *Nature* **614**, 762–766 (2023).

850. Konstantinov, I. E. & Alexi-Meskishvili, V. V. Sergei S. Brukhonenko: the development of the first heart-lung machine for total body perfusion. *The Annals of Thoracic Surgery* **69**, 962–966 (2000).

851. Lamba, N., Holsgrove, D. & Broekman, M. L. The history of head transplantation: a review. *Acta Neurochir (Wien)* **158**, 2239–2247 (2016).

852. Barker, J. H., Frank, J. M. & Leppik, L. Head Transplantation: Editorial Commentary. *CNS Neurosci Ther* **21**, 613–614 (2015).

853. White, R. J., Wolin, L. R., Massopust, L. C., Taslitz, N. & Verdura, J. Primate cephalic transplantation: neurogenic separation, vascular association. *Transplant Proc* **3**, 602–604 (1971).

854. White, R. J., Wolin, L. R., Massopust, L. C., Taslitz, N. & Verdura, J. Cephalic exchange transplantation in the monkey. *Surgery* **70**, 135–139 (1971).

855. Konstantinov, I. E. At the Cutting Edge of the Impossible. *Tex Heart Inst J* **36**, 453–458 (2009).

856. Langer, R. M. Vladimir P. Demikhov, a pioneer of organ transplantation. *Transplant Proc* **43**, 1221–1222 (2011).

857. Canavero, S. HEAVEN: The head anastomosis venture Project outline for the first human head transplantation with spinal linkage (GEMINI). *Surg Neurol Int* **4**, S335–S342 (2013).

858. Schneeberger, S. *et al.* 20-Year Follow-up of Two Cases of Bilateral Hand Transplantation. *N Engl J Med* **383**, 1791–1792 (2020).

859. Suskin, Z. D. & Giordano, J. J. Body –to-head transplant; a 'caputal' crime? Examining the corpus of ethical and legal issues. *Philos Ethics Humanit Med* **13**, 10, s13010-018-0063–2 (2018).

860. Furr, A., Hardy, M. A., Barret, J. P. & Barker, J. H. Surgical, ethical, and psychosocial considerations in human head transplantation. *Int J Surg* **41**, 190–195 (2017).

861. Ren, X. *et al.* First cephalosomatic anastomosis in a human model. *Surg Neurol Int* **8**, 276 (2017).

862. Capogrosso, M. *et al.* A brain–spine interface alleviating gait deficits after spinal cord injury in primates. *Nature* **539**, 284–288 (2016).

863. Velliste, M., Perel, S., Spalding, M. C., Whitford, A. S. & Schwartz, A. B. Cortical control of a prosthetic arm for self-feeding. *Nature* **453**, 1098–1101 (2008).

864. Flesher, S. N. *et al.* A brain-computer interface that evokes tactile sensations improves robotic arm control. *Science* **372**, 831–836 (2021).

REFERENCES

865. Land, B. B., Brayton, C. E., Furman, K. E., LaPalombara, Z. & DiLeone, R. J. Optogenetic inhibition of neurons by internal light production. *Front Behav Neurosci* **8**, 108 (2014).

866. Boyden, E. S., Zhang, F., Bamberg, E., Nagel, G. & Deisseroth, K. Millisecond-timescale, genetically targeted optical control of neural activity. *Nat Neurosci* **8**, 1263–1268 (2005).

867. Lima, S. Q. & Miesenböck, G. Remote control of behavior through genetically targeted photostimulation of neurons. *Cell* **121**, 141–152 (2005).

868. Hernandez, V. H. *et al.* Optogenetic stimulation of the auditory pathway. *J Clin Invest* **124**, 1114–1129 (2014).

869. Wrobel, C. *et al.* Optogenetic stimulation of cochlear neurons activates the auditory pathway and restores auditory-driven behavior in deaf adult gerbils. *Sci Transl Med* **10**, eaao0540 (2018).

870. Adamczyk, A. K. & Zawadzki, P. The Memory-Modifying Potential of Optogenetics and the Need for Neuroethics. *Nanoethics* **14**, 207–225 (2020).

871. Liu, X. *et al.* Optogenetic stimulation of a hippocampal engram activates fear memory recall. *Nature* **484**, 381–385 (2012).

872. Nyns, E. C. A. *et al.* Optogenetic termination of ventricular arrhythmias in the whole heart: towards biological cardiac rhythm management. *Eur Heart J* **38**, 2132–2136 (2017).

873. Park, E. & Lee, K.-S. A new approach to urinary bladder control with optogenetics. *Investig Clin Urol* **60**, 61–63 (2019).

874. Kim, T., Folcher, M., Doaud-El Baba, M. & Fussenegger, M. A synthetic erectile optogenetic stimulator enabling blue-light-inducible penile erection. *Angew Chem Int Ed Engl* **54**, 5933–5938 (2015).

875. Method of the Year 2010. *Nat Methods* **8**, 1–1 (2011).

876. Su, Y. *et al.* An optimized bioluminescent substrate for non-invasive imaging in the brain. *Nat Chem Biol* (2023) doi:10.1038/s41589-023-01265-x.

877. Oba, Y. & Hosaka, K. The Luminous Fungi of Japan. *J Fungi (Basel)* **9**, 615 (2023).

878. Arshavsky, Y. I. *et al.* Analysis of the central pattern generator for swimming in the mollusk Clione. *Ann N Y Acad Sci* **860**, 51–69 (1998).

879. Mühlethaler, M., de Curtis, M., Walton, K. & Llinás, R. The isolated and perfused brain of the guinea-pig in vitro. *Eur J Neurosci* **5**, 915–926 (1993).

REFERENCES

880. de Curtis, M., Librizzi, L. & Uva, L. The in vitro isolated whole guinea pig brain as a model to study epileptiform activity patterns. *J Neurosci Methods* **260**, 83–90 (2016).

881. Could We Transport Our Consciousness Into Robots? (Big Think, 2011).

882. Szigeti, B. et al. OpenWorm: an open-science approach to modeling Caenorhabditis elegans. *Front Comput Neurosci* **8**, 137 (2014).

883. Sarma, G. P. et al. OpenWorm: overview and recent advances in integrative biological simulation of Caenorhabditis elegans. *Philos Trans R Soc Lond B Biol Sci* **373**, 20170382 (2018).

884. Winding, M. et al. The connectome of an insect brain. *Science* **379**, eadd9330 (2023).

885. Billeh, Y. N. et al. Systematic Integration of Structural and Functional Data into Multi-scale Models of Mouse Primary Visual Cortex. *Neuron* **106**, 388-403.e18 (2020).

886. Shapson-Coe, A. et al. A connectomic study of a petascale fragment of human cerebral cortex. 2021.05.29.446289 Preprint at https://doi.org/10.1101/2021.05.29.446289 (2021).

887. Justinia, T. Blockchain Technologies: Opportunities for Solving Real-World Problems in Healthcare and Biomedical Sciences. *Acta Inform Med* **27**, 284–291 (2019).

888. Leritz, E. C., McGlinchey, R. E., Kellison, I., Rudolph, J. L. & Milberg, W. P. Cardiovascular Disease Risk Factors and Cognition in the Elderly. *Curr Cardiovasc Risk Rep* **5**, 407–412 (2011).

889. Sofi, F. et al. Physical activity and risk of cognitive decline: a meta-analysis of prospective studies: Physical activity and risk of cognitive decline. *Journal of Internal Medicine* **269**, 107–117 (2011).

890. Kennedy, G., Hardman, R. J., Macpherson, H., Scholey, A. B. & Pipingas, A. How Does Exercise Reduce the Rate of Age-Associated Cognitive Decline? A Review of Potential Mechanisms. *J Alzheimers Dis* **55**, 1–18 (2017).

891. Rasmussen, P. et al. Evidence for a release of brain-derived neurotrophic factor from the brain during exercise. *Exp Physiol* **94**, 1062–1069 (2009).

892. Cohen-Cory, S., Kidane, A. H., Shirkey, N. J. & Marshak, S. Brain-Derived Neurotrophic Factor and the Development of Structural Neuronal Connectivity. *Dev Neurobiol* **70**, 271–288 (2010).

893. Miranda, M., Morici, J. F., Zanoni, M. B. & Bekinschtein, P. Brain-Derived Neurotrophic Factor: A Key Molecule for Memory

in the Healthy and the Pathological Brain. *Front Cell Neurosci* **13**, 363 (2019).

894. Bueller, J. A. *et al.* BDNF Val66Met allele is associated with reduced hippocampal volume in healthy subjects. *Biol Psychiatry* **59**, 812–815 (2006).

895. Hariri, A. R. *et al.* Brain-derived neurotrophic factor val-66met polymorphism affects human memory-related hippocampal activity and predicts memory performance. *J Neurosci* **23**, 6690–6694 (2003).

896. Miyajima, F. *et al.* Brain-derived neurotrophic factor polymorphism Val66Met influences cognitive abilities in the elderly. *Genes Brain Behav* **7**, 411–417 (2008).

897. Sanchez, M. M. *et al.* BDNF polymorphism predicts the rate of decline in skilled task performance and hippocampal volume in healthy individuals. *Transl Psychiatry* **1**, e51–e51 (2011).

898. Magrassi, L., Leto, K. & Rossi, F. Lifespan of neurons is uncoupled from organismal lifespan. *Proc. Natl. Acad. Sci. U.S.A.* **110**, 4374–4379 (2013).

899. Ludwig, F. C. & Elashoff, R. M. Mortality in syngeneic rat parabionts of different chronological age. *Trans N Y Acad Sci* **34**, 582–587 (1972).

900. Yankova, T., Dubiley, T., Shytikov, D. & Pishel, I. Three-Month Heterochronic Parabiosis Has a Deleterious Effect on the Lifespan of Young Animals, Without a Positive Effect for Old Animals. *Rejuvenation Research* **25**, 191–199 (2022).

901. Katsimpardi, L. *et al.* Vascular and Neurogenic Rejuvenation of the Aging Mouse Brain by Young Systemic Factors. *Science* **344**, 630–634 (2014).

902. Ozek, C., Krolewski, R. C., Buchanan, S. M. & Rubin, L. L. Growth Differentiation Factor 11 treatment leads to neuronal and vascular improvements in the hippocampus of aged mice. *Sci Rep* **8**, 17293 (2018).

903. Mayweather, B. A., Buchanan, S. M. & Rubin, L. L. GDF11 expressed in the adult brain negatively regulates hippocampal neurogenesis. *Mol Brain* **14**, 134 (2021).

904. Kiss, T. *et al.* Circulating anti-geronic factors from heterochonic parabionts promote vascular rejuvenation in aged mice: transcriptional footprint of mitochondrial protection, attenuation of oxidative stress, and rescue of endothelial function by young blood. *Geroscience* **42**, 727–748 (2020).

905. Horowitz, A. M. *et al.* Blood factors transfer beneficial effects of exercise on neurogenesis and cognition to the aged brain. *Science* **369**, 167–173 (2020).

906. Shytikov, D., Balva, O., Debonneuil, E., Glukhovskiy, P. & Pishel, I. Aged Mice Repeatedly Injected with Plasma from Young Mice: A Survival Study. *Biores Open Access* **3**, 226–232 (2014).

907. Villeda, S. A. *et al.* Young blood reverses age-related impairments in cognitive function and synaptic plasticity in mice. *Nat Med* **20**, 659–663 (2014).

908. Danon, D., Kowatch, M. A. & Roth, G. S. Promotion of wound repair in old mice by local injection of macrophages. *Proc Natl Acad Sci U S A* **86**, 2018–2020 (1989).

909. Chen, D. *et al.* Demyelinating processes in aging and stroke in the central nervous system and the prospect of treatment strategy. *CNS Neurosci Ther* **26**, 1219–1229 (2020).

910. Ruckh, J. M. *et al.* Rejuvenation of regeneration in the aging central nervous system. *Cell Stem Cell* **10**, 96–103 (2012).

911. Iram, T. *et al.* Young CSF restores oligodendrogenesis and memory in aged mice via Fgf17. *Nature* **605**, 509–515 (2022).

912. Villeda, S. A. *et al.* The aging systemic milieu negatively regulates neurogenesis and cognitive function. *Nature* **477**, 90–94 (2011).

913. Lieschke, S. *et al.* CCL11 Differentially Affects Post-Stroke Brain Injury and Neuroregeneration in Mice Depending on Age. *Cells* **9**, 66 (2019).

914. Minhas, P. S. *et al.* Restoring metabolism of myeloid cells reverses cognitive decline in ageing. *Nature* **590**, 122–128 (2021).

915. Mehdipour, M. *et al.* Plasma dilution improves cognition and attenuates neuroinflammation in old mice. *GeroScience* **43**, 1–18 (2020).

916. Mehdipour, M. *et al.* Rejuvenation of three germ layers tissues by exchanging old blood plasma with saline-albumin. *Aging (Albany NY)* **12**, 8790–8819 (2020).

917. Sha, S. J. *et al.* Safety, Tolerability, and Feasibility of Young Plasma Infusion in the Plasma for Alzheimer Symptom Amelioration Study: A Randomized Clinical Trial. *JAMA Neurol* **76**, 35 (2019).

918. Vacher, M. C. *et al.* Alzheimer's disease-like neuropathology in three species of oceanic dolphin. *Eur J of Neuroscience* **57**, 1161–1179 (2023).

919. Cao, Q. *et al.* The Prevalence of Dementia: A Systematic Review and Meta-Analysis. *JAD* **73**, 1157–1166 (2020).

REFERENCES

920. Farfel, J. M. *et al.* Alzheimer's disease frequency peaks in the tenth decade and is lower afterwards. *acta neuropathol commun* **7**, 104 (2019).

921. GBD 2019 Dementia Forecasting Collaborators. Estimation of the global prevalence of dementia in 2019 and forecasted prevalence in 2050: an analysis for the Global Burden of Disease Study 2019. *Lancet Public Health* **7**, e105–e125 (2022).

922. Cummings, J., Lee, G., Ritter, A., Sabbagh, M. & Zhong, K. Alzheimer's disease drug development pipeline: 2020. *A&D Transl Res & Clin Interv* **6**, e12050 (2020).

923. Cummings, J. L., Morstorf, T. & Zhong, K. Alzheimer's disease drug-development pipeline: few candidates, frequent failures. *Alzheimers Res Ther* **6**, 37 (2014).

924. Temp, A. G. M. *et al.* A Bayesian perspective on Biogen's aducanumab trial. *Alzheimers Dement* **18**, 2341–2351 (2022).

925. Tampi, R. R., Forester, B. P. & Agronin, M. Aducanumab: evidence from clinical trial data and controversies. *Drugs Context* **10**, 2021-7-3 (2021).

926. Sims, J. R. *et al.* Donanemab in Early Symptomatic Alzheimer Disease: The TRAILBLAZER-ALZ 2 Randomized Clinical Trial. *JAMA* **330**, 512 (2023).

927. Raji, C. A., Lopez, O. L., Kuller, L. H., Carmichael, O. T. & Becker, J. T. Age, Alzheimer disease, and brain structure. *Neurology* **73**, 1899–1905 (2009).

928. Hardy, J. & Allsop, D. Amyloid deposition as the central event in the aetiology of Alzheimer's disease. *Trends Pharmacol Sci* **12**, 383–388 (1991).

929. Awasthi, A., Matsunaga, Y. & Yamada, T. Amyloid-beta causes apoptosis of neuronal cells via caspase cascade, which can be prevented by amyloid-beta-derived short peptides. *Experimental Neurology* **196**, 282–289 (2005).

930. APP protein expression summary - The Human Protein Atlas.

931. Ishiura, S. *et al.* APP α-Secretase, a Novel Target for Alzheimer Drug Therapy. in *Madame Curie Bioscience Database [Internet]* (Landes Bioscience, 2013).

932. Fahrenholz, F. & Postina, R. Alpha-secretase activation--an approach to Alzheimer's disease therapy. *Neurodegener Dis* **3**, 255–261 (2006).

933. Barage, S. H. & Sonawane, K. D. Amyloid cascade hypothesis: Pathogenesis and therapeutic strategies in Alzheimer's disease. *Neuropeptides* **52**, 1–18 (2015).

REFERENCES

934. Lanoiselée, H.-M. *et al.* APP, PSEN1, and PSEN2 mutations in early-onset Alzheimer disease: A genetic screening study of familial and sporadic cases. *PLoS Med* **14**, e1002270 (2017).

935. Cacace, R., Sleegers, K. & Van Broeckhoven, C. Molecular genetics of early-onset Alzheimer's disease revisited. *Alzheimers Dement* **12**, 733–748 (2016).

936. Giau, V. V. *et al.* Genetic analyses of early-onset Alzheimer's disease using next generation sequencing. *Sci Rep* **9**, 8368 (2019).

937. Liu, C.-C., Kanekiyo, T., Xu, H. & Bu, G. Apolipoprotein E and Alzheimer disease: risk, mechanisms, and therapy. *Nat Rev Neurol* **9**, 106–118 (2013).

938. Farrer, L. A. *et al.* Effects of age, sex, and ethnicity on the association between apolipoprotein E genotype and Alzheimer disease. A meta-analysis. APOE and Alzheimer Disease Meta Analysis Consortium. *JAMA* **278**, 1349–1356 (1997).

939. Yamazaki, Y., Zhao, N., Caulfield, T. R., Liu, C.-C. & Bu, G. Apolipoprotein E and Alzheimer disease: pathobiology and targeting strategies. *Nat Rev Neurol* **15**, 501–518 (2019).

940. Silvius, J. R. Role of cholesterol in lipid raft formation: lessons from lipid model systems. *Biochimica et Biophysica Acta (BBA) — Biomembranes* **1610**, 174–183 (2003).

941. Wang, H. *et al.* Regulation of beta-amyloid production in neurons by astrocyte-derived cholesterol. *Proceedings of the National Academy of Sciences* **118**, e2102191118 (2021).

942. Blanchard, J. W. *et al.* APOE4 impairs myelination via cholesterol dysregulation in oligodendrocytes. *Nature* **611**, 769–779 (2022).

943. Lesné, S. *et al.* RETRACTED ARTICLE: A specific amyloid-β protein assembly in the brain impairs memory. *Nature* **440**, 352–357 (2006).

944. Potential fabrication in research images threatens key theory of Alzheimer's disease. (2022) doi:10.1126/science.ade0209.

945. Schmid, S. *et al.* Intracerebroventricular injection of beta-amyloid in mice is associated with long-term cognitive impairment in the modified hole-board test. *Behavioural Brain Research* **324**, 15–20 (2017).

946. Faucher, P., Mons, N., Micheau, J., Louis, C. & Beracochea, D. J. Hippocampal Injections of Oligomeric Amyloid β-peptide (1–42) Induce Selective Working Memory Deficits and Long-lasting Alterations of ERK Signaling Pathway. *Front. Aging Neurosci.* **7**, (2016).

REFERENCES

947. Kim, H. Y., Lee, D. K., Chung, B.-R., Kim, H. V. & Kim, Y. Intracerebroventricular Injection of Amyloid-β Peptides in Normal Mice to Acutely Induce Alzheimer-like Cognitive Deficits. *J Vis Exp* 53308 (2016) doi:10.3791/53308.

948. Sun, Y. *et al.* Intra-gastrointestinal amyloid-β1–42 oligomers perturb enteric function and induce Alzheimer's disease pathology. *The Journal of Physiology* **598**, 4209–4223 (2020).

949. Maniv, I. *et al.* Altered ubiquitin signaling induces Alzheimer's disease-like hallmarks in a three-dimensional human neural cell culture model. *Nat Commun* **14**, 5922 (2023).

950. Arrasate, M. & Finkbeiner, S. Protein aggregates in Huntington's disease. *Exp Neurol* **238**, 1–11 (2012).

951. Chiti, F. & Dobson, C. M. Protein Misfolding, Amyloid Formation, and Human Disease: A Summary of Progress Over the Last Decade. *Annu Rev Biochem* **86**, 27–68 (2017).

952. Stefanis, L. α-Synuclein in Parkinson's Disease. *Cold Spring Harb Perspect Med* **2**, a009399 (2012).

953. Zabel, M. D. & Reid, C. A brief history of prions. *Pathog Dis* **73**, ftv087 (2015).

954. Prusiner, S. B. Prions. *Proc Natl Acad Sci U S A* **95**, 13363–13383 (1998).

955. Soto, C. & Satani, N. The intricate mechanisms of neurodegeneration in prion diseases. *Trends Mol Med* **17**, 14–24 (2011).

956. Han, X.-J. *et al.* Amyloid β-42 induces neuronal apoptosis by targeting mitochondria. *Mol Med Rep* **16**, 4521–4528 (2017).

957. Tharp, W. G. & Sarkar, I. Origins of amyloid-β. *BMC Genomics* **14**, 290 (2013).

958. Walsh, D. M. *et al.* The APP family of proteins: similarities and differences. *Biochem Soc Trans* **35**, 416–420 (2007).

959. Koike, M. A. *et al.* APP knockout mice experience acute mortality as the result of ischemia. *PLoS One* **7**, e42665 (2012).

960. Fulop, T. *et al.* Can an Infection Hypothesis Explain the Beta Amyloid Hypothesis of Alzheimer's Disease? *Front Aging Neurosci* **10**, 224 (2018).

961. Brothers, H. M., Gosztyla, M. L. & Robinson, S. R. The Physiological Roles of Amyloid-β Peptide Hint at New Ways to Treat Alzheimer's Disease. *Front Aging Neurosci* **10**, 118 (2018).

962. Revised clinical trial form for Alzheimer's antibody warned of fatal brain bleeds. (2022) doi:10.1126/science.adg4937.

963. Mahase, E. Aducanumab: 4 in 10 high dose trial participants experienced brain swelling or bleeding. *BMJ* n2975 (2021) doi:10.1136/bmj.n2975.

964. Thal, D. R., Ghebremedhin, E., Orantes, M. & Wiestler, O. D. Vascular pathology in Alzheimer disease: correlation of cerebral amyloid angiopathy and arteriosclerosis/lipohyalinosis with cognitive decline. *J Neuropathol Exp Neurol* **62**, 1287–1301 (2003).

965. Ng, T. K. S., Ho, C. S. H., Tam, W. W. S., Kua, E. H. & Ho, R. C.-M. Decreased Serum Brain-Derived Neurotrophic Factor (BDNF) Levels in Patients with Alzheimer's Disease (AD): A Systematic Review and Meta-Analysis. *Int J Mol Sci* **20**, 257 (2019).

966. Tian, Y. *et al.* Alcohol consumption and all-cause and cause-specific mortality among US adults: prospective cohort study. *BMC Med* **21**, 208 (2023).

967. Hendriks, S. *et al.* Risk Factors for Young-Onset Dementia in the UK Biobank. *JAMA Neurol* **81**, 134 (2024).

968. Ramos-Cejudo, J. *et al.* Traumatic Brain Injury and Alzheimer's Disease: The Cerebrovascular Link. *EBioMedicine* **28**, 21–30 (2018).

969. Oh, H. S.-H. *et al.* Organ aging signatures in the plasma proteome track health and disease. *Nature* **624**, 164–172 (2023).

970. Gao, L., Zhang, Y., Sterling, K. & Song, W. Brain-derived neurotrophic factor in Alzheimer's disease and its pharmaceutical potential. *Transl Neurodegener* **11**, 4 (2022).

971. Vijaya Kumar, D. K. *et al.* Amyloid-β Peptide Protects Against Microbial Infection In Mouse and Worm Models of Alzheimer's Disease. *Sci Transl Med* **8**, 340ra72 (2016).

972. Wozniak, M. A., Itzhaki, R. F., Shipley, S. J. & Dobson, C. B. Herpes simplex virus infection causes cellular beta-amyloid accumulation and secretase upregulation. *Neurosci Lett* **429**, 95–100 (2007).

973. Wozniak, M. A., Frost, A. L., Preston, C. M. & Itzhaki, R. F. Antivirals reduce the formation of key Alzheimer's disease molecules in cell cultures acutely infected with herpes simplex virus type 1. *PLoS One* **6**, e25152 (2011).

974. Steel, A. J. & Eslick, G. D. Herpes Viruses Increase the Risk of Alzheimer's Disease: A Meta-Analysis. *J Alzheimers Dis* **47**, 351–364 (2015).

975. Ou, Y.-N. *et al.* Associations of Infectious Agents with Alzheimer's Disease: A Systematic Review and Meta-Analysis. *J Alzheimers Dis* **75**, 299–309 (2020).

REFERENCES

976. Harris, K. *et al.* The Impact of Routine Vaccinations on Alzheimer's Disease Risk in Persons 65 Years and Older: A Claims-Based Cohort Study using Propensity Score Matching. *J Alzheimers Dis* **95**, 703–718 (2023).

977. Beydoun, M. A. *et al.* Clinical and Bacterial Markers of Periodontitis and Their Association with Incident All-Cause and Alzheimer's Disease Dementia in a Large National Survey. *J Alzheimers Dis* **75**, 157–172 (2020).

978. Konradt, C. *et al.* Endothelial cells are a replicative niche for entry of Toxoplasma gondii to the central nervous system. *Nat Microbiol* **1**, 16001 (2016).

979. Nayeri Chegeni, T. *et al.* Is Toxoplasma gondii a potential risk factor for Alzheimer's disease? A systematic review and meta-analysis. *Microb Pathog* **137**, 103751 (2019).

980. Itzhaki, R. F. Overwhelming Evidence for a Major Role for Herpes Simplex Virus Type 1 (HSV1) in Alzheimer's Disease (AD); Underwhelming Evidence against. *Vaccines (Basel)* **9**, 679 (2021).

981. Wang, C. & Holtzman, D. M. Bidirectional relationship between sleep and Alzheimer's disease: role of amyloid, tau, and other factors. *Neuropsychopharmacology* **45**, 104–120 (2020).

982. Ma, Y. *et al.* Association Between Sleep Duration and Cognitive Decline. *JAMA Netw Open* **3**, e2013573 (2020).

983. Panchin, Y. & Kovalzon, V. M. Total Wake: Natural, Pathological, and Experimental Limits to Sleep Reduction. *Front Neurosci* **15**, 643496 (2021).

984. Xie, L. *et al.* Sleep Drives Metabolite Clearance from the Adult Brain. *Science* **342**, 373–377 (2013).

985. Fultz, N. E. *et al.* Coupled electrophysiological, hemodynamic, and cerebrospinal fluid oscillations in human sleep. *Science* **366**, 628–631 (2019).

986. Lewis, L. D. The interconnected causes and consequences of sleep in the brain. *Science* **374**, 564–568 (2021).

987. Eide, P. K., Vinje, V., Pripp, A. H., Mardal, K.-A. & Ringstad, G. Sleep deprivation impairs molecular clearance from the human brain. *Brain* **144**, 863–874 (2021).

988. Nous, A., Engelborghs, S. & Smolders, I. Melatonin levels in the Alzheimer's disease continuum: a systematic review. *Alz Res Therapy* **13**, 52 (2021).

989. Roy, J. *et al.* Role of melatonin in Alzheimer's disease: From preclinical studies to novel melatonin-based therapies. *Frontiers in Neuroendocrinology* **65**, 100986 (2022).

REFERENCES

990. Spira, A. P., Chen-Edinboro, L. P., Wu, M. N. & Yaffe, K. Impact of Sleep on the Risk of Cognitive Decline and Dementia. *Curr Opin Psychiatry* **27**, 478–483 (2014).

991. Dimitrova-Paternoga, L. *et al.* Molecular basis of mRNA transport by a kinesin-1–atypical tropomyosin complex. *Genes Dev* **35**, 976–991 (2021).

992. Yildiz, A., Tomishige, M., Vale, R. D. & Selvin, P. R. Kinesin Walks Hand-Over-Hand. *Science* **303**, 676–678 (2004).

993. Carter, N. J. & Cross, R. A. Mechanics of the kinesin step. *Nature* **435**, 308–312 (2005).

994. Taniguchi, Y. & Yanagida, T. The forward and backward stepping processes of kinesin are gated by ATP binding. *Biophysics (Nagoya-shi)* **4**, 11–18 (2008).

995. Guillaud, L., El-Agamy, S. E., Otsuki, M. & Terenzio, M. Anterograde Axonal Transport in Neuronal Homeostasis and Disease. *Front. Mol. Neurosci.* **13**, (2020).

996. MacGibeny, M. A., Koyuncu, O. O., Wirblich, C., Schnell, M. J. & Enquist, L. W. Retrograde axonal transport of rabies virus is unaffected by interferon treatment but blocked by emetine locally in axons. *PLoS Pathog* **14**, e1007188 (2018).

997. Kadavath, H. *et al.* Tau stabilizes microtubules by binding at the interface between tubulin heterodimers. *Proc Natl Acad Sci U S A* **112**, 7501–7506 (2015).

998. Ikegami, S., Harada, A. & Hirokawa, N. Muscle weakness, hyperactivity, and impairment in fear conditioning in tau-deficient mice. *Neurosci Lett* **279**, 129–132 (2000).

999. Ke, Y. D. *et al.* Lessons from Tau-Deficient Mice. *Int J Alzheimers Dis* **2012**, 873270 (2012).

1000. Stamer, K., Vogel, R., Thies, E., Mandelkow, E. & Mandelkow, E.-M. Tau blocks traffic of organelles, neurofilaments, and APP vesicles in neurons and enhances oxidative stress. *J Cell Biol* **156**, 1051–1063 (2002).

1001. Noble, W., Hanger, D. P., Miller, C. C. J. & Lovestone, S. The Importance of Tau Phosphorylation for Neurodegenerative Diseases. *Front Neurol* **4**, 83 (2013).

1002. Guha, S., Johnson, G. V. & Nehrke, K. The crosstalk between pathological tau phosphorylation and mitochondrial dysfunction as a key to understanding and treating Alzheimer's disease. *Mol Neurobiol* **57**, 5103–5120 (2020).

1003. Abouna, G. M. Organ Shortage Crisis: Problems and Possible Solutions. *Transplantation Proceedings* **40**, 34–38 (2008).

REFERENCES

1004. Vanholder, R. *et al.* Organ donation and transplantation: a multi-stakeholder call to action. *Nat Rev Nephrol* **17**, 554–568 (2021).

1005. Nordham, K. D. & Ninokawa, S. The history of organ transplantation. *Proc (Bayl Univ Med Cent)* **35**, 124–128.

1006. Ayala García, M. A., González Yebra, B., López Flores, A. L. & Guaní Guerra, E. The Major Histocompatibility Complex in Transplantation. *J Transplant* **2012**, 842141 (2012).

1007. Sommer, S. The importance of immune gene variability (MHC) in evolutionary ecology and conservation. *Front Zool* **2**, 16 (2005).

1008. Yamazaki, K. & Beauchamp, G. K. Genetic basis for MHC-dependent mate choice. *Adv Genet* **59**, 129–145 (2007).

1009. Santos, P. S. C., Schinemann, J. A., Gabardo, J. & Bicalho, M. da G. New evidence that the MHC influences odor perception in humans: a study with 58 Southern Brazilian students. *Horm Behav* **47**, 384–388 (2005).

1010. Havlíček, J., Winternitz, J. & Roberts, S. C. Major histocompatibility complex-associated odour preferences and human mate choice: near and far horizons. *Philosophical Transactions of the Royal Society B: Biological Sciences* **375**, 20190260 (2020).

1011. Qiao, Z., Powell, J. & Evans, D. MHC-Dependent Mate Selection within 872 Spousal Pairs of European Ancestry from the Health and Retirement Study. *Genes* **9**, 53 (2018).

1012. Lu, Y., Zhou, Y., Ju, R. & Chen, J. Human-animal chimeras for autologous organ transplantation: technological advances and future perspectives. *Ann Transl Med* **7**, 576 (2019)

1013. Murphy, S. V. & Atala, A. 3D bioprinting of tissues and organs. *Nat Biotechnol* **32**, 773–785 (2014).

1014. Agarwal, S. *et al.* Current Developments in 3D Bioprinting for Tissue and Organ Regeneration–A Review. *Front. Mech. Eng.* **6**, 589171 (2020).

1015. Isakov, N. Histocompatibility and Reproduction: Lessons from the Anglerfish. *Life (Basel)* **12**, 113 (2022).

1016. Lipsker, D., Flory, E., Wiesel, M.-L., Hanau, D. & De La Salle, H. Between Light and Dark, the Chimera Comes Out. *Arch Dermatol* **144**, (2008).

1017. Wolinsky, H. A mythical beast. Increased attention highlights the hidden wonders of chimeras. *EMBO Rep* **8**, 212–214 (2007).

1018. Chen, J., Lansford, R., Stewart, V., Young, F. & Alt, F. W. RAG-2-deficient blastocyst complementation: an assay of gene function in lymphocyte development. *Proc Natl Acad Sci U S A* **90**, 4528–4532 (1993).

1019. Usui, J. *et al.* Generation of kidney from pluripotent stem cells via blastocyst complementation. *Am J Pathol* **180**, 2417–2426 (2012).

1020. Mori, M. *et al.* Generation of functional lungs via conditional blastocyst complementation using pluripotent stem cells. *Nat Med* **25**, 1691–1698 (2019).

1021. Kobayashi, T. *et al.* Generation of rat pancreas in mouse by interspecific blastocyst injection of pluripotent stem cells. *Cell* **142**, 787–799 (2010).

1022. Manji, R. A., Menkis, A. H., Ekser, B. & Cooper, D. K. C. Porcine bioprosthetic heart valves: The next generation. *Am Heart J* **164**, 177–185 (2012).

1023. Kostyunin, A. E. *et al.* Degeneration of Bioprosthetic Heart Valves: Update 2020. *JAHA* **9**, e018506 (2020).

1024. Wu, J. *et al.* Interspecies Chimerism with Mammalian Pluripotent Stem Cells. *Cell* **168**, 473-486.e15 (2017).

1025. Das, S. *et al.* Generation of human endothelium in pig embryos deficient in ETV2. *Nat Biotechnol* **38**, 297–302 (2020).

1026. Zheng, C. *et al.* Cell competition constitutes a barrier for interspecies chimerism. *Nature* **592**, 272–276 (2021).

1027. Maeng, G. *et al.* Humanized skeletal muscle in MYF5/MYOD/MYF6-null pig embryos. *Nat Biomed Eng* **5**, 805–814 (2021).

1028. Nelson, E. D. *et al.* Limited Expansion of Human Hepatocytes in FAH/ *RAG2* -Deficient Swine. *Tissue Engineering Part A* **28**, 150–160 (2022).

1029. Griffith, B. P. *et al.* Genetically Modified Porcine-to-Human Cardiac Xenotransplantation. *N Engl J Med* **387**, 35–44 (2022).

1030. Wang, W., He, W., Ruan, Y. & Geng, Q. First pig-to-human heart transplantation. *Innovation (Camb)* **3**, 100223 (2022).

1031. Reardon, S. First pig-to-human heart transplant: what can scientists learn? *Nature* **601**, 305–306 (2022).

1032. Wolf, E., Reichart, B., Moretti, A. & Laugwitz, K.-L. Designer pigs for xenogeneic heart transplantation and beyond. *Dis Model Mech* **16**, dmm050177 (2023).

1033. Sykes, M. & Sachs, D. H. Transplanting organs from pigs to humans. *Sci Immunol* **4**, eaau6298 (2019).

1034. Moazami, N. *et al.* Pig-to-human heart xenotransplantation in two recently deceased human recipients. *Nat Med* **29**, 1989–1997 (2023).

REFERENCES

1035. Montgomery, R. A. *et al.* Results of Two Cases of Pig-to-Human Kidney Xenotransplantation. *N Engl J Med* **386**, 1889–1898 (2022).

1036. Mohiuddin, M. M. *et al.* Chimeric 2C10R4 anti-CD40 antibody therapy is critical for long-term survival of GTKO.hCD46.hTBM pig-to-primate cardiac xenograft. *Nat Commun* **7**, 11138 (2016).

1037. Yamamoto, T. *et al.* Life-supporting Kidney Xenotransplantation From Genetically Engineered Pigs in Baboons: A Comparison of Two Immunosuppressive Regimens. *Transplantation* **103**, 2090 (2019).

1038. Adams, A. B. *et al.* Xenoantigen Deletion and Chemical Immunosuppression Can Prolong Renal Xenograft Survival. *Annals of Surgery* **268**, 564–573 (2018).

1039. Denner, J. Why was PERV not transmitted during preclinical and clinical xenotransplantation trials and after inoculation of animals? *Retrovirology* **15**, 28 (2018).

1040. Niu, D. *et al.* Inactivation of porcine endogenous retrovirus in pigs using CRISPR-Cas9. *Science* **357**, 1303–1307 (2017).

1041. Marshall, J., Molloy, R., Moss, G. W., Howe, J. R. & Hughes, T. E. The jellyfish green fluorescent protein: a new tool for studying ion channel expression and function. *Neuron* **14**, 211–215 (1995).

1042. Lundstrom, K. Viral Vectors in Gene Therapy. *Diseases* **6**, 42 (2018).

1043. Logunov, D. Y. *et al.* Safety and efficacy of an rAd26 and rAd5 vector-based heterologous prime-boost COVID-19 vaccine: an interim analysis of a randomised controlled phase 3 trial in Russia. *Lancet* **397**, 671–681 (2021).

1044. Arroyo-Olarte, R. D., Bravo Rodríguez, R. & Morales-Ríos, E. Genome Editing in Bacteria: CRISPR-Cas and Beyond. *Microorganisms* **9**, 844 (2021).

1045. Karginov, F. V. & Hannon, G. J. The CRISPR system: small RNA-guided defense in bacteria and archaea. *Mol Cell* **37**, 7 (2010).

1046. Xiao, Q., Guo, D. & Chen, S. Application of CRISPR/Cas9-Based Gene Editing in HIV-1/AIDS Therapy. *Front Cell Infect Microbiol* **9**, 69 (2019).

1047. Cao, Y., Zhou, H., Zhou, X. & Li, F. Control of Plant Viruses by CRISPR/Cas System-Mediated Adaptive Immunity. *Front. Microbiol.* **11**, (2020).

1048. Cyranoski, D. What CRISPR-baby prison sentences mean for research. *Nature* **577**, 154–155 (2020).

1049. Zhang, X.-H., Tee, L. Y., Wang, X.-G., Huang, Q.-S. & Yang, S.-H. Off-target Effects in CRISPR/Cas9-mediated Genome Engineering. *Mol Ther Nucleic Acids* **4**, e264 (2015).

1050. Jamal, M. *et al.* Improving CRISPR-Cas9 On-Target Specificity. *Curr Issues Mol Biol* **26**, 65–80 (2018).

1051. Slaymaker, I. M. *et al.* Rationally engineered Cas9 nucleases with improved specificity. *Science* **351**, 84–88 (2016).

1052. Blond, J. L. *et al.* An envelope glycoprotein of the human endogenous retrovirus HERV-W is expressed in the human placenta and fuses cells expressing the type D mammalian retrovirus receptor. *J Virol* **74**, 3321–3329 (2000).

1053. Pastuzyn, E. D. *et al.* The Neuronal Gene Arc Encodes a Repurposed Retrotransposon Gag Protein that Mediates Intercellular RNA Transfer. *Cell* **172**, 275-288.e18 (2018).

1054. Plath, N. *et al.* Arc/Arg3.1 is essential for the consolidation of synaptic plasticity and memories. *Neuron* **52**, 437–444 (2006).

1055. Korb, E. & Finkbeiner, S. Arc in synaptic plasticity: from gene to behavior. *Trends Neurosci* **34**, 591–598 (2011).

1056. Li, M. *et al.* Downregulation of Human Endogenous Retrovirus Type K (HERV-K) Viral *env* RNA in Pancreatic Cancer Cells Decreases Cell Proliferation and Tumor Growth. *Clinical Cancer Research* **23**, 5892–5911 (2017).

1057. Ko, E.-J. *et al.* Human Endogenous Retrovirus (HERV)-K env Gene Knockout Affects Tumorigenic Characteristics of nupr1 Gene in DLD-1 Colorectal Cancer Cells. *Int J Mol Sci* **22**, 3941 (2021).

1058. Grow, E. J. *et al.* Intrinsic retroviral reactivation in human preimplantation embryos and pluripotent cells. *Nature* **522**, 221–225 (2015).

1059. Lancaster, M. A. & Knoblich, J. A. Organogenesis in a dish: modeling development and disease using organoid technologies. *Science* **345**, 1247125 (2014).

1060. Eiraku, M. *et al.* Self-Organized Formation of Polarized Cortical Tissues from ESCs and Its Active Manipulation by Extrinsic Signals. *Cell Stem Cell* **3**, 519–532 (2008).

1061. Rungarunlert, S., Techakumphu, M., Pirity, M. K. & Dinnyes, A. Embryoid body formation from embryonic and induced pluripotent stem cells: Benefits of bioreactors. *World J Stem Cells* **1**, 11–21 (2009).

REFERENCES

1062. Hughes, C. S., Postovit, L. M. & Lajoie, G. A. Matrigel: A complex protein mixture required for optimal growth of cell culture. *Proteomics* **10**, 1886–1890 (2010).

1063. Adibi, J. J., Marques, E. T. A., Cartus, A. & Beigi, R. H. Teratogenic effects of the Zika virus and the role of the placenta. *Lancet* **387**, 1587–1590 (2016).

1064. Garcez, P. P. *et al.* Zika virus impairs growth in human neurospheres and brain organoids. *Science* **352**, 816–818 (2016).

1065. Dang, J. *et al.* Zika Virus Depletes Neural Progenitors in Human Cerebral Organoids through Activation of the Innate Immune Receptor TLR3. *Cell Stem Cell* **19**, 258–265 (2016).

1066. Zhou, J. *et al.* Differentiated human airway organoids to assess infectivity of emerging influenza virus. *Proceedings of the National Academy of Sciences* **115**, 6822–6827 (2018).

1067. Heo, I. *et al.* Modelling Cryptosporidium infection in human small intestinal and lung organoids. *Nat Microbiol* **3**, 814–823 (2018).

1068. Driehuis, E. & Clevers, H. CRISPR/Cas 9 genome editing and its applications in organoids. *American Journal of Physiology-Gastrointestinal and Liver Physiology* **312**, G257–G265 (2017).

1069. Berkers, G. *et al.* Rectal Organoids Enable Personalized Treatment of Cystic Fibrosis. *Cell Rep* **26**, 1701-1708.e3 (2019).

1070. Vlachogiannis, G. *et al.* Patient-derived organoids model treatment response of metastatic gastrointestinal cancers. *Science* **359**, 920–926 (2018).

1071. Sidhaye, J. & Knoblich, J. A. Brain organoids: an ensemble of bioassays to investigate human neurodevelopment and disease. *Cell Death Differ* **28**, 52–67 (2021).

1072. Miller, A. J. *et al.* Generation of lung organoids from human pluripotent stem cells in vitro. *Nat Protoc* **14**, 518–540 (2019).

1073. Prior, N., Inacio, P. & Huch, M. Liver organoids: from basic research to therapeutic applications. *Gut* **68**, 2228–2237 (2019).

1074. Takasato, M. & Little, M. H. Making a Kidney Organoid Using the Directed Differentiation of Human Pluripotent Stem Cells. *Methods Mol Biol* **1597**, 195–206 (2017).

1075. Balak, J. R. A., Juksar, J., Carlotti, F., Lo Nigro, A. & de Koning, E. J. P. Organoids from the Human Fetal and Adult Pancreas. *Curr Diab Rep* **19**, 160 (2019).

1076. Date, S. & Sato, T. Mini-gut organoids: reconstitution of the stem cell niche. *Annu Rev Cell Dev Biol* **31**, 269–289 (2015).

REFERENCES

1077. Vasyutin, I., Zerihun, L., Ivan, C. & Atala, A. Bladder Organoids and Spheroids: Potential Tools for Normal and Diseased Tissue Modelling. *Anticancer Res* **39**, 1105–1118 (2019).

1078. Kim, J., Koo, B.-K. & Knoblich, J. A. Human organoids: model systems for human biology and medicine. *Nat Rev Mol Cell Biol* **21**, 571–584 (2020).

1079. Wimmer, R. A. *et al.* Human blood vessel organoids as a model of diabetic vasculopathy. *Nature* **565**, 505–510 (2019).

1080. Watanabe, S. *et al.* Transplantation of intestinal organoids into a mouse model of colitis. *Nat Protoc* **17**, 649–671 (2022).

1081. Lin, B. *et al.* Retina Organoid Transplants Develop Photoreceptors and Improve Visual Function in RCS Rats With RPE Dysfunction. *Invest. Ophthalmol. Vis. Sci.* **61**, 34 (2020).

1082. Mandai, M. *et al.* Autologous Induced Stem-Cell–Derived Retinal Cells for Macular Degeneration. *N Engl J Med* **376**, 1038–1046 (2017).

1083. Nam, S. A. *et al.* Graft immaturity and safety concerns in transplanted human kidney organoids. *Exp Mol Med* **51**, 1–13 (2019).

1084. Sağraç, D. *et al.* Organoids in Tissue Transplantation. *Adv Exp Med Biol* **1347**, 45–64 (2021).

1085. Atala, A., Bauer, S. B., Soker, S., Yoo, J. J. & Retik, A. B. Tissue-engineered autologous bladders for patients needing cystoplasty. *Lancet* **367**, 1241–1246 (2006).

1086. Di Nicola, V. Omentum a powerful biological source in regenerative surgery. *Regenerative Therapy* **11**, 182–191 (2019).

1087. Irwin, R. M. *et al.* The clot thickens: Autologous and allogeneic fibrin sealants are mechanically equivalent in an ex vivo model of cartilage repair. *PLoS One* **14**, e0224756 (2019).

1088. Adamowicz, J., Pokrywczynska, M., Van Breda, S. V., Kloskowski, T. & Drewa, T. Concise Review: Tissue Engineering of Urinary Bladder; We Still Have a Long Way to Go? *Stem Cells Translational Medicine* **6**, 2033–2043 (2017).

1089. Gomez-Amaya, S. M. *et al.* Neural reconstruction methods of restoring bladder function. *Nat Rev Urol* **12**, 100–118 (2015).

1090. Das, S. *et al.* Innervation: the missing link for biofabricated tissues and organs. *npj Regen Med* **5**, 11 (2020).

1091. A tissue-engineered artificial human thymus from human iPSCs to study T cell immunity. *Nat Methods* **19**, 1191–1192 (2022).

1092. Bredenkamp, N. *et al.* An organized and functional thymus generated from FOXN1-reprogrammed fibroblasts. *Nat Cell Biol* **16**, 902–908 (2014).

REFERENCES

1093. Noor, N. *et al.* 3D Printing of Personalized Thick and Perfusable Cardiac Patches and Hearts. *Advanced Science* **6**, 1900344 (2019).

1094. Kang, H.-W. *et al.* A 3D bioprinting system to produce human-scale tissue constructs with structural integrity. *Nat Biotechnol* **34**, 312–319 (2016).

1095. Matai, I., Kaur, G., Seyedsalehi, A., McClinton, A. & Laurencin, C. T. Progress in 3D bioprinting technology for tissue/organ regenerative engineering. *Biomaterials* **226**, 119536 (2020).

1096. Barczyński, M., Gołkowski, F. & Nawrot, I. Parathyroid transplantation in thyroid surgery. *Gland Surg* **6**, 530–536 (2017).

1097. Lo, C. Y. Parathyroid Autotransplantation During Thyroidectomy: Documentation of Graft Function. *Arch Surg* **136**, 1381 (2001).

1098. Faleo, G. *et al.* Co-Transplant of Parathyroid Gland and Stem Cell-Derived Insulin-Producing Cells Enhances Graft Survival through Release of Pro-Angiogenic and Pro-Survival Factors. *Transplantation* **102**, S350 (2018).

1099. UCSF. UCSF Diabetes Trial: Pancreatic Islets and Parathyroid Gland Co-transplantation for Treatment of Type 1 Diabetes. https://clinicaltrials.ucsf.edu/trial/NCT03977662 (2019).

1100. Kano, M. *et al.* Functional calcium-responsive parathyroid glands generated using single-step blastocyst complementation. *Proc Natl Acad Sci U S A* **120**, e2216564120 (2023).

1101. Tashman, J. W., Shiwarski, D. J. & Feinberg, A. W. Development of a high-performance open-source 3D bioprinter. *Sci Rep* **12**, 22652 (2022).

1102. Bartke, A. *et al.* Extending the lifespan of long-lived mice. *Nature* **414**, 412–412 (2001).

1103. Mattison, J. A. *et al.* Studies of aging in ames dwarf mice: Effects of caloric restriction. *J Am Aging Assoc* **23**, 9–16 (2000).

1104. Sun, L. Y. *et al.* Growth hormone-releasing hormone disruption extends lifespan and regulates response to caloric restriction in mice. *eLife* **2**, e01098 (2013).

1105. Bonkowski, M. S., Rocha, J. S., Masternak, M. M., Al Regaiey, K. A. & Bartke, A. Targeted disruption of growth hormone receptor interferes with the beneficial actions of calorie restriction. *Proc. Natl. Acad. Sci. U.S.A.* **103**, 7901–7905 (2006).

1106. Rybina, O. Y., Symonenko, A. V. & Pasyukova, E. G. Compound combinations targeting longevity: Challenges and perspectives. *Ageing Res Rev* **85**, 101851 (2023).

1107. Bunu, G. *et al.* SynergyAge, a curated database for synergistic and antagonistic interactions of longevity-associated genes. *Sci Data* **7**, 366 (2020).

1108. Chen, D. *et al.* Germline signaling mediates the synergistically prolonged longevity produced by double mutations in daf-2 and rsks-1 in C. elegans. *Cell Rep* **5**, 1600–1610 (2013).

1109. Hansen, M. *et al.* Lifespan extension by conditions that inhibit translation in Caenorhabditis elegans. *Aging Cell* **6**, 95–110 (2007).

1110. Houthoofd, K., Braeckman, B. P., Johnson, T. E. & Vanfleteren, J. R. Life extension via dietary restriction is independent of the Ins/IGF-1 signalling pathway in Caenorhabditis elegans. *Exp Gerontol* **38**, 947–954 (2003).

1111. Castillo-Quan, J. I. *et al.* A triple drug combination targeting components of the nutrient-sensing network maximizes longevity. *Proceedings of the National Academy of Sciences* **116**, 20817–20819 (2019).

1112. Shaposhnikov, M. V. *et al.* Molecular mechanisms of exceptional lifespan increase of Drosophila melanogaster with different genotypes after combinations of pro-longevity interventions. *Commun Biol* **5**, 566 (2022).

1113. Kaur, P. *et al.* Combining stem cell rejuvenation and senescence targeting to synergistically extend lifespan. *Aging (Albany NY)* **14**, 8270–8291 (2022).

1114. Kimsey, I. J. *et al.* Dynamic basis for dG•dT misincorporation via tautomerization and ionization. *Nature* **554**, 195–201 (2018).

1115. Chen, L. *et al.* WRN, the protein deficient in Werner syndrome, plays a critical structural role in optimizing DNA repair. *Aging Cell* **2**, 191–199 (2003).

1116. Cagan, A. *et al.* Somatic mutation rates scale with lifespan across mammals. *Nature* **604**, 517–524 (2022).

1117. Belshaw, R. *et al.* Long-term reinfection of the human genome by endogenous retroviruses. *Proc Natl Acad Sci U S A* **101**, 4894–4899 (2004).

1118. Gorbunova, V. *et al.* The role of retrotransposable elements in ageing and age-associated diseases. *Nature* **596**, 43–53 (2021).

1119. Della Valle, F. *et al.* *LINE-1* RNA causes heterochromatin erosion and is a target for amelioration of senescent phenotypes in progeroid syndromes. *Sci. Transl. Med.* **14**, eabl6057 (2022).

1120. Simon, M. *et al.* LINE1 Derepression in Aged Wild-Type and SIRT6-Deficient Mice Drives Inflammation. *Cell Metabolism* **29**, 871-885.e5 (2019).

REFERENCES

1121. Chatterjee, N. & Walker, G. C. Mechanisms of DNA damage, repair and mutagenesis. *Environ Mol Mutagen* **58**, 235–263 (2017).

1122. Caratero, A., Courtade, M., Bonnet, L., Planel, H. & Caratero, C. Effect of a continuous gamma irradiation at a very low dose on the life span of mice. *Gerontology* **44**, 272–276 (1998).

1123. Sponsler, R. & Cameron, J. R. Nuclear shipyard worker study (1980–1988): a large cohort exposed to low-dose-rate gamma radiation. *International Journal of Low Radiation* **1**, 463–478 (2005).

1124. Lacoste-Collin, L. *et al.* Effect of Continuous Irradiation with a Very Low Dose of Gamma Rays on Life Span and the Immune System in SJL Mice Prone to B-Cell Lymphoma. *Radiation Research* **168**, 725–732 (2007).

1125. Tanaka, S. *et al.* No Lengthening of Life Span in Mice Continuously Exposed to Gamma Rays at Very Low Dose Rates. *Radiation Research* **160**, 376–379 (2003).

1126. Dalke, C. *et al.* Lifetime study in mice after acute low-dose ionizing radiation: a multifactorial study with special focus on cataract risk. *Radiat Environ Biophys* **57**, 99–113 (2018).

1127. Kristiani, L., Kim, M. & Kim, Y. Role of the Nuclear Lamina in Age-Associated Nuclear Reorganization and Inflammation. *Cells* **9**, 718 (2020).

1128. Hackett, J. A., Feldser, D. M. & Greider, C. W. Telomere Dysfunction Increases Mutation Rate and Genomic Instability. *Cell* **106**, 275–286 (2001).

1129. Froidure, A. *et al.* Short telomeres increase the risk of severe COVID-19. *Aging (Albany NY)* **12**, 19911 (2020).

1130. M'kacher, R. *et al.* Telomere aberrations, including telomere loss, doublets, and extreme shortening, are increased in patients with infertility. *Fertility and Sterility* **115**, 164–173 (2021).

1131. Tian, X. *et al.* SIRT6 Is Responsible for More Efficient DNA Double-Strand Break Repair in Long-Lived Species. *Cell* **177**, 622-638. e22 (2019).

1132. Klein, M. A. & Denu, J. M. Biological and catalytic functions of sirtuin 6 as targets for small-molecule modulators. *Journal of Biological Chemistry* **295**, 11021–11041 (2020).

1133. Meter, M. V. *et al.* SIRT6 represses LINE1 retrotransposons by ribosylating KAP1 but this repression fails with stress and age. *Nature communications* **5**, 5011 (2014).

1134. Mostoslavsky, R. *et al.* Genomic instability and aging-like phenotype in the absence of mammalian SIRT6. *Cell* **124**, 315–329 (2006).

REFERENCES

1135. Kanfi, Y. *et al.* The sirtuin SIRT6 regulates lifespan in male mice. *Nature* **483**, 218–221 (2012).

1136. Zhang, Y. *et al.* SIRT6, a novel direct transcriptional target of FoxO3a, mediates colon cancer therapy. *Theranostics* **9**, 2380 (2019).

1137. Chen, J. *et al.* SIRT6 enhances telomerase activity to protect against DNA damage and senescence in hypertrophic ligamentum flavum cells from lumbar spinal stenosis patients. *Aging (Albany NY)* **13**, 6025–6040 (2021).

1138. Zhang, Z. *et al.* Nicotine induces senescence in spermatogonia stem cells by disrupting homeostasis between circadian oscillation and rhythmic mitochondrial dynamics via the SIRT6/Bmal1 pathway. *Life Sciences* **352**, 122860 (2024).

1139. Kulaksiz, D. *et al.* Sperm concentration and semen volume increase after smoking cessation in infertile men. *Int J Impot Res* **34**, 614–619 (2022).

1140. Simon, M. *et al.* A rare human centenarian variant of SIRT6 enhances genome stability and interaction with Lamin A. *The EMBO Journal* **41**, e110393 (2022).

1141. Calderwood, S. K., Murshid, A. & Prince, T. The Shock of Aging: Molecular Chaperones and the Heat Shock Response in Longevity and Aging — A Mini-Review. *Gerontology* **55**, 550–558 (2009).

1142. Hipp, M. S., Kasturi, P. & Hartl, F. U. The proteostasis network and its decline in ageing. *Nat Rev Mol Cell Biol* **20**, 421–435 (2019).

1143. Waudby, C. A., Dobson, C. M. & Christodoulou, J. Nature and Regulation of Protein Folding on the Ribosome. *Trends in Biochemical Sciences* **44**, 914–926 (2019).

1144. Santoro, M. G. Heat shock factors and the control of the stress response. *Biochemical Pharmacology* **59**, 55–63 (2000).

1145. Fernández-Fernández, M. R., Gragera, M., Ochoa-Ibarrola, L., Quintana-Gallardo, L. & Valpuesta, J. M. Hsp70 — a master regulator in protein degradation. *FEBS Letters* **591**, 2648–2660 (2017).

1146. Hsu, A.-L., Murphy, C. T. & Kenyon, C. Regulation of Aging and Age-Related Disease by DAF-16 and Heat-Shock Factor. *Science* **300**, 1142–1145 (2003).

1147. Steele, A. D. *et al.* Heat shock factor 1 regulates lifespan as distinct from disease onset in prion disease. *Proc. Natl. Acad. Sci. U.S.A.* **105**, 13626–13631 (2008).

REFERENCES

1148. Servello, F. A. & Apfeld, J. The heat shock transcription factor HSF-1 protects *Caenorhabditis elegans* from peroxide stress. *Translational Medicine of Aging* **4**, 88–92 (2020).

1149. Grossi, V. *et al.* The longevity SNP rs2802292 uncovered: HSF1 activates stress-dependent expression of FOXO3 through an intronic enhancer. *Nucleic Acids Research* **46**, 5587 (2018).

1150. Baird, N. A. *et al.* HSF-1-mediated cytoskeletal integrity determines thermotolerance and life span. *Science* **346**, 360–363 (2014).

1151. Trivedi, R. & Jurivich, D. A. A molecular perspective on age-dependent changes to the heat shock axis. *Experimental Gerontology* **137**, 110969 (2020).

1152. Pignatti, C., D'Adamo, S., Stefanelli, C., Flamigni, F. & Cetrullo, S. Nutrients and Pathways that Regulate Health Span and Life Span. *Geriatrics* **5**, 95 (2020).

1153. Burkewitz, K., Zhang, Y. & Mair, W. B. AMPK at the Nexus of Energetics and Aging. *Cell metabolism* **20**, 10 (2014).

1154. Hajer, G. R., Van Haeften, T. W. & Visseren, F. L. J. Adipose tissue dysfunction in obesity, diabetes, and vascular diseases. *European Heart Journal* **29**, 2959–2971 (2008).

1155. Ali, M. & Bracko, O. VEGF Paradoxically Reduces Cerebral Blood Flow in Alzheimer's Disease Mice. *Neuroscience Insights* **17**, 26331055221109254 (2022).

1156. Fazeli, P. K., Lee, H. & Steinhauser, M. L. Aging Is a Powerful Risk Factor for Type 2 Diabetes Mellitus Independent of Body Mass Index. *Gerontology* **66**, 209–210 (2020).

1157. Junnila, R. K., List, E. O., Berryman, D. E., Murrey, J. W. & Kopchick, J. J. The GH/IGF-1 axis in ageing and longevity. *Nature reviews. Endocrinology* **9**, 366 (2013).

1158. Amorim, J. A. *et al.* Mitochondrial and metabolic dysfunction in ageing and age-related diseases. *Nat Rev Endocrinol* **18**, 243–258 (2022).

1159. Borsche, M., Pereira, S. L., Klein, C. & Grünewald, A. Mitochondria and Parkinson's Disease: Clinical, Molecular, and Translational Aspects. *J Parkinsons Dis* **11**, 45–60 (2021).

1160. Hwang, A. B., Jeong, D.-E. & Lee, S.-J. Mitochondria and Organismal Longevity. *Current Genomics* **13**, 519 (2012).

1161. Schriner, S. E. *et al.* Extension of murine life span by overexpression of catalase targeted to mitochondria. *Science* **308**, 1909–1911 (2005).

1162. Pérez-Estrada, J. R. *et al.* Reduced lifespan of mice lacking catalase correlates with altered lipid metabolism without oxidative damage or premature aging. *Free Radic Biol Med* **135**, 102–115 (2019).

1163. Lapointe, J. & Hekimi, S. Early Mitochondrial Dysfunction in Long-lived Mclk1+/- Mice *. *Journal of Biological Chemistry* **283**, 26217–26227 (2008).

1164. Lionaki, E. *et al.* Mitochondrial protein import determines lifespan through metabolic reprogramming and de novo serine biosynthesis. *Nat Commun* **13**, 651 (2022).

1165. Liu, X. *et al.* Evolutionary conservation of the clk-1-dependent mechanism of longevity: loss of mclk1 increases cellular fitness and lifespan in mice. *Genes & Development* **19**, 2424 (2005).

1166. Heidler, T., Hartwig, K., Daniel, H. & Wenzel, U. Caenorhabditis elegans lifespan extension caused by treatment with an orally active ROS-generator is dependent on DAF-16 and SIR-2.1. *Biogerontology* **11**, 183–195 (2010).

1167. Raamsdonk, J. M. V. & Hekimi, S. Superoxide dismutase is dispensable for normal animal lifespan. *Proceedings of the National Academy of Sciences of the United States of America* **109**, 5785 (2012).

1168. Raamsdonk, J. M. V. & Hekimi, S. Deletion of the Mitochondrial Superoxide Dismutase sod-2 Extends Lifespan in Caenorhabditis elegans. *PLoS Genetics* **5**, e1000361 (2009).

1169. Demyanenko, I. A. *et al.* Mitochondria-Targeted Antioxidant SkQ1 Improves Dermal Wound Healing in Genetically Diabetic Mice. *Oxidative Medicine and Cellular Longevity* **2017**, 6408278 (2017).

1170. Skulachev, V. P. *et al.* Mitochondrion-targeted antioxidant SkQ1 prevents rapid animal death caused by highly diverse shocks. *Scientific Reports* **13**, 4326 (2023).

1171. Ryu, D. *et al.* Urolithin A induces mitophagy and prolongs lifespan in C. elegans and increases muscle function in rodents. *Nat Med* **22**, 879–888 (2016).

1172. Geng, L. *et al.* Low-dose quercetin positively regulates mouse healthspan. *Protein Cell* **10**, 770–775 (2019).

1173. Y, Z. *et al.* The Achilles' heel of senescent cells: from transcriptome to senolytic drugs. *Aging cell* **14**, (2015).

1174. Clevers, H. & Watt, F. M. Defining Adult Stem Cells by Function, not by Phenotype. *Annu. Rev. Biochem.* **87**, 1015–1027 (2018).

1175. Dos Santos, G. A., Magdaleno, G. D. V. & de Magalhães, J. P. Evidence of a pan-tissue decline in stemness during human aging. *Aging (Albany NY)* **16**, 5796–5810 (2024).

REFERENCES

1176. Kollman, C. *et al.* The effect of donor characteristics on survival after unrelated donor transplantation for hematologic malignancy. *Blood* **127**, 260 (2015).

1177. Bonnet, D. Biology of human bone marrow stem cells. *Clin Exp Med* **3**, 140–149 (2003).

1178. Gao, Q. *et al.* Bone Marrow Mesenchymal Stromal Cells: Identification, Classification, and Differentiation. *Front. Cell Dev. Biol.* **9**, (2022).

1179. Søraas, A. *et al.* Epigenetic age is a cell-intrinsic property in transplanted human hematopoietic cells. *Aging Cell* **18**, e12897 (2019).

1180. Kovina, M. V. *et al.* Extension of Maximal Lifespan and High Bone Marrow Chimerism After Nonmyeloablative Syngeneic Transplantation of Bone Marrow From Young to Old Mice. *Front. Genet.* **10**, (2019).

1181. Guderyon, M. J. *et al.* Mobilization-based transplantation of young-donor hematopoietic stem cells extends lifespan in mice. *Aging Cell* **19**, e13110 (2020).

1182. Rudnitsky, E. *et al.* Stem cell-derived extracellular vesicles as senotherapeutics. *Ageing Res Rev* **99**, 102391 (2024).

1183. Puzianowska-Kuźnicka, M. *et al.* Interleukin-6 and C-reactive protein, successful aging, and mortality: the PolSenior study. *Immun Ageing* **13**, 21 (2016).

1184. Oh, H., Lewis, D. A. & Sibille, E. The Role of BDNF in Age-Dependent Changes of Excitatory and Inhibitory Synaptic Markers in the Human Prefrontal Cortex. *Neuropsychopharmacol* **41**, 3080–3091 (2016).

1185. Grunewald, M. *et al.* Counteracting age-related VEGF signaling insufficiency promotes healthy aging and extends life span. *Science* **373**, eabc8479 (2021).

1186. Harris, R., Miners, J. S., Allen, S. & Love, S. VEGFR1 and VEGFR2 in Alzheimer's Disease. *J Alzheimers Dis* **61**, 741–752 (2018).

1187. Ray, S. K. & Manz, S. N. BRAIN HEALTH ASSESSMENT IN MACULAR DEGENERATION PATIENTS UNDERGOING INTRAVITREAL ANTI-VASCULAR ENDOTHELIAL GROWTH FACTOR INJECTIONS (THE BHAM STUDY): An Interim Analysis. *Retina* **41**, 1748–1753 (2021).

1188. Carmeliet, P. VEGF as a key mediator of angiogenesis in cancer. *Oncology* **69 Suppl 3**, 4–10 (2005).

1189. Duda, D. G., Batchelor, T. T., Willett, C. G. & Jain, R. K. VEGF-targeted cancer therapy strategies: current progress, hurdles and future prospects. *Trends in molecular medicine* **13**, 223 (2007).

1190. Haibe, Y. *et al.* Resistance Mechanisms to Anti-angiogenic Therapies in Cancer. *Front. Oncol.* **10**, (2020).

1191. López-Otín, C., Blasco, M. A., Partridge, L., Serrano, M. & Kroemer, G. Hallmarks of aging: An expanding universe. *Cell* **186**, 243–278 (2023).

1192. Franceschi, C. *et al.* Inflamm-aging. An evolutionary perspective on immunosenescence. *Ann N Y Acad Sci* **908**, 244–254 (2000).

1193. Hirata, T. *et al.* Associations of cardiovascular biomarkers and plasma albumin with exceptional survival to the highest ages. *Nat Commun* **11**, 3820 (2020).

1194. Fish-Trotter, H. *et al.* Inflammation and Circulating Natriuretic Peptide Levels. *Circ: Heart Failure* **13**, e006570 (2020).

1195. Muslimovic, A., Tulumovic, D., Hasanspahic, S., Hamzic-Mehmedbasic, A. & Temimovi, R. Serum Cystatin C — Marker of Inflammation and Cardiovascular Morbidity in Chronic Kidney Disease Stages 1-4. *Materia Socio-Medica* **27**, 75 (2015).

1196. Alfaddagh, A. *et al.* Inflammation and cardiovascular disease: From mechanisms to therapeutics. *American Journal of Preventive Cardiology* **4**, 100130 (2020).

1197. Liu, T., Zhang, L., Joo, D. & Sun, S.-C. NF-κB signaling in inflammation. *Sig Transduct Target Ther* **2**, 17023 (2017).

1198. Zhang, G. *et al.* Hypothalamic programming of systemic ageing involving IKK-β, NF-κB and GnRH. *Nature* **497**, 211–216 (2013).

1199. Widjaja, A. A. *et al.* IL11 Stimulates IL33 Expression and Proinflammatory Fibroblast Activation across Tissues. *International Journal of Molecular Sciences* **23**, 8900 (2022).

1200. Cook, S. A. Understanding interleukin 11 as a disease gene and therapeutic target. *Biochemical Journal* **480**, 1987–2008 (2023).

1201. Venter, J. C. *et al.* Environmental genome shotgun sequencing of the Sargasso Sea. *Science* **304**, 66–74 (2004).

1202. Quince, C., Walker, A. W., Simpson, J. T., Loman, N. J. & Segata, N. Shotgun metagenomics, from sampling to analysis. *Nat Biotechnol* **35**, 833–844 (2017).

1203. Hibbing, M. E., Fuqua, C., Parsek, M. R. & Peterson, S. B. Bacterial competition: surviving and thriving in the microbial jungle. *Nat Rev Microbiol* **8**, 15–25 (2010).

1204. Yang, J. *et al.* Species-Level Analysis of Human Gut Microbiota With Metataxonomics. *Front. Microbiol.* **11**, (2020).

1205. Wilmanski, T. *et al.* Gut microbiome pattern reflects healthy aging and predicts survival in humans. *Nature metabolism* **3**, 274 (2021).

1206. Cronin, P., Joyce, S. A., O'Toole, P. W. & O'Connor, E. M. Dietary Fibre Modulates the Gut Microbiota. *Nutrients* **13**, 1655 (2021).

1207. Sonestedt, E., Borné, Y., Wirfält, E. & Ericson, U. Dairy Consumption, Lactase Persistence, and Mortality Risk in a Cohort From Southern Sweden. *Frontiers in Nutrition* **8**, 779034 (2021).

1208. Makino, H. Bifidobacterial strains in the intestines of newborns originate from their mothers. *Bioscience of Microbiota, Food and Health* **37**, 79 (2018).

1209. Odamaki, T. *et al.* Age-related changes in gut microbiota composition from newborn to centenarian: a cross-sectional study. *BMC Microbiology* **16**, 90 (2016).

1210. Claesson, M. J. *et al.* Composition, variability, and temporal stability of the intestinal microbiota of the elderly. *Proc Natl Acad Sci U S A* **108 Suppl 1**, 4586–4591 (2011).

1211. Sharif, S., Meader, N., Oddie, S. J., Rojas-Reyes, M. X. & McGuire, W. Probiotics to prevent necrotising enterocolitis in very preterm or very low birth weight infants. *The Cochrane Database of Systematic Reviews* **2020**, CD005496 (2020).

1212. Morgan, R. L. *et al.* Probiotics Reduce Mortality and Morbidity in Preterm, Low-Birth-Weight Infants: A Systematic Review and Network Meta-analysis of Randomized Trials. *Gastroenterology* **159**, 467 (2020).

1213. Lievin, V. *et al.* Bifidobacterium strains from resident infant human gastrointestinal microflora exert antimicrobial activity. *Gut* **47**, 646 (2000).

1214. Narula, N. *et al.* Systematic Review and Meta-analysis: Fecal Microbiota Transplantation for Treatment of Active Ulcerative Colitis. *Inflamm Bowel Dis* **23**, 1702–1709 (2017).

1215. Xu, L. *et al.* Fecal microbiota transplantation from young donor mice improves ovarian function in aged mice. *J Genet Genomics* **49**, 1042–1052 (2022).

1216. Chen, Y. *et al.* Transplant of microbiota from long-living people to mice reduces aging-related indices and transfers beneficial bacteria. *Aging (Albany NY)* **12**, 4778 (2020).

1217. Ridaura, V. K. *et al.* Cultured gut microbiota from twins discordant for obesity modulate adiposity and metabolic phenotypes in mice. *Science (New York, N.Y.)* **341**, 10.1126/science.1241214 (2013).

REFERENCES

1218. Agus, A., Clément, K. & Sokol, H. Gut microbiota-derived metabolites as central regulators in metabolic disorders. *Gut* **70**, 1174 (2021).

1219. Gordon, H. A., Bruckner-Kardoss, E. & Wostmann, B. S. Aging in germ-free mice: life tables and lesions observed at natural death. *J Gerontol* **21**, 380–387 (1966).

1220. Tazume, S. *et al.* Effects of germfree status and food restriction on longevity and growth of mice. *Jikken Dobutsu* **40**, 517–522 (1991).

1221. Kuma, A. *et al.* The role of autophagy during the early neonatal starvation period. *Nature* **432**, 1032–1036 (2004).

1222. Hara, T. *et al.* Suppression of basal autophagy in neural cells causes neurodegenerative disease in mice. *Nature* **441**, 885–889 (2006).

1223. Pyo, J.-O. *et al.* Overexpression of Atg5 in mice activates autophagy and extends lifespan. *Nature Communications* **4**, 2300 (2013).

1224. Schmauck-Medina, T. *et al.* New hallmarks of ageing: a 2022 Copenhagen ageing meeting summary. *Aging (Albany NY)* **14**, 6829–6839 (2022).

1225. Neverov, A. *et al.* Alternative splicing and protein function. *BMC Bioinformatics* **6**, 266 (2005).

1226. Hattori, D., Millard, S. S., Wojtowicz, W. M. & Zipursky, S. L. Dscam-Mediated Cell Recognition Regulates Neural Circuit Formation. *Annual review of cell and developmental biology* **24**, 597 (2008).

1227. Holly, A. C. *et al.* Changes in splicing factor expression are associated with advancing age in man. *Mechanisms of ageing and development* **134**, 356 (2013).

1228. Gyenis, A. *et al.* Genome-wide RNA polymerase stalling shapes the transcriptome during aging. *Nat Genet* **55**, 268–279 (2023).

1229. Saito, T. *et al.* VEGF-A induces its negative regulator, soluble form of VEGFR-1, by modulating its alternative splicing. *FEBS Lett* **587**, 2179–2185 (2013).

1230. Mamer, S. B., Wittenkeller, A. & Imoukhuede, P. I. VEGF-A splice variants bind VEGFRs with differential affinities. *Sci Rep* **10**, 14413 (2020).

1231. Panchin, A. Y. *et al.* Targeting multiple hallmarks of mammalian aging with combinations of interventions. *Aging (Albany NY)* **16**, 12073–12100 (2024).

1232. Matheu, A. *et al.* Delayed ageing through damage protection by the Arf/p53 pathway. *Nature* **448**, 375–379 (2007).

1233. Spindler, S. R., Mote, P. L. & Flegal, J. M. Combined statin and angiotensin-converting enzyme (ACE) inhibitor treatment increases the lifespan of long-lived F1 male mice. *AGE* **38**, 379–391 (2016).

1234. Davidsohn, N. *et al.* A single combination gene therapy treats multiple age-related diseases. *Proceedings of the National Academy of Sciences* **116**, 23505–23511 (2019).

1235. Hui, H. *et al.* Klotho suppresses the inflammatory responses and ameliorates cardiac dysfunction in aging endotoxemic mice. *Oncotarget* **8**, 15663 (2017).

1236. Kuro-o, M. *et al.* Mutation of the mouse klotho gene leads to a syndrome resembling ageing. *Nature* **390**, 45–51 (1997).

1237. Kurosu, H. *et al.* Suppression of Aging in Mice by the Hormone Klotho. *Science (New York, N.Y.)* **309**, 1829 (2005).

1238. Gaertner, R. F. *et al.* Reduced brain tissue perfusion in TGF-beta 1 transgenic mice showing Alzheimer's disease-like cerebrovascular abnormalities. *Neurobiol Dis* **19**, 38–46 (2005).

1239. Ladiges, W. & Liggitt, D. Testing drug combinations to slow aging. *Pathobiology of Aging & Age-related Diseases* **8**, 1407203 (2018).

1240. Tao, X., Zhang, Z., Yang, Z. & Rao, B. The effects of taurine supplementation on diabetes mellitus in humans: A systematic review and meta-analysis. *Food Chemistry: Molecular Sciences* **4**, 100106 (2022).

1241. Huang, H. *et al.* Dietary resources shape the adaptive changes of cyanide detoxification function in giant panda (Ailuropoda melanoleuca). *Sci Rep* **6**, 34700 (2016).

1242. Kalk, E. *et al.* Safety and Effectiveness of Isoniazid Preventive Therapy in Pregnant Women Living with Human Immunodeficiency Virus on Antiretroviral Therapy: An Observational Study Using Linked Population Data. *Clin Infect Dis* **71**, e351–e358 (2020).

1243. Schmid, D. R., Lee, J. A., Wismer, T. A., Diniz, P. P. V. P. & Murtaugh, R. J. Isoniazid toxicosis in dogs: 137 cases (2004-2014). *J Am Vet Med Assoc* **251**, 689–695 (2017).

1244. Bostrom, N. The fable of the dragon tyrant. *J Med Ethics* **31**, 273–277 (2005).

1245. *Fable of the Dragon-Tyrant.* (2018).

1246. About - Hevolution Foundation. https://hevolution.com/about.

REFERENCES

1247. Overview | XPRIZE Healthspan | XPRIZE Foundation. https://www.xprize.org/prizes/healthspan.

1248. ARPA-H Home | ARPA-H. https://arpa-h.gov/ (2024).

1249. Keane, M. *et al.* Insights into the Evolution of Longevity from the Bowhead Whale Genome. *Cell Reports* **10**, 112 (2015).

1250. Hébert, J. M. & Vijg, J. Cell Replacement to Reverse Brain Aging: Challenges, Pitfalls, and Opportunities. *Trends Neurosci* **41**, 267–279 (2018).

1251. Quezada, A. *et al.* An In Vivo Platform for Rebuilding Functional Neocortical Tissue. *Bioengineering* **10**, 263 (2023).

1252. Longevity Impetus Grants. *Longevity Impetus Grants* https://impetusgrants.org.

1253. Norn Group. *Norn Group* https://norn.group.

1254. Berry, B. J. *et al.* Optogenetic rejuvenation of mitochondrial membrane potential extends C. elegans lifespan. *Nat Aging* **3**, 157–161 (2022).

1255. Bottlenecks of Aging. *Amaranth Foundation* https://amaranth.foundation/bottlenecks-of-aging.

1256. TIME Initiative. https://www.timeinitiative.org/.

1257. age1. https://www.age1.com/.

1258. The Longevity Fund. https://www.longevity.vc/.

1259. https://www.afar.org/. *American Federation for Aging Research* https://www.afar.org/.

1260. AbbVie. *A Randomized, Double-Blind, Placebo-Controlled Study to Assess Safety, Tolerability, and Pharmacokinetics Following Multiple Doses of Fosigotifator in Subjects With Amyotrophic Lateral Sclerosis Followed by an Active Treatment Extension.* https://clinicaltrials.gov/study/NCT04948645 (2024).

1261. Baumgartner, C. K. *et al.* The PTPN2/PTPN1 inhibitor ABBV-CLS-484 unleashes potent anti-tumour immunity. *Nature* **622**, 850–862 (2023).

1262. Cudkowicz, M. E. *HEALEY ALS Platform Trial - Regimen F ABBV-CLS-7262.* https://clinicaltrials.gov/study/NCT05740813 (2023).

1263. Sridar, J. *et al.* Cryo-EM structure of human PAPP-A2 and mechanism of substrate recognition. *Commun Chem* **6**, 234 (2023).

1264. Sam Altman invested $180 million into a company trying to delay death | MIT Technology Review. https://www.technologyreview.com/2023/03/08/1069523/sam-altman-investment-180-million-retro-biosciences-longevity-death/.

1265. Bottlenecks. https://www.longbiofellowship.org/bottlenecks.

REFERENCES

1266. Say Forever! *Say Forever!* https://sayforever.org.

1267. Longevity Biotech Fellowship. https://www.longbiofellowship.org/.

1268. Vitalism - Unlimited lifespans in peak health. https://www.vitalism.io/.

1269. The latest longevity, life extension, and rejuvenation news. https://www.lifespan.io/.

1270. Fiscal Year 2024 Budget. *National Institute on Aging* https://www.nia.nih.gov/about/budget/fiscal-year-2024-budget.

1271. Deaths in Wars and Conflicts in the 20th Century | Center for International and Security Studies at Maryland. https://cissm.umd.edu/research-impact/publications/deaths-wars-and-conflicts-20th-century (2006).

1272. How Many People Have Ever Lived on Earth? | PRB. https://www.prb.org/articles/how-many-people-have-ever-lived-on-earth/.

1273. Dublin Longevity Declaration. *Dublin Longevity Declaration* https://dublinlongevitydeclaration.org.

1274. Dog Aging Project. *Dog Aging Project* https://dogagingproject.org/.

1275. Bray, E. E. *et al.* Once-daily feeding is associated with better health in companion dogs: results from the Dog Aging Project. *Geroscience* **44**, 1779–1790 (2022).

1276. Barnett, B. G. *et al.* A masked, placebo-controlled, randomized clinical trial evaluating safety and the effect on cardiac function of low-dose rapamycin in 17 healthy client-owned dogs. *Front Vet Sci* **10**, 1168711 (2023).

1277. Kent, M. S. *et al.* Longevity and mortality in cats: A single institution necropsy study of 3108 cases (1989-2019). *PLoS One* **17**, e0278199 (2022).

1278. Sugisawa, R. *et al.* Impact of feline AIM on the susceptibility of cats to renal disease. *Sci Rep* **6**, 35251 (2016).

1279. Miyazaki, T., Yamazaki, T., Sugisawa, R., Gershwin, M. E. & Arai, S. AIM associated with the IgM pentamer: attackers on stand-by at aircraft carrier. *Cell Mol Immunol* **15**, 563–574 (2018).

1280. Sugisawa, R. *et al.* Independent modes of disease repair by AIM protein distinguished in AIM-felinized mice. *Sci Rep* **8**, 13157 (2018).

1281. Feline generous: Japan cat lovers give US$2m to kidney research | Malay Mail. https://www.malaymail.com/news/life/2021/09/14/feline-generous-japan-cat-lovers-give-us2m-to-kidney-research/2005569.

1282. Aiming to double cats' lifespan. *The University of Tokyo* https://www.u-tokyo.ac.jp/focus/en/features/z1304_00039.html.

1283. Runacres, A., Mackintosh, K. A. & McNarry, M. A. Health Consequences of an Elite Sporting Career: Long-Term Detriment or Long-Term Gain? A Meta-Analysis of 165,000 Former Athletes. *Sports Med* **51**, 289–301 (2021).

1284. McKee, A. C. *et al.* Neuropathologic and Clinical Findings in Young Contact Sport Athletes Exposed to Repetitive Head Impacts. *JAMA Neurol* **80**, 1037 (2023).

1285. Szigeti, B. *et al.* Self-blinding citizen science to explore psychedelic microdosing. *eLife* **10**, e62878 (2021).

1286. Szigeti, B., Nutt, D., Carhart-Harris, R. & Erritzoe, D. The difference between 'placebo group' and 'placebo control': a case study in psychedelic microdosing. *Sci Rep* **13**, 12107 (2023).

1287. Newman, S. J. Supercentenarian and remarkable age records exhibit patterns indicative of clerical errors and pension fraud. 704080 Preprint at https://doi.org/10.1101/704080 (2024).

1288. Sho, H. History and characteristics of Okinawan longevity food. *Asia Pac J Clin Nutr* **10**, 159–164 (2001).

1289. Kreouzi, M., Theodorakis, N. & Constantinou, C. Lessons Learned From Blue Zones, Lifestyle Medicine Pillars and Beyond: An Update on the Contributions of Behavior and Genetics to Well-being and Longevity. *American Journal of Lifestyle Medicine* **18**, 750–765 (2024).

1290. Murder Victims|Statistics Japan : Prefecture Comparisons. *Murder Victims/Statistics Japan : Prefecture Comparisons* https://stats-japan.com/t/kiji/10567.

1291. Baker, M. 1,500 scientists lift the lid on reproducibility. *Nature* **533**, 452–454 (2016).

1292. Spiridonova, O., Kriukov, D., Nemirovich-Danchenko, N. & Peshkin, L. On standardization of controls in lifespan studies. *Aging* **16**, 3047–3055 (2024).

1293. First Approval. https://intro.dev.firstapproval.io/.

1294. Thng, C. E., Lim-Ashworth, N. S., Poh, B. Z. & Lim, C. G. Recent developments in the intervention of specific phobia among adults: a rapid review. *F1000Research* **9**, F1000 Faculty Rev (2020).

1295. Homepage - a4li.org. https://a4li.org/ (2024).

1296. Jellyfish DAO Landing. https://www.jellyfishdao.org/.

1297. VitaDAO | The Longevity DAO. https://www.vitadao.com/.

1298. LessWrong. https://www.lesswrong.com/ (2024).

REFERENCES

1299. Moqri, M. *et al.* Validation of biomarkers of aging. *Nat Med* **30**, 360–372 (2024).

1300. Moqri, M. *et al.* Biomarkers of aging for the identification and evaluation of longevity interventions. *Cell* **186**, 3758–3775 (2023).

1301. Glaberman, S., Bulls, S. E., Vazquez, J. M., Chiari, Y. & Lynch, V. J. Concurrent Evolution of Antiaging Gene Duplications and Cellular Phenotypes in Long-Lived Turtles. *Genome Biology and Evolution* **13**, evab244 (2021).

1302. Huang, Z. *et al.* Duplications of Human Longevity-Associated Genes Across Placental Mammals. *Genome Biology and Evolution* **15**, evad186 (2023).

1303. Korotkova, D. D. *et al.* Bioinformatics Screening of Genes Specific for Well-Regenerating Vertebrates Reveals c-answer, a Regulator of Brain Development and Regeneration. *Cell reports* **29**, 1027 (2019).

1304. Harris, M. P., Hasso, S. M., Ferguson, M. W. J. & Fallon, J. F. The development of archosaurian first-generation teeth in a chicken mutant. *Curr Biol* **16**, 371–377 (2006).

1305. Ravi, V. *et al.* Advances in tooth agenesis and tooth regeneration. *Regenerative Therapy* **22**, 160 (2023).

1306. Rafikova, E. *et al.* Open Genes-a new comprehensive database of human genes associated with aging and longevity. *Nucleic Acids Res* **52**, D950–D962 (2024).

1307. Tacutu, R. *et al.* Human Ageing Genomic Resources: new and updated databases. *Nucleic Acids Res* **46**, D1083–D1090 (2018).

1308. Lewin, H. A. *et al.* Earth BioGenome Project: Sequencing life for the future of life. *Proceedings of the National Academy of Sciences of the United States of America* **115**, 4325 (2018).

1309. Schröder, K.-P. & Connon Smith, R. Distant future of the Sun and Earth revisited. *Monthly Notices of the Royal Astronomical Society* **386**, 155–163 (2008).

Popular science publication

Alexander Panchin

IMMORTALITY OR DEATH

From Aging Research to Real-Life Solutions

Illustrations by *Olga Posukh*
Book cover by *Kira Severinova*

An Imprint of Open Longevity
openlongevity.org

Font type *«Libre Caslon Text»*

Made in the USA
Columbia, SC
07 June 2025

b477a6a1-2bb9-4b69-a6c9-1d2db69c513aR02